Communicating
Technical
Information

COMMUNICATING TECHNICAL INFORMATION

A Guide for the Electronic Age

Donald Pattow

William Wresch

University of Wisconsin,
Stevens Point

A BLAIR PRESS BOOK

PRENTICE HALL, ENGLEWOOD CLIFFS, NJ 07632

Library of Congress Cataloging-in-Publication Data

Pattow, Donald.
 Communicating technical information : a guide for the electronic age / Donald Pattow,
William Wresch.
  p.   cm.
 "A Blair Press book."
 Includes index.
 ISBN 0-13-898669-X
 1. Communication of technical information. 2. Technical writing.
I. Wresch, William. II. Title.
T10.5.P3785 1993
808'.0666--dc20

92-31805
CIP

Cover designer: Richard Stalzer Associates, Ltd.
Cover art: Bob Forrest
Prepress buyer: Herb Klein
Manufacturing buyer: Robert Anderson/Patrice Fraccio

Blair Press
The Statler Building
20 Park Plaza, Suite 1113
Boston, MA 02116-4399

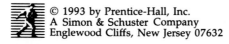 © 1993 by Prentice-Hall, Inc.
A Simon & Schuster Company
Englewood Cliffs, New Jersey 07632

Printed in the United States of America
10 9 8 7 6 5 4 3 2 1

ISBN 0-13-898669-X

Prentice-Hall International (UK) Limited, *London*
Prentice-Hall of Australia Pty. Limited, *Sydney*
Prentice-Hall Canada Inc., *Toronto*
Prentice-Hall Hispanoamericana, S.A., *Mexico*
Prentice-Hall of India Private Limited, *New Delhi*
Prentice-Hall of Japan, Inc., *Tokyo*
Simon & Schuster Asia Pte. Ltd., *Singapore*
Editora Prentice-Hall do Brasil, Ltda., *Rio de Janeiro*

Preface

The field of technical communication has changed dramatically in the past few years. Perhaps the most obvious change has been in who produces it: technical communication is no longer limited to those trained specifically in technical writing or to a small group of engineers and experts. The number of people employed in highly technical fields continues to rise. At the same time, nontechnical occupations now involve technology to a degree that would have been unthinkable only ten years ago. Office managers, political activists, small business owners, salespeople, and journalists, among many others, have all seen new technologies transform their day-to-day activities —and when these people communicate, chances are that at least some of what they communicate will be about technology. In other words, virtually every person in every field must now learn to communicate technical information well.

An equally important change is the higher standards readers have for today's technical documents. This change is no doubt due to the recent explosion in the amount and variety of technical documents, combined with the fact that they are being read by a broader audience than ever before. Whether to make important business or personal decisions or merely to stay informed, the average person today must read a wide variety of surprisingly technical material. Computer tutorials, instruction manuals for home appliances and office machines, news reports on chemical hazards, economic forecasts, and articles on medical advances—the constant flow of information that sustains our society—are just as much technical documents as feasibility studies and progress reports. As soon as general readers began to encounter technical documents on a regular basis, they made known their dissatisfactions with indecipherable instructions, unending and overly technical proposals, and apparently contradictory medical bulletins. Today's businesspeople and consumers alike demand documents that are absolutely accurate, thorough but not overwhelming, and easy to navigate.

These higher standards have led to another change, the increased responsibility of the technical writer. It used to be that a technical writer was just that: a writer. Someone else was expected to pay attention to a document's design, production, and quality control—if they were given much attention at all. But with reader satisfaction such an important consideration today, these matters can no longer be pushed to the bottom of the list. Someone has to take full responsibility for them, and that person is increasingly the technical writer.

Finally, one thing links all these changes and brings new changes of its own: the computer. This single piece of technology has transformed every aspect of technical communication. The computer is now an important subject for technical communication, primarily in computer manuals, tutorials, and reference guides. The possibilities for routine research and analysis have been enlarged by on-line indexes and databases, spreadsheets, and statistics packages. The process of drafting and revision has been transformed by word processors, by group-writing software, and by spelling, grammar, and style checkers. The way documents are produced has been changed by desktop publishing programs. Even the transmission of information has been affected by the computer, since electronic mail and hypertext can create and disseminate documents without using a single piece of paper.

Approach and Features

To address the changing nature of technical communication and give students the background and skills they will need to be good technical communicators, this book emphasizes the broader scope of contemporary technical communication, the importance of the writing process, the need to understand the social context of writing, and the growing influence and importance of new technologies.

Rather than define technical communication as a series of rhetorical modes, we emphasize the importance of technical communication as a professional activity in the workplace. Rather than discussing standard documents as rigid forms to be filled, we present them as tasks with typical (but not unvarying) audiences, purposes, writing processes, and methods of organization. Because we believe that effective technical writing means making effective choices, this book focuses on the choices writers make. And in order to put those choices into a meaningful context, we offer extended descriptions of realistic situations that require technical communication.

This philosophy can be seen at work in the strong emphasis we place on the following topics, many of which are treated only superficially by other technical writing textbooks:

- **The Writing Process.** We devote a chapter early in the book to the typical technical writing process—evaluating the situation, conducting research, writing, reviewing and revising, and publishing—and then use it as the framework for each technical writing task covered later on.

- **Group Writing.** Because group writing has long been a reality in the workplace, we give a full chapter to a practical discussion of group dynamics and the challenges of group writing. Collaborative projects are included in every set of end-of-chapter exercises.
- **The Computer.** The growing importance of the computer as a tool for communication, as a subject of communication, and as a medium for communication is thoroughly addressed in our book. We devote three chapters to computer-related topics, and we discuss other aspects of the computer's role in technical communication as needed.
- **Graphics and Document Design.** These two components of visual presentation, increasingly recognized as a significant method of communication, are given a strong emphasis in our book. We cover each in its own chapter and in addition offer suggestions for using specific graphics and design elements for each type of writing.
- **Quality Reviews.** To prepare students to meet the higher standards placed on today's technical documents, we give a full chapter in our book to discussing the goals, principles, and techniques of document testing and revision. Reviewing for quality is presented as a standard step in the writing process for every technical writing task we cover.
- **Argumentation.** In the increasingly competitive marketplace, an effective technical document must often be an effective technical argument. We devote an entire chapter to argumentation as a writing technique indispensable to today's technical communicator.
- **Nonsexist Language.** The higher standards for technical communication have led to a new awareness of the importance of nonsexist language. In a special section in our chapter on style, we thoroughly discuss the rhetorical and ethical significance of exclusive language and give specific recommendations for avoiding it.

In a further effort to address the practical needs of communicators in today's workplace, we give full coverage to writing tasks that have assumed new importance as a result of the information explosion and the computer revolution:

- **Abstracts**
- **Literature Reviews**
- **Computer Manuals**
- **Electronic Mail**
- **Hypertext**

Organization

The straightforward organization of our book leads students through the material in a logical, easily understood sequence.

- **Part I, "Basic Principles and Concepts,"** introduces students to the fundamentals of technical communication. The major emphasis here is on the writing process and the social context of communication.

- **Part II, "Visual Presentation,"** includes a chapter on graphics and a chapter on document design; both emphasize the various computer options available today.
- **Part III, "Technical Writing Techniques,"** consists of six chapters that introduce students to the range of techniques they will need in order to be successful writers and speakers: definition, description, process description, summary, analysis, and argumentation.
- **Part IV, "Technical Research,"** comprises four chapters that lead students through the typical technical research process: collecting information, analyzing information, organizing information, and using and citing technical sources. Special attention is given to the computer tools available for each stage of the writing process.
- **Part V, "Technical Writing Tasks,"** is the heart of the book, pulling together the principles and techniques students have learned thus far and showing how these culminate in effective technical documents. Eight chapters cover typical writing tasks: literature reviews, research reports, proposals, feasibility studies, progress reports, instructions, computer manuals, and electronic documents.
- **Part VI, "Other Technical Communication Tasks,"** covers other skills that students need in order to become good technical communicators. The section includes one chapter on oral presentations, one on correspondence, and one on job applications.

Acknowledgments

We would like to express our thanks to our colleague Jim Gifford for initiating this project. We would also like to thank the following instructors around the country for providing helpful reviews of the book at its various stages of development: Laura E. Casari, University of Nebraska, Lincoln; Patricia E. Connors, Memphis State University; David H. Covington, North Carolina State University; Robert Cullen, San Jose State University; Sam Dragga, Texas Tech University; Irene F. Gale, University of South Florida; Judith Kaufman, Eastern Washington University; Gloria Kitto Lewis, Wayne State University; Shirley W. Logan, University of Maryland; Fred Reynolds, Old Dominion University; and Dene Kay Thomas, University of Idaho. At Blair Press, our appreciation to Nancy Perry, for encouraging us all along the way, and to Denise Branch Wydra, for her silver tongue and barbed whip. Finally, this book is dedicated to Rebecca Pattow and Sue Wresch, who kept us on an even keel throughout.

Donald Pattow
William Wresch

Brief Contents

Preface   v

Part I   **Basic Principles and Concepts**   **1**

1   Introduction to Technical Communication   3
2   The Process of Technical Writing   17
3   Purpose and Audience in Technical Writing   35
4   Technical Style   54
5   Quality Reviews   74
6   Group Writing   101
7   Writing with a Computer   115

Part II   **Visual Presentation**   **131**

8   Graphics   133
9   Document Design   172

Part III   **Technical Writing Techniques**   **199**

10   Technical Definition   201
11   Technical Description   215
12   Technical Process Description   228
13   Technical Summary   246
14   Technical Analysis   264
15   Technical Argumentation   280

Part IV   Technical Research   295

16   Collecting Information   297
17   Analyzing Information   322
18   Organizing Information   338
19   Using and Citing Technical Sources   354

Part V   Technical Writing Tasks   375

20   Literature Reviews   377
21   Research Reports   394
22   Proposals   410
23   Feasibility Studies   432
24   Progress Reports   453
25   Instructions   470
26   Computer Manuals   495
27   Electronic Documents   513

Part VI   Other Technical Communication Tasks   529

28   Oral Presentations   531
29   Correspondence   552
30   Job Applications   567

References   585
Index   591

Contents

Preface   v

Part I   Basic Principles and Concepts   1

1   Introduction to Technical Communication   3

The *Challenger* Disaster   4
  The Breakdown in Technical Communication   7
  Lessons of the *Challenger*   11

Technical Communication and Your Career   11
  The Increasing Amount of Technical Communication   11
  The Growing Dissatisfaction with Current Technical
    Communication   12
  The Growing Complexity of the Technical Communication
    Process   13

An Overview of This Book   14

Activities   16

2   The Process of Technical Writing   17

Evaluating the Writing Situation   18
  What Is the Purpose of This Document?   19
  Who Is the Audience for This Document?   19
  What Is the Assignment?   20
  How Does This Communication Fit into the Larger Context?   20
  Will This Be Written in a Group?   21
  What Are the Available Resources?   21
  What Are the Time Constraints?   22
  Case Study   22

Conducting Research   23
   What Do I Need to Find Out?   23
   How Can I Find This Out?   24
   What Does It Mean?   25
   What Is the Best Way to Present It?   25
   Case Study   25

Writing the Document   26
   What Would Be the Best Writing Style?   26
   What Computer Writing Techniques Can I Use?   27
   What Kind of Graphics Should This Document Have?   27
   What Would Be the Best Design?   27
   Case Study   27

Reviewing and Revising   30
   What Do I Need to Find Out?   30
   How Can I Determine This?   30
   Case Study   31

Publishing   32
   What Would Be Most Suitable?   32
   What Are the Available Options?   32
   Case Study   33

Activities   33

3   Purpose and Audience in Technical Writing   35

What Is My Purpose?   37
   Explicit Purposes   37
   Implicit Purposes   41

Who Is My Audience?   42
   Experts   42
   Technicians   43
   Executives and Managers   43
   Novices   43
   The Mixed Audience   44

What Are My Audience's Needs and Wants?   44
   What Is Their Purpose in Reading?   44
   How Much Do They Know?   45
   How Receptive Will They Be?   46

How Can I Adapt My Document to This Audience and Achieve My
   Purpose?   47
   Employ Characteristics of All Good Technical Writing   47
   Vary Content and Presentation According to Audience   48

Examples   49

Checklist for Assessing Purpose and Audience   52

Activities   52

4 **Technical Style** **54**

Style Conventions 55

Specialized Conventions of Technical Style 56
Company Style Guides 56
Professional Style Manuals 56

General Conventions of Technical Style 57
Principle 1: Help Readers Feel Involved in What They Are
Reading 57
Principle 2: Select Words Carefully 59
Principle 3: Write Clear Sentences 62
Principle 4: Create Documents That Are Easy to Navigate 66

Checklist for Technical Writing Style 70

Activities 71

5 **Quality Reviews** **74**

Determining Quality 78
Reasons to Review for Quality 78
Common Problems in Determining Quality 78
Strategies for Determining Quality 80

Writer-Based Reviews 80
Review for Company Standards 80
Review for Accuracy and Organization 81
Review for Correctness, Format, and Style 82
Guidelines for Writer-Based Reviews 88

Checklist for Writer-Based Reviews 89

Colleague-Based Reviews 90
Review for Organization and Format 90
Review for Effectiveness 90
Review for Accuracy 90
Guidelines for Colleague-Based Reviews 91

User Testing 92
Comprehension Tests 92
Performance Tests 95
Evaluation Surveys 96
Guidelines for User Testing 98

Activities 99

6 **Group Writing** **101**

Eight Management Tasks for Group Writing 102
Team Building 103
Task Analysis 104
Task Management 105
Time Management 106

Document Management   108
Style and Format Management   108
Technology Management   110
Conflict Management   110
Case Study   111

Activities   113

7 Writing with a Computer   115

Computer Tools   116
Information-Access Tools   116
Spreadsheets   118
Graphics Programs   118
Word Processors   119
Desktop Publishing Programs   119
Group Writing Software   120

Computer Writing Techniques   120
Reasons to Use Shortcuts   120
Scavenging   122
Templates   123
Structured Document Processors   124
Boilerplate   126
Macros   127
Search and Replace   127

Activities   128

Part II Visual Presentation   131

8 Graphics   133

When Should You Use Graphics?   134
To Condense the Text   135
To Emphasize Patterns   136
To Clarify Relationships   137

Common Graphics   138
Tables   138
Graphs   140
Illustrations   150
Diagrams   153

Producing Graphics   156
Graphics from Sources   157
Manual Methods   158
Word Processors   158
Spreadsheets   160
Computer Scanners   160

Computer-Aided Design Programs   161

Using Graphics   162
Integrating Graphics and Text   162
Adding Titles and Labels   163
Placing the Graphic in the Paper   163
The Danger of Novelty   165

Graphics and Ethics   166
Guidelines for Ethical Graphics   168

Activities   170

9   ## Document Design   172

Principles of Design   173
Design and Information Processing   174
Design and Visual Coherence   175

Visual Aspects of Documents   175
Page Layout   175
Text Blocking   179
Type Size, Font, and Style   181
Headings   183
Marginal Glosses   185
Lists and Bullets   186
Lines and Boxes   187
Color   189
Icons   189
Graphic Balance   191

Tools of Design   192
Word Processors   193
Desktop Publishing Programs   194

Activities   196

Part III   Technical Writing Techniques   199

10   ## Technical Definition   201

The Role of Technical Definition   202

When to Use Technical Definition   203

Types of Technical Definition   204
Informal Definitions   204
Formal Definitions   205
Stipulative Definitions   205

Writing a Technical Definition   206
Writing a Single-Word Definition   206
Writing a Phrase Definition   207

Writing a Sentence Definition   207
Writing an Expanded Definition   207

Guidelines for Technical Definition   209
1. Use the Least Intrusive Method Possible   209
2. Place Definitions Carefully   210

Examples of Technical Definition   212

Activities   214

11 Technical Description   215

The Role of Technical Description   216

When to Use Technical Description   216

Types of Technical Description   216

Writing a Technical Description   217
Introduction   217
Part-by-Part Description   218
Graphics   219

Guidelines for Technical Description   222
1. Use an Appropriate Level of Detail   222
2. Use Figurative Language   222
3. Indicate Subdivisions Clearly   223

Examples of Technical Description   223

Activities   226

12 Technical Process Description   228

The Role of Technical Process Description   229

When to Use Technical Process Description   229

Types of Technical Process Description   229
Process Descriptions That Inform   230
Process Descriptions That Instruct   230

Writing a Technical Process Description   231
Accurate Information   231
Strong Organization   233
Appropriate Level of Detail   234
Description of the Purpose of the Process   234
Description of Observation Methods   235
Description of Equipment   235
Graphics   237

Guidelines for Technical Process Description   239
1. Use an Appropriate Level of Detail   239
2. Describe Time Spans as Needed   239

Example of a Technical Process Description   240

Activities   243

13   **Technical Summary   246**

The Role of Technical Summary   247

When to Use Technical Summary   247
  To Help the Reader Find Documents of Interest   247
  To Provide Main Points of a Document   248
  To Guide the Reading of a Document   248

Types of Technical Summary   249
  Descriptive Abstracts   249
  Informative Abstracts   250
  Executive Summaries   250
  Glosses   251

Writing a Technical Summary   253
  Writing a Descriptive Abstract   253
  Writing an Informative Abstract   253
  Writing an Executive Summary   256
  Writing Glosses   256

Guidelines for Technical Summary   257
  1. Keep All Summaries Concise   257
  2. Do Not Use a Summary to Evaluate the Document   258
  3. Use a Logical Order   258
  4. Do Not Add Information   258
  5. Do Not Omit Important Points   259
  6. Make Glosses Visually Distinctive   259

Examples of Technical Summary   259

Activities   260

14   **Technical Analysis   264**

The Role of Technical Analysis   265

When to Use Technical Analysis   265

Types of Technical Analysis   265
  Classifications   266
  Partitions   267

Writing a Technical Analysis   267
  Writing a Classification   267
  Writing a Partition   269

Guidelines for Technical Analysis   272
  1. Use a Justifiable Principle of Analysis   272
  2. Don't Use the Same Approach in Every Document   272
  3. Name the Approach You Are Using   272
  4. Use Labels for Each Group or Part   272
  5. Keep the Number of Groups or Parts Reasonable   272

Examples of Technical Analysis   273

Activities   274

15 **Technical Argumentation 280**

The Role of Technical Argumentation 281

When to Use Technical Argumentation 281

Types of Argumentation 281
  Logical Appeal 282
  Emotional Appeal 282
  Ethical Appeal 282

Writing a Technical Argument 282
  Thesis Statement 283
  Background Information 284
  Logical Argument 284
  Evidence 286
  Anticipation of Objections 287
  Clear Conclusion 287

Guidelines for Technical Argumentation 287
  1. Project a Credible Image 287
  2. Anticipate Objections Early 288
  3. Do Not Argue Matters of Taste 288
  4. Do Not Preach to the Converted 288

Examples of Technical Argumentation 288

Activities 293

Part IV Technical Research **295**

16 **Collecting Information 297**

Determining What Information You Need 298
  Using Tagmemics 299
  Using Burke's Pentad 300
  Framing a Research Question 302

Sources of Information 302

Yourself 303
  Brainstorming 303
  Free Writing 304
  Invisible Writing 304

The Library 305
  The Card Catalog 306
  Periodical Indexes 306
  Reference Books 306
  The On-Line Catalog 306
  Computerized Indexes 308
  On-Line Databases 311

The Field 313

Interviews   313
Letters of Inquiry   314
Questionnaires   315
On-Site Visits   319
Observations   319
Controlled Experiments   320

Activities   320

17 Analyzing Information   322

Checking Completeness   323
Conventional Standards   323
Heuristic Techniques   324
Statistical Significance   325

Developing Conclusions   325
Standard Procedures   325
Spreadsheets   326
Spreadsheet-Based Graphics   330

Checking Conclusions for Reasonableness   335
Evaluating Sources   335
Checking Procedures   335
Checking Data Sufficiency   336
Considering Alternative Conclusions   336

Activities   336

18 Organizing Information   338

Content-Driven Organization   339
Chronological   339
Spatial   340
Order of Importance   341
General to Specific   342
Specific to General   342
Problem – Cause – Solution   344

Format-Driven Organization   345

Using Templates   348

Activities   350

19 Using and Citing Technical Sources   354

Incorporating Source Material in Your Writing   355
Quotations   355
Paraphrase and Summary   357
Guidelines for Incorporating Source Material   359

Documenting Your Sources   361
In-Text Citations and Reference Lists   362

APA Style for In-Text Citations   363
CBE Style for In-Text Citations   365
APA Style for Reference Lists   366
CBE Style for Reference Lists   368
Number System   371
Guidelines for Documenting Sources   372

Activities   373

Part V   Technical Writing Tasks   375

20   Literature Reviews   377

Situation   378

Form   378

Writing the Literature Review   379
Collecting Information   379
Organizing the First Draft   380
Reviewing to Assure Quality   385
Publishing the Literature Review   385

Examples of Literature Reviews   386

Activities   392

21   Research Reports   394

Situation   395

Form   397
Title and Author   398
Abstract   399
Objective   399
Literature Review   400
Procedures   400
Results   401
Conclusions   403
Appendixes   404

Writing the Research Report   404
Collecting Information   404
Organizing the First Draft   405
Reviewing to Assure Quality   406
Publishing the Research Report   406

Example of a Research Report   407

Activities   407

22   Proposals   410

Situation   411

Solicited Proposals   411
Unsolicited Proposals   412
Internal Proposals   413
Audience   414
Common Constraints   415

Form   415
Summary   416
Introduction   417
Description of the Solution   418
Organizational Qualifications   419
Evaluation   419
Budget   419
Appendixes   420

Writing the Proposal   421
Setting a Schedule   421
Collecting Information   421
Organizing the First Draft   422
Reviewing to Assure Quality   426
Publishing the Proposal   427

Example of a Proposal   428

Activities   428

23   Feasibility Studies   432

What Is Feasibility?   433

Situation   435

Form   436
Recommendations   437
Original Situation   437
Background of the Investigation   437
Comparison of Alternatives   437
Conclusion   438
Appendixes   438

Writing a Feasibility Study   438
Collecting Information   439
Organizing the First Draft   442
Reviewing to Assure Quality   446
Publishing the Feasibility Study   446

Example of a Feasibility Study   446

Activities   451

24   Progress Reports   453

Situation   454

Form   455

Introduction   456
Summary   457
Body   458
Conclusion   458
Appendixes   459

Writing the Progress Report   459
Collecting Information   459
Organizing the First Draft   459
Reviewing to Assure Quality   461
Publishing the Progress Report   462

Example of a Progress Report   462

Activities   469

25 | Instructions   470

Situation   471

Form   473
Parts List   474
Warnings   475
General Background   475
Preparatory Steps   475
Sequential Instructions   475
Trouble-Shooting Guide   475

Writing Instructions   477
Collecting Information   477
Organizing the First Draft   479
Reviewing to Assure Quality   482
Publishing the Instructions   485

Example of Instructions   486

Activities   486

26 | Computer Manuals   495

Tutorials   498
Situation   499
Audience   499
Guidelines for Tutorials   501

Reference Guides   502
Situation   505
Audience   505
Guidelines for Reference Guides   505

On-Line Help   507
Situation   508
Audience   508

Guidelines for On-Line Help   508

Activities   511

27   **Electronic Documents**   513

Features of Electronic Documents   514
Speed   514
Larger Audiences   515
Media Integration   515
Flexibility   515
Authorship Ambiguity   516
Increased Access   516
Technical Limitations   516

Electronic Mail   517
Situation   518
Audience   518
Guidelines for Electronic Mail   519

Hypertext   521
Situation   521
Audience   525
Guidelines for Hypertext   526

Activities   526

Part VI   **Other Technical Communication Tasks**   529

28   **Oral Presentations**   531

Situation   532
Work Group Reports   532
Customer Presentations   533
Professional Meetings   534

Preparing for an Oral Presentation   535
Adapting to a Location   535
Creating an Organization   535
Creating Note Cards or Prompts   537
Choosing Information to Include in Visual Aids   538
Selecting the Kind of Visual Aid to Use   540
Practicing the Presentation   545

Giving an Oral Presentation   546
Oral Presentations Are Flexible   546
Oral Presentations Are Interactive   547
Oral Presentations Present You   547
People Don't Listen Well   548

Following Up the Presentation   549

Activities   549

29   Correspondence   552

Memorandums   553
  Types of Memorandums   553
  Elements of a Memorandum   554

Letters   555
  Types of Letters   556
  Elements of a Letter   560

The Writing Process   561
  Assess the Writing Situation   563
  Determine the Solution   563
  Collect Information   564
  Organize the Information   564
  Evaluate the Communication   565
  Distribute the Communication   565

Activities   565

30   Job Applications   567

Securing an Interview   568

The Résumé   569
  Types of Résumés   570
  Elements of a Résumé   573
  Guidelines for Résumés   575

The Letter of Application   577
  Elements of a Letter of Application   577
  Guidelines for Letters of Application   578

The Interview   580
  How You Look   580
  How You Communicate   581
  Follow-Up   582

Getting the Right Job   582

Activities   584

References   585

Index   591

I

BASIC PRINCIPLES
AND CONCEPTS

Introduction to Technical Communication

The *Challenger* Disaster
  The Breakdown in Technical Communication
  Lessons of the *Challenger*

Technical Communication and Your Career
  The Increasing Amount of Technical Communication
  The Growing Dissatisfaction with Current Technical Communication
  The Growing Complexity of the Technical Communication Process

An Overview of This Book
Activities

Our world is more technologically driven than ever before. We marvel at the latest computers, bring ever more complex systems into our homes and job sites, agonize over the legal and ethical consequences of medical technology, and debate the environmental impact of agricultural technology. We are surrounded by technology in many forms. As a consequence, we need to communicate about technology. We need to discuss it, debate it, instruct others in its use. No matter what your chosen career, you will be involved with technology, and you will have to communicate information about it.

There was a time when technical information was communicated by specialists — technical writers. Such people are still around, but there is far too much technical information around, far too many decisions, far too much instruction, to leave such communication to specialists. We have all become technical writers. At its most basic level, technical communication is just people talking about the tools around them. Those tools need to be used wisely and their limitations understood. The tools we now have available give us increasing power, but also increasing responsibility. The space shuttle *Challenger* disaster demonstrates in chilling detail just how important clear communication is. The people involved were engineers, not technical writers, yet it was their inability to communicate about the technology they worked with every day that led to disaster.

The *Challenger* Disaster

At 11:38 AM on January 28, 1986, shuttle flight 51-L — the *Challenger* — lifted off from Cape Canaveral. It was a cold morning, thirty-six degrees Fahrenheit, far colder than for any previous shuttle launch. On board were seven astronauts: Francis R. Scobee, commander; Michael John Smith, pilot; Ellison S. Onizuka, mission specialist; Judith Arlene Resnick, mission specialist; Ronald Erwin McNair, mission specialist; S. Christa McAuliffe, payload specialist; and Gregory Bruce Jarvis, payload specialist. In seventy-three seconds, all seven would be dead.

Within milliseconds of ignition, gases from the right solid rocket booster began pouring through a loose joint between sections of the booster. Smoke from this area was already visible less than one second into the flight. At 58 seconds flames from the hot gases were burning visibly. For about 5 seconds the plume grew in size, with flames aimed directly at the huge external fuel tank. At 64.660 seconds the hydrogen tank portion of the external tank was breached. Cryogenic liquid hydrogen rushed from the tank directly onto the burning gases from the booster rocket. At 72.201 seconds the shuttle began to break up. The lower attaching strut between the right solid rocket booster and the external tank broke. The solid rocket booster began twisting counterclockwise. At 73.124 seconds the hydrogen tank exploded, causing disintegration of the intertank and the liquid oxygen tank. The *Challenger* exploded in a huge white fireball (Presidential Commission, 1986).

Millions of schoolchildren all over the country were watching the launch because the first teacher-astronaut, Christa McAuliffe, was aboard. Her mission assignment was to broadcast lessons from space. The children sat shocked in school auditoriums as they watched the pieces of the shuttle arc down into the sea. The lesson they learned was that the shuttle was dangerous and that people could die in an instant.

There were more lessons for adults. Most were shocked to learn that space flight could not be considered "routine." Business and military groups had been told that the shuttle represented a reliable vehicle for putting payloads in space. Shuttle flights were to be the equivalent of commercial airlines, with 24 flights scheduled each year (NASA, 1987). The shuttle would demonstrate America's mastery of space and compete with commercial challengers such as Europe's Ariane rocket.

But some of the most troubling lessons came as the President's Commission (also referred to as the Rogers' Commission, after its chair William Rogers) reviewed engineering documents. They discovered that the leak between sections of the solid rocket booster—the leak that led to the *Challenger* explosion—had been a source of concern for years. As early as 1977 engineers at the Marshall Space Flight Center were highly critical of the design for the solid rocket booster. Charged with supervising the work of independent contractors, the engineers at Marshall were especially worried about the joint being designed by Morton-Thiokol to connect sections of the solid rocket booster. Two Marshall engineers put their objections in writing. Leon Ray called the joint "unacceptable" and recommended redesign. John Q. Miller wrote a memo stating

> We find the Thiokol position regarding design adequacy of the clevis joint to be completely unacceptable for the following reasons: (1) the gap created by excess tang-clevis movement causes the primary O-ring seal to function in a way that violates industry and government O-ring application practices; (2) excessive tang-clevis movement allows the secondary O-ring to become completely disengaged from its sealing surface on the tang (Lewis, 1988).

These findings notwithstanding, Morton-Thiokol received the contract for the solid rocket boosters and used their joint design. But the engineers at Marshall were correct in their estimate of the joint problem, and not just on flight 51-L. Damage to O-rings was found on most flights, beginning with flights in 1984. Just as the Marshall engineers had predicted, the seals needed to keep hot gases from leaking were not seating properly and were being damaged in flight. Additional pressure tests were initiated to help identify the problem, but these tests may have worsened things by damaging a layer of putty placed over the joint and also used to prevent leaks. In all flights now the O-rings were leaking.

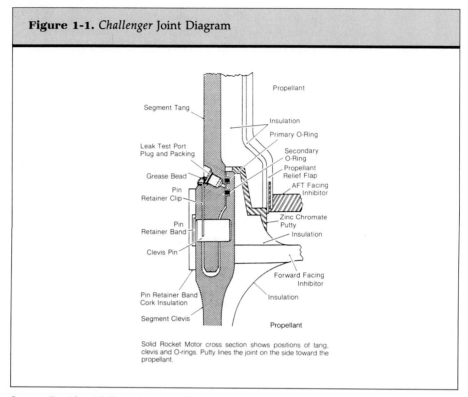

Figure 1-1. *Challenger* Joint Diagram

Propellant

Segment Tang

Insulation

Primary O-Ring

Leak Test Port
Plug and Packing

Secondary
O-Ring

Propellant
Relief Flap

Grease Bead

AFT Facing
Inhibitor

Pin
Retainer Clip

Zinc Chromate
Putty

Pin
Retainer Band

Insulation

Clevis Pin

Forward Facing
Inhibitor

Pin Retainer Band
Cork Insulation

Insulation

Segment Clevis

Propellant

Solid Rocket Motor cross section shows positions of tang,
clevis and O-rings. Putty lines the joint on the side toward the
propellant.

Source: Presidential Commission. (1986). *Report to the President on the space shuttle Challenger accident* (Vol. 1, p. 57). Washington, DC: U.S. Government Printing Office.

Then came the first low-temperature launch. Flight 51-C was launched January 24, 1985, when the temperature was fifty-three degrees Fahrenheit. This was the lowest temperature to date for a shuttle launch, and demonstrated a new danger. Stiffened with cold, the O-rings sealed even worse than on previous flights, and there was "blow-by"—evidence that hot gases had passed the primary O-ring and charred the grease between the two seals. The engineers at Morton-Thiokol now knew that cold was a problem.

For that reason, engineers were especially worried the following January about flight 51-L, the *Challenger*. Temperatures at the launch site had dropped into the low twenties, far below the temperatures of any previous shuttle flight. A group of Morton-Thiokol engineers convened a meeting on January 27 to discuss the problems caused by such unusual cold. They called the Kennedy Space Center at Cape Canaveral to get more information about the weather and to warn their liaison there that they considered the cold a problem for the O-rings.

Engineers at the Kennedy Space Center were so concerned that they set up a conference call for that afternoon between the Marshall Space Flight Center, Kennedy Space Center, and Morton-Thiokol. That conference began at 5:45 P.M., was broken off, reconvened at about 9:00 P.M., was recessed at 10:30, and reconvened in its final session at 11:00. The dialogue of that extended conference was later examined in detail by the President's Commission which examined the *Challenger* disaster.

Much information was presented in the course of the conference calls. Morton-Thiokol engineer Roger Boisjoly transmitted two charts to the other locations (see Fig. 1-2) explaining the problem with the O-rings. In their later testimony, engineers at Morton-Thiokol asserted that they opposed the upcoming launch of the *Challenger*. But Morton-Thiokol's management recommended going ahead with the launch and signed an assessment of the problem stating that "if the primary seal does not seat, the secondary seal will seat" (Presidential Commission, 1986). With this memo in hand, the Kennedy Space Center and Marshall Space Flight Center also recommended launch of flight 51-L. The secondary seal did not seat and the *Challenger* disaster resulted.

The Breakdown in Technical Communication

Who was responsible for the disaster? The Presidential Commission investigated every aspect of the flight. Testimony revealed that many people and systems contributed to the final unfortunate outcome. There were political and economic pressures to launch; there was pressure to launch because of media attention to the teacher in space; there was considerable human fatigue from overwork immediately prior to the flight. But a major focus of the Presidential Commission was the engineers at Morton-Thiokol. In the face of substantial warnings from the engineers, why was the shuttle launched anyway? How did communications break down so badly?

Testimony made it clear that several communication problems were involved.

1. *The engineers were unable to present their case convincingly to their managers.* Most of the teleconference the evening before the flight was devoted to engineers explaining the problems presented by the cold, and the history of problems with the O-rings. Here is part of the testimony to the Presidential Commission by Roger Boisjoly, an engineer at Morton-Thiokol, about what happened during the conference:

> Those of us who opposed the launch continued to speak out, and I am specifically speaking of Mr. Thompson and myself because in my recollection he and I were the only ones who vigorously continued to oppose the launch. And we were attempting to go back and rereview and try to make clear what we were trying to get across, and we couldn't understand why it was going to be reversed.

Figure 1-2. Charts Presented at *Challenger* Meeting

Primary Concerns

Field Joint - Highest Concern

- Erosion penetration of primary seal requires reliable secondary seal for pressure integrity
  - Ignition transient - (0-600 MS)
    - (0-170 MS) high probability of reliable secondary seal
    - (170-330 MS) reduced probability of reliable secondary seal
    - (330-600 MS) high probability of no secondary seal capability
- Steady state - (600 MS − 2 minutes)
  - If erosion penetrates primary O-ring seal - high probability of no secondary seal capability
    - Bench testing showed O-ring not capable of maintaining contact with metal parts gap opening rate to IEOP
    - Bench testing showed capability to maintain O-ring contact during initial phase (0-170 MS) of transient

Joint Primary Concerns   SRM 25

- A temperature lower than current data base results in changing primary O-ring sealing timing function
- SRM 15A-80° ARC black grease between O-rings
  SRM 15B-110° ARC black grease between O-rings
- Lower O-ring squeeze due to lower temp.
- Higher O-ring shore hardness
- Thicker grease viscosity
- Higher O-ring pressure actuation time
- If actuation time increases, threshold of secondary seal pressurization capability is approached
- If threshold is reached then secondary seal may not be capable of being pressurized

The first chart summarizes the Thiokol's primary concerns with the field joint and its O-ring seals on the boosters. The second chart stresses the concern about the effect of temperature on seal actuation time. Both charts were presented by Thiokol's Roger Boisjoly.

Source: Presidential Commission. (1986). *Report to the President on the space shuttle Challenger accident* (Vol. 1, p. 89). Washington, DC: U.S. Government Printing Office.

So we spoke out and tried to explain once again the effects of low temperature. Arnie actually got up from his position which was down the table, and walked up the table and put a quarter pad down in front of the table, in front of the management folks, and tried to sketch out once again what his concern was with the joint, and when he realized he wasn't getting through, he just stopped.

I tried once more with the photos. I grabbed the photos and I went up and discussed the photos once again and tried to make the point that it was my opinion from actual observations that temperature was indeed a discriminator and we should not ignore the physical evidence that we had observed.

And again I brought up the point that SRM-15 [Flight 51-C, January 1985] had 110 degree arc of black grease while SRM-22 [Flight 61-A, October 1985] had a relatively different amount, which was less and wasn't quite as black. I also stopped when it was apparent that I couldn't get anybody to listen" (Presidential Commission, 1986).

Both engineers used visual aids, both attempted to be as persuasive as possible. And both failed to convince management of their position. The engineers had a case to make but were unable to do so.

2. *Communication between organizations was a problem.* Multiple contractors built different parts of the shuttle with the overall project coordinated by a team within NASA. Morton-Thiokol normally reported to a group within the Marshall Space Flight Center, a part of NASA. Communication between the two organizations was based on specific roles: Morton-Thiokol was the producer and the Marshall Space Flight Center/NASA was the consumer. Based on these conventional roles, communication between these two organizations followed a typical pattern: Morton-Thiokol was the insistent seller, urging the project forward; Marshall Space Flight Center was the conservative buyer that wanted to check quality at each stage along the way. Yet on the night the launch decision was made, the roles were reversed, with a significant effect on communication — one that engineers only felt in hindsight. Here is how one of the Morton-Thiokol managers felt about this change in communication pattern:

Mr. Lund: We have dealt with Marshall for a long time and have always been in the position of defending our position to make sure that we were ready to fly, and I guess I didn't realize until after that meeting and after several days that we had absolutely changed our position from what we had been before. But that evening I guess I never had those kinds of things come from the people at Marshall. We had to prove to them that we weren't ready, and so we got ourselves in the thought process that we were trying to find some way to prove to them that it wouldn't work, and we were unable to do that. We couldn't prove absolutely the motor wouldn't work.

Chairman Rogers: In other words, you honestly believed that you had a duty to prove that it would not work?

Mr. Lund: Well, that is kind of the mode we got ourselves into that evening. It seems like we have always been in the opposite mode. I should have detected that, but I did not, but the roles kind of switched . . . (Presidential Commission, 1986).

No one was aware of it at the time, but the rules of communication had been changed. Only days later did participants begin to see how unusual that conversation had been. Only days later did they begin to see why things had gone wrong. They had spent hours on cross-country conference calls making a life-and-death decision without ever understanding the basis of their communication.

3. *Communication between levels within NASA was a problem.* There are always administrative levels within an organization and decisions always have to be made about which concerns to take to senior management. In the case of the shuttle, there were specific rules about what areas were critical to safety and so had to be taken up NASA's chain of command. The O-rings were known to be a problem and were nominally on that list. Yet there was still a sense among many engineers that because there was a second ring, the loss of the primary O-ring was not really critical. Consequently, O-ring problems often were not reported up the levels.

For this flight, Stanley Reinartz, Shuttle Project Manager, decided he would not pass along possible problems with the O-rings to the launch directors, or even describe the nature of the discussion that had occurred the previous evening. No one at the top two levels of NASA knew that the teleconference had even been held.

Would it have made a difference? Here is the pertinent testimony before the Presidential Commission by J. A. (Gene) Thomas, Mission Launch Director:

Mr. Hotz: . . . Mr. Thomas, you are familiar with the testimony that this Commission has taken in the last several days on the relationship of temperature to the seals on the Solid Rocket Booster?

Mr. Thomas: Yes, sir, I have been here all week.

Mr. Hotz: Is this the type of information that you feel you should have as Launch Director to make a launch decision?

Mr. Thomas: If you refer to the fact that the temperature according to the Launch Commit Criteria should have been 53 degrees, as has been testified, rather than 31, yes, I expect that to be in the LCC. That is a controlling document that we use in most cases to make a decision for launch.

Mr. Hotz: But you are not really very happy about not having this information before the launch?

Mr. Thomas: No, sir, I can assure you that if we had had that information, we wouldn't have launched if it hadn't been 53 degrees (Presidential Commission, 1986).

Determining what information to send to whom is a critical decision. Too much information is just clutter. But too little means that decisions are being made with insufficient information or misleading information. Determining when to communicate can be as important as determining what to communicate. The middle managers of NASA chose wrong. The *Challenger* disaster resulted.

Lessons of the *Challenger*

The *Challenger* disaster has been portrayed elsewhere as an example of weak American engineering, of an inability to develop complex systems correctly. It was nothing of the sort. The shuttle flew successfully 24 times prior to flight 51-L and would have flown successfully again. The engineers knew what the problem was and how the problem could be resolved. The *Challenger* was not an engineering disaster; it was a communication disaster.

As we have seen, vital communication was hindered by three problems: the engineers were unable to be persuasive, the communication situation was misunderstood, and conventions prevented the necessary flow of information. The *Challenger* incident shows that good communication—especially for technically trained people—is essential.

The *Challenger* disaster was a spectacular and very public communications disaster. As a consequence, changes were made in NASA, mostly in the area of communications. But technical communication is not limited to rocket engineers. Larger and larger numbers of people are being employed in technical careers, and all of these people have to communicate. Whether it is a matter of product design, medical information, or environmental considerations, decisions are being made and information communicated that affect people's lives.

Technical Communication and Your Career

The engineers working on the *Challenger* project probably never thought of themselves as technical communicators. They were just engineers doing their job. Communication was a part of the job they didn't think about until later. That is the point: technical communication is a crucial part of many jobs, but is almost never thought of. In careers as diverse as environmental studies, engineering, computer programming, medical services, and manufacturing, technical communication is frequent, and sometimes crucial. And it is changing.

The Increasing Amount of Technical Communication

According to one source, from March 1985 to March 1989, 73% of the new jobs created in the United States fell into the three highest paying and most responsible job categories: professional administrative, sales and technical, and precision crafts. This same source predicted that during the 1990s, more than 2 million new administrative, managerial, and technical jobs would be created each year (Naisbitt, 1982). Clearly, workplace communication will become increasingly technical, because more people in the work force will be technically trained.

Technical communication affects the home as well. Consider the gas grill, the staple of suburban cookouts. A recent study showed that the average time needed for consumers to assemble a grill was four hours (*Consumer Reports*, 1991). This means four hours of following exacting instructions to build an appliance that is highly flammable. One false step and disaster may result. Yet we do this sort of thing without ever thinking how odd it is. Imagine if you bought a kitchen stove or refrigerator and then discovered that you had to assemble it before you could use it. This hasn't happened yet, but consumers are given more and more responsibility for the products they buy.

Ordinary people are also being given more responsibility for the air they breathe and the water they drink. According to a recent Supreme Court decision, every municipality can determine for itself what pollution standards it wants. That means that local experts, from county agricultural agents to fertilizer salespeople, must learn for themselves what the dangers are, and then learn how to present that information to common people. If they communicate well, communities will have safe environments. If they don't, civic leaders will either regulate farming out of the community or allow pollutants to injure the population. We will no longer be able to leave such decisions to officials in Washington. The technology of agriculture is now a political issue that nonspecialists must understand and discuss.

In short, we live in a more technical world. Whether we are at our place of employment, making a routine purchase at the mall, or filling a glass of water at the kitchen sink, we are surrounded by technical issues and challenges. Our jobs, our leisure, and our personal health all depend on technology. To communicate about the world is to communicate about technology.

The Growing Dissatisfaction with Current Technical Communication

As technical communication becomes more commonplace for larger numbers of people, there is increased interest in the quality of that communication. We understand that we are making life and death discussions based on technical information. But it also becomes a quality of life issue. How much of what we read can we understand? How much of our time is given to decoding badly written reports?

Consider the VCR. We may laugh about the number of people who can't program their VCRs, but consider what a huge portion of our population owns what is essentially a highly evolved electronic device. They are doing their best to find the on/off switch. Whose fault is it that they often cannot use the other capabilities of their machines? As technology moves out of the workplace and into the home, information explaining that technology must be adjusted to a new audience.

Technical communication is being held to higher standards these days.

Consumer Reports, for instance now evaluates and rates instructions as part of product evaluations. It is not well disposed toward products that include strange instructions. A recent article on gas grills explicitly described one reviewer's experience: "As you dig through the carton, you begin to turn up cryptic instructions: 'Position shelf boards to handles and align predrilled holes, using #10×1 ¼ THMS, hole closest to edges of board goes on rear handle'" (*Consumer Reports,* 1991). The article names particular brands that are especially difficult to assemble. One grill received this scathing comment: "Good points: I didn't spend my own money to buy this grill." This isn't the sort of thing most writers like to hear about their writing.

This objection to the current quality of technical communication is all too frequent. The pressure is on to do better. Whether it is a consumer appliance, a medical report, an insurance form, or an environmental impact statement, we expect technical information to be presented clearly. And readers are more ready than ever before to complain if the communication is not clear. Anyone going into a technical profession will be held to a higher communication standard than existed in the past.

The Growing Complexity of the Technical Communication Process

There was a time when people presented technical information in a rather rudimentary fashion. Reports were almost totally text—words—with few illustrations. This was because creating illustrations was expensive, as was printing them. Even the text was all the same size, with only an occasional use of underlining or capitalization to highlight particular words or phrases. That was the technology of the time. And while it didn't create inspiring documents, at least it was simple.

The publishing technology of today is far different. It gives the writer far more power—and far more responsibility. With a simple desktop publishing program a writer can set many sizes of type and several type styles, insert numerous illustrations, and print a document with typeset quality. The power to produce beautiful and clear documents is now in the writer's hands. And so is the total responsibility. In the past, there were others around who would do much of our work. Why worry about spelling?—the secretary will catch it. Why worry about appearance?—the folks in graphics will put it right. Most of those folks are gone now. In their place today's technical writer has a desktop computer with a good word processor, illustration software, desktop publishing package, and presentation graphics. The writer not only writes the report but makes sure it is spelled correctly, laid out effectively, placed in the right type size and font, illustrated adequately, and printed on a laser printer. The media team consists of one person: you.

Technical communication will also be occurring at a greater distance. The globalization of the economy has already had dramatic impact on communication. In 1977 Americans made 580 million minutes of overseas calls. By 1987 the number was 4.7 billion minutes—an eightfold increase (Naisbitt, 1982). Increased optical fiber capacity will enable even faster growth in this area in coming years. At the same time, plane travel is accelerating. It is estimated that every day, 3 million people travel by plane (Naisbitt, 1982). The people we will be working with will no longer be down the hall; they may be anywhere in the world.

All of this creates a challenge for anyone entering a technical field. So much of the world's economy is technical, and so many more people are involved in technical fields, that increased communication of technical information is certain. At the same time, that communication is being held to an ever higher standard. The tools of technical communication are changing, giving writers more power and more direct responsibility for the quality of their communication. The skills of technical communication will occupy a growing place in the education of any person anticipating a technical career.

An Overview of This Book

Communicating Technical Information was built around four basic ideas.

1. *It is important to understand why technical communication takes the form it does.* The best communicating will be done by people who have thought about what they are doing, whom they are addressing, and how the best results can be achieved.

The first two sections of the book cover these basic principles. Part I, "Basic Principles and Concepts," examines the most important ideas behind good technical writing, including writing as a process, the importance of audience and purpose, and technical writing as a distinct category of communication with conventions and concerns all its own. Part II, "Visual Presentation," touches on topics once considered peripheral to a discussion of writing but that are now understood as central to communication. Prior to desktop publishing, most of the information in this section could have been left to artists; now it is the responsibility of all writers.

2. *The general principles of technical writing become most useful when they are examined in specific situations.* Knowing the general rules is fine, but it is also important to discover how these rules are followed or broken in real-life writing situations.

Many of the topics and discussions in the first two parts of the book reflect this concern with real-life situations. The chapter "Group Writing," for example, addresses a basic fact about technical writing today: much of it is done in collaboration with other people. Part III, "Technical Writing Techniques," concentrates on the specific writing techniques that are most useful in everyday technical writing. Part IV, "Technical Research," covers the research process as it is commonly applied in technical writing. Part V, "Technical Writing Tasks," focuses on specific documents at the core of technical writing in the workplace today: literature reviews, research reports, proposals, feasibility studies, progress reports, instructions, computer manuals, and electronic documents.

3. *Technical writing's new higher standards and broader scope are very real.* In a world where so much of our personal and professional lives is dependent on technology, and where there is so much competition among technologies and technology producers, every piece of technical writing is expected to be informative, accurate, and well-suited to its readers. At the same time, more and more communication is "technical": not only must writers do a better job of communicating technical information, there is more technical information to communicate.

One of the best ways to assure quality documents is to review, test, and revise them thoroughly. This is the subject of Chapter 5, "Quality Reviews," and is a topic discussed consistently throughout the second half of the book. A wide variety of writing situations and documents are examined in this book, some of which would not even have been considered germane to technical writing a few years ago. Similarly, the shift from "writing" to the broader idea of "communication" means that today's technical writer not only must create words for a document but must also determine how those words should look on the page and what other visual elements should appear in the document; these issues are examined in Part II, "Visual Presentation."

4. *The computer is now an integral part of all areas of communication.* It is both an unavoidable part of the communication process and an important subject of communication itself. Those who use computers well will have much improved documents. Those who are unclear of the role of computers will be at a significant disadvantage.

Chapter 7, "Writing with a Computer," gives a basic overview of the electronic options available to today's technical writer and then suggests some strategies for taking advantage of this powerful tool. Two of the chapters in Part V are devoted to the new documents of the electronic age. Throughout the book, computer tools and strategies are discussed wherever they are relevant, whether for document design, graphics, research, or analysis.

This chapter began with a description of what can happen when technical communication breaks down. It then discussed three of the major forces affecting technical communication: the growth in volume, the higher standards, and the latest tools. All of these are challenges for technical fields.

It may seem impossible to communicate technical information well. Without question it is difficult, but it is not impossible. In time you may have the pleasure of seeing someone using an instruction manual of yours and following it well, or of writing a feasibility report that is accepted on the first try, or of persuasively presenting a technical position and having it understood by others. None of this will come easily, but it can be done, and you can start preparing for it with this course.

Activities

1. What techniques did the Morton-Thiokol engineers rely on in their presentations? Why did those techniques not appear to work? How could the engineers have presented their case more effectively? List five suggestions, and compare your list with the lists of your classmates.

2. This chapter identified three communication problems that led to the *Challenger* disaster. As a group, discuss what recommendations you would make in each of these areas.

   a. What techniques could have been used to improve the teleconference?

   b. What techniques would enable organizations like NASA and Morton-Thiokol to communicate better?

   c. What procedures could be established to improve supervision by upper management at NASA?

3. To see how much technical information is now available, select one area of technology, such as pollution control, computer graphics, or satellite communications, and visit your library reference room. Compare the number of articles published on the subject annually twenty years ago, ten years ago, and today. For some subjects the numbers are so large that you may want to count the number of journals in the field rather than the number of articles.

4. Photocopy the first page of any instruction manual for a consumer product. In a study group, estimate the reaction of consumers to the manual. As a class, identify the best manual and the worst. List the features that make them the best or worst.

5. Bring to class one good example of technical writing (computer manuals, technical reports, laboratory instructions, field guides, and the like) you have used as a student. Explain what features made the document easy to use.

6. Review any position descriptions for positions you might apply for after graduation. How much communication is mentioned? What kinds of communication are most common? Do a survey of recent graduates in your major. What kind of communication do they do? How many hours each day are devoted to technical communication? How important is communication for promotion? Write a summary of your findings.

The Process of Technical Writing

Evaluating the Writing Situation
What Is the Purpose of This Document?
Who Is the Audience for This Document?
What Is the Assignment?
How Does This Communication Fit into the Larger Context?
Will This Be Written in a Group?
What Are the Available Resources?
What Are the Time Constraints?
Case Study

Conducting Research
What Do I Need to Find Out?
How Can I Find This Out?
What Does It Mean?
What Is the Best Way to Present It?
Case Study

Writing the Document
What Would Be the Best Writing Style?
What Computer Writing Techniques Can I Use?
What Kind of Graphics Should This Document Have?
What Would Be the Best Design?
Case Study

Reviewing and Revising
What Do I Need to Find Out?
How Can I Determine This?
Case Study

Publishing
What Would Be Most Suitable?
What Are the Available Options?
Case Study

Activities

Susan is an intern working in the administrative computing area of a small liberal arts college. The college has recently started a major renovation of its computer communications system. As part of its overall plan, the college has introduced a variety of interactive programs to be run over an Information Systems Network (ISN). Because Susan indicated on her résumé that she had taken a course in technical writing, her immediate supervisor asked her to prepare a manual describing the new ISN interactive programs.

If you had been given this typical assignment, how might you have begun? Where might you have begun? What would you have done? What would have been your choices? Or, to put the question more broadly, how would you have begun to solve the problems this assignment poses? Many technical writers claim that technical writing is, in fact, problem-solving. Because Susan is like many writers, she hadn't thought about *how* to proceed; for example, she began by immediately drafting a response. When it was clear that she had no immediate response, and because somewhere she remembered a teacher telling her not to write before she had an outline, she decided to make a rough outline for the manual. However, because she had too little information, she couldn't make an outline. Finally she looked for someone who knew something about ISN interactive programs. Like many novice writers, she stopped and started in her search for a way to proceed with her writing task.

People who study writing practices in a wide variety of technical communities know that successful writers of technical information are systematic in their approach to their writing tasks. Successful technical writers know where to begin, and they do not attempt to deal with all writing problems at once: they treat appropriate concerns at appropriate stages of the writing process. For example, they define their audience before determining which details they will need, and they identify any constraints before attempting to set a schedule.

The technical writing process can be thought of as a problem-solving task, a series of questions to be considered and answered. Technical writers might not all use the same approach, and they might not use the same approach with each of their writing tasks, but they almost always use some variation of the process described in this chapter.

Evaluating the Writing Situation ─────────────────

Before writers can turn out a good document they must first consider what the document is about. This means evaluating certain standard features of the writing situation: the purpose, the audience, the type of document. It also means considering the specific conditions under which *this* document is being written: the larger organizational context, whether it is being written by a group, time and budget constraints, and so on.

What Is the Purpose of This Document?

If the purpose of a document is not clear, the choices a writer makes will also not be clear. Because audience and purpose typically determine both the content and the presentation of a technical document, it is important to determine its purpose as soon as possible.

Generally there are three purposes in technical writing: writing to inform, writing to help a reader make a decision, and writing to help a reader learn how to do something. For example, you might be asked to report to managers the results of an experiment, or to submit a proposal, or to give instructions. It is important to determine the purpose of a document before you begin writing; for example, if you write only an informative report, but your audience is expected to base a decision on the report, you have lost an opportunity to be persuasive. (See chapter 3, "Purpose and Audience in Technical Writing," for further discussion.)

Who Is the Audience for This Document?

Within the past ten years more has been written about audience than about any other aspect of writing. The reason for this is clear: even if all your sentences are clear, all your paragraphs unified and well-organized, your language precise, and your words spelled correctly, determining and evaluating the audience for a specific piece of writing is a *crucial* step of the writing process. Your document is a failure if it doesn't communicate effectively with your specific audience. Many features of the final document, such as format, amount of detail, and tone, will be determined by the audience for the document.

While it is difficult to generalize about the audiences you might have to write for, there are essentially five audiences for most technical communication. (See chapter 3, "Purpose and Audience in Technical Writing," for further discussion.)

1. *Experts*. Experts are highly knowledgeable about their subject area and its theory. These readers seldom need details such as background information or definitions.

2. *Technicians*. Technicians are those who fix and operate the equipment. Because they have a strong technical understanding of the subject, a document written for this group will include technical language, details, and step-by-step instructions.

3. *Executives*. Executives, who are often managers, read primarily to make decisions based on what they read. They may not need theory; they always need to understand the potential effects of their decisions.

4. *Novices*. Novices know little about the subject you are writing about. They often read not because they have to, but because they are interested in

the subject. On the other hand, they may need to learn new information quickly about a completely unfamiliar subject. They benefit from relatively nontechnical language, little or no theory, and interesting writing.

5. *Mixed audience*. These are people with varying degrees of knowledge about the subject you are writing about. Because each group reads the document for different purposes, documents written for mixed audiences must be carefully planned so that the needs of each subgroup, from novice to technician, expert to executive, are met.

What Is the Assignment?

People communicating technical information rarely write documents just because they want to; more often than not, the writing process begins when someone asks the writer to produce something. But exactly what sort of document is the writer being asked to write? A feasibility study? A report? A proposal? While this is usually absolutely apparent — and thus not much of a problem — it's still a question whose answer will have a greater influence on the writing process. It is important because each writing task has its own set of techniques, conventions, and strategies.

Occasionally assignments merely ask the writer to accomplish a certain purpose, with the choice of document type left up to the writer: "We might need to upgrade our computers — come up with something that will help us get a fix on the problem" could result in a research report, a feasibility study, or even a proposal. However, except for brief, informal, and internal documents, this is rarely the case. Each important technical writing task has a conventional type of document that has been developed to accomplish this task, so the type of document is usually predetermined by the assignment itself. If the type of document you are being asked to prepare isn't absolutely clear, ask.

How Does This Communication Fit into the Larger Context?

According to a comprehensive survey of on-the-job technical communication conducted by Laura Casari and Joyce Povlacs (1988), two researchers from the University of Nebraska–Lincoln, "what professionals write depends upon the role the writer is expected to play or upon the role the writer holds in the structure of the firm or agency." According to these researchers, successful writers "knew when they were writing up or down the hierarchy."

Awareness of one's role does not mean that writers should be overbearing to subordinates and subservient to their bosses. It means being sensitive to the particular demands of the social context that exists in every piece of writing. This is especially important when you are writing for external readers, because here you are functioning as a representative of the company.

Is the document one of a series of documents that have already been prepared or that are about to be prepared? If it is, you will want to know what came before and possibly what is still to come. If you are explaining company policy, interpreting laboratory analysis, describing a policy, or giving instructions, you will want to know the format others used in presenting their information. For example, there may be a company format that you should be aware of.

There are many other contextual concerns you will need to be aware of: Does the company or organization have a style guide? Is there a particular image that the company wants to project? What is the writer's position or role in the company?

Do not be shy about asking for information, advice, or clarification. Managers and supervisors claim that employees ask too few questions rather than too many.

Will This Be Written in a Group?

At some point in your career, it is likely that you will be working with other people in the preparation of a document. Either everyone will have a hand in the entire document or each person will be responsible for different portions of the document. You may even be able to include in your document something that was previously written by another person, or your document may later be used by someone else. In any case, the document will not be complete until each person has contributed his or her section. Writing in a group has a great influence on the writing process, so it is important to know the extent of collaboration before you get too far into the project. (See chapter 6, "Group Writing," for further discussion.)

What Are the Available Resources?

What, other than your own mind, can help you write this document? First, figure out what tools are available to you. Will you have a personal computer with a graphics package and a laser printer, or only a typewriter and a ruler for sketching out some graphs? Are on-line library searches feasible, or will the reference librarian be your main source of information?

Budgets can greatly affect the writing process. For example, if money is available, you might be able to hire research assistants, make on-site visits, plan focus groups, and implement extensive user testing. If not, many of these options will be out of reach.

The availability of information is another concern that shouldn't be overlooked at this stage. If you are asked to write about something you know well, you will not have to worry about getting information — you, after all, are the source. However, it is more likely that you will have to go to others to get information. A good rule to follow is, never assume that information will be readily available. If the information is readily available, you might have some time to spare; if the information is not readily available, you might find yourself wishing you had a few more hours, days, or even weeks.

What Are the Time Constraints?

Of the various constraints on every writing task, the most common constraint is a deadline. You have likely had deadlines imposed upon you from the time you can remember: parents asking you to be in by a certain time, friends needing to know by Wednesday if you are going to the movies on Friday, teachers wanting homework in *on time*. Though parents, friends, and even teachers, were often lenient, especially if you had a good reason, employers are usually not lenient: a bid submitted a day late could jeopardize an entire contract; a late memo to engineering requesting information on a project could hold up production for several days and cost considerable time and money.

Time constraints will determine how much time can be devoted to each step in the writing process, which in turn will determine what sorts of tasks can be performed for each task. For example, you may have enough time to conduct interviews but not to send out questionnaires. Many experienced technical writers report that failing to find out whether there was a deadline for their document caused sleepless nights and rushed jobs.

Part of the solution to the problem of time constraints is to set a schedule. As part of a systematic writing process, a schedule can make your life easier and less hectic. A schedule can help you organize your tasks into manageable segments. If your writing project is very short (a memo to maintenance recommending a new procedure, an interpretation of a laboratory analysis, a request for information), you might not need a schedule. However, for much technical writing, such as proposals, feasibility studies, manuals, you will likely need a schedule.

Case Study

Susan thought about the assignment to prepare a manual describing the new ISN interactive programs. She realized that she could both *inform* (describe new services) and *help a reader learn how to do something* (offer instructions). Susan wasn't sure whether she was to write a purely informative document or one that also gave readers instructions about how to use the various programs; the assignment seemed vague. So she asked her boss, who told her the finished text should do both.

At this point Susan understood that she would basically be writing an instruction manual: although the primary purpose of instructions is to tell the reader how to perform a task, they often include a good amount of background information.

Because it was difficult for Susan to know exactly who her audience was going to be, but because she knew her readers would not be experts, she decided to aim for a novice audience: she reasoned that while most readers, as members of the university community, would have had some familiarity with computers, they would not have much familiarity with the specific programs she was describing.

Susan was the only person working on the project, so group writing concerns were not an issue. But she didn't know how much time and money were available. Her supervisor told her that there wasn't any specific deadline but that she shouldn't take too much time—as soon as the manual was ready, people could begin using the system. Although considerable money had been invested in the new computer system, there was little left over to spend on extensive research and testing for the manual or on a glitzy design; even a demonstration program or an on-line tutorial was out of the question. Susan knew that all she really needed to write a simple instruction manual was time, a word processor, access to the ISN itself, and a few volunteers to test out her first draft—all of which she had available.

At this point Susan looked at the other instruction manuals that had been written at the college. She noticed a few consistencies of format and style and decided to use them in her own manual, since they would make her document seem somewhat familiar and therefore more accessible to readers.

Conducting Research

After determining the purpose and audience for a document, the next step is to collect, analyze, and organize the information you will need to write the document. Again, taking a systematic approach to the task saves time and energy.

What Do I Need to Find Out?

The first step is to figure out what it is you need to know in order to write the document. Sometimes the writer immediately knows exactly what information is needed: when it's time to write the sixteenth progress report in a series, it's pretty clear what pieces of information are needed. Sometimes it is a matter of identifying the different types of data needed to make a strong argument in a proposal. Occasionally it is less clearcut. If part of your assignment is to describe the social interaction of chimpanzees, you might have to determine what exactly is meant by "social interaction."

How Can I Find This Out?

Basically, there are three sources for information: yourself, the library, and the field. (See chapter 16, "Collecting Information," for further discussion.)

YOURSELF

Don't underestimate how much you know. Although you may not know a lot about the subject, it is unlikely that you know absolutely nothing. Therefore, find out how much you know before looking elsewhere.

While there are no guarantees that you will come up with anything, there is a good chance that thinking about (and possibly writing about) a system, a program, a method, a service, a problem, a procedure, or a policy for just ten minutes will dredge up quite a bit that you had forgotten you knew.

THE LIBRARY

Secondary sources record the information gathered by other people. The single most important resource for this sort of information is the library. Public libraries (city, state, and regional), academic libraries (college and university), and professional libraries (business, industrial, hospital) contain an astounding amount of information.

The electronic age has brought many new forms of storing and accessing information—so many that the idea of a library as a building with books in it is quite misleading. At some time you will probably have occasion to use computerized indexes, on-line catalogs, on-line databases, databases on compact disks, or electronic bulletin boards, all of which expand the possibilities for researching secondary sources.

THE FIELD

Although for some writing tasks you need only figure out what you already know, and then use the library to collect information from secondary sources, very often you will also need to use primary sources, that is, first-hand data that you collect yourself. Typically, collecting first-hand data for technical writing means finding out what other people think, going into the field to observe what is actually going on, or conducting a controlled experiment.

What Does It Mean?

In order to form a conclusion about the information you have gathered, you will have to analyze it. Sometimes this entails only performing certain calculations and coming up with a final figure. Often, however, the process of analysis is far more subtle and complex. Several tools, such as standard questioning techniques and computer programs that can easily manipulate data, are available to help writers develop reasonable conclusions. It is important at this stage to re-evaluate both the information-collecting process and the analysis process to make sure they are reliable: faulty procedures or errors in the data can invalidate even the most carefully constructed conclusions. (See chapter 17, "Analyzing Information," for further discussion.)

What Is the Best Way to Present It?

Once you have gathered and analyzed the information you need, the next step is to select and arrange the details in a way that makes the reader understand—and in some cases agree with—the information you are presenting.

Whether you are writing a proposal, a memo, or a manual, you will have to include details. However, some details or facts are not necessary in the final document; the wise writer chooses the important facts from among the many that were gathered during research.

There are as many ways to organize and arrange information within a document as there are documents. In general, though, the organization may be either content-driven (determined mainly by the information to be presented) or format-driven (determined mainly by a conventional format). When communicating technical information, part of your job is to discover what would be most effective—which often means discovering what is expected—and then arranging your information according to this standard. (See chapter 18, "Organizing Information," for further discussion.)

Case Study

Because Susan was already an expert user of the interactive programs she was asked to write about, she decided to rely on herself as the principal source of information. She realized that observing other people performing a task is usually an important step in writing instructions. But given her time and budget constraints, she decided to rely on her own knowledge for the writing stage of the project and then bring in other people for user testing in the revision stage.

Because Susan anticipated that most readers of the ISN manual would be novices who knew little about ISN interactive programs, she considered organizing her manual in the order she felt readers would be using the manual. She decided to begin with a section introducing the subject and giving the user general information, such as an overview of the document and what needs to be done before running the programs. This section she titled "General User Information."

Susan wanted the next section to follow the process of a typical user. Therefore, since the programs only appeared on a computer screen and in a particular sequence after the user logged on, the next section, "ISN Program Listing," would list and describe the programs available.

Next, Susan thought about listing in a separate section the programs available, since these would be what the user saw next as a menu pick. Readers had to know what programs were available and how to use and access them. Because the instructions for using the programs would not be displayed on the computer, Susan had two choices in organizing the information: she could list the programs separately from instructions on how to use them, or she could list each program along with instructions for using that particular program. How to choose? Because one principle of writing is brevity, Susan asked herself which method would take up the least amount of space and thus make it easier for the reader to follow.

If the instructions for all the programs were similar, the answer would have been easy: Susan would have listed all the programs and then offered the general directions with one or two examples. That structure would use the least amount of paper. However, in this case Susan needed to describe the operating procedure for each one of the programs available. Therefore, because she knew that not everyone would use all the programs, a reasonable choice was to first describe the programs available, and then describe the procedures.

Writing the Document

You've evaluated the writing situation. You've conducted research. Now it's time to write the document. Here are four important questions to ask as you prepare your document.

What Would Be the Best Writing Style?

Style is the way in which we say or write something. A concern for style means helping readers feel involved in what they are reading, choosing the right level of detail, selecting words carefully, writing clear sentences, and creating documents that are easy to navigate. Keep your readers in mind when you choose a writing style: the same no-nonsense paragraph that would be appropriate for a group of technicians might alienate a group of novices, even if all the necessary information were included. (See chapter 4, "Technical Style," for further discussion.)

What Computer Writing Techniques Can I Use?

Computers have transformed the writing process. Any basic word processing program offers the writer a number of convenient shortcuts that would be unthinkable with paper and pencil. These shortcuts not only allow writers to produce documents more quickly, but often create *better* documents because they rely on the computer's capacity to store information accurately. Boiler-plate, macros, templates, search-and-replace routines, and the structured document processor are among the most common of these computer writing shortcuts. Whenever you can, take advantage of them. (See chapter 7, "Writing with a Computer," for further discussion.)

What Kind of Graphics Should This Document Have?

Graphics are an important element of technical writing, and all technical writers should consider the use of graphics as part of their standard arsenal. When preparing a document, consider whether or not a graphic would help the reader more easily understand the material. Tables, drawings, photographs, and graphs are excellent communication tools, but writers have to carefully consider them in the context of the audience, purpose, and nature of their document. (See chapter 8, "Graphics," for further discussion.)

What Would Be the Best Design?

Because technical documents communicate visually as well as verbally, a number of visual considerations are part of document design. Some of the more important aspects of appearance are type size, font and style, page design, lists and bullets, lines and boxes, icons, text blocking, graphic balance, section heads and labels, glosses, and color. Successful technical writers are familiar with the large number of design options available to them and know how to choose among them. (See chapter 9, "Document Design," for further discussion.)

Case Study

If you were preparing the ISN manual, how detailed would you make the instructions? For example, in the section "Logging On to the Burroughs," Susan wasn't sure whether she had to tell her readers that they had to log on to the university's Burroughs computer before running any of the programs. Initially, Susan assumed that the readers might already know that. However, Susan also knew that her assumption might be erroneous, that there might be readers who did not know that; therefore, because she knew it is generally better to have too much detail rather than too little, she added the following sentence:

> In order to run DARGAL programs over the ISN you must first log on to the Burroughs.

The question of how much information to give is difficult to answer. For example, if there is a reader who does not know *how* to log on to the Burroughs, Susan has not given enough information. All she could do at this point, though, was make her best guess based on her knowledge of the audience.

Because Susan's instruction manual was being written from scratch, there were no computer files she could pick up and use in her document. She did decide, though, that once she'd written the instructions for one program, she could use the same text as the basis for other instructions. Therefore, she saved the first set of instructions she wrote in two separate files so that she could use one of them to write the second set of instructions.

Given the budget constraints and the relatively simple nature of her document, Susan decided that she didn't really need to use any graphics in her document. Document design was another matter—every writer who produces the finished version of a document has to consider design. Susan knew all about the standard conventions for instructions: the headings should be clear and visible, each step should start on its own line, the steps should be numbered, and so on. Looking at some of the other computer instruction manuals that the college produced, she got some other ideas for a good format and design for her document.

Here are two excerpts from Susan's descriptions of the ISN interactive programs available at her college. The first excerpt is a description of one of the ISN programs. The second excerpt is the corresponding description of the program commands used to run the ISN program.

ISN Program Listing

ISN/STUDENT/COURSES

This program displays a student's entire course history, broken down by semester. For each course the student has taken there will be a letter grade and the number of credits received. At the end of the display you will see the student's current cumulative grade point average, total credits earned, and classification.

Program Commands

ISN/STUDENT/COURSES

1. Type RUN ISN/STUDENT/COURSES. Press [return].
2. When prompted to enter the student's number, a total of nine digits must be entered. If the student's number is less than nine digits, precede it with the correct number of zeros.
3. When prompted to enter the student's last semester enrolled, the date must be entered in the form, 8520, which represents the school year and the semester.
   EXAMPLE:
   8520 = Second semester of the 1985–86 school year
   8510 = First semester of the 1985–86 school year
   8430 = Summer session of the 1984–85 school year
   8310 = First semester of the 1983–84 school year

Although the description of each program could have been written in one sentence, simply listing all the kinds of information available on the display, Susan chose to write it in three sentences to make it easier for her readers to follow the description.

In the description of the actual commands, Susan decided to oversimplify rather than undersimplify. For example, even though most readers would likely know what to do after the initial command, Susan did not want to take any chances on a novice reader typing in the command

RUN ISN/STUDENT/COURSES

and then waiting (perhaps forever) for the computer to do something. Therefore, she included the instruction

Press [return].

to prevent even the slightest possibility that some readers might not know what to do.

Although it is true that good examples can save the reader (and the writer) a lot of time, Susan decided not to give an example illustrating the need for the student's number to be nine digits. She was rightfully concerned with the reader's ego and decided that that was so obvious that the reader might feel insulted. However, Susan did choose to offer multiple examples to illustrate the procedure to follow when entering the student's last semester enrolled. The choice was either to write out in great detail what each digit represented, or simply to give enough examples to represent each possible situation.

Reviewing and Revising ────────────────

As discussed in chapter 5, the quality of a technical document is a matter of how well the writer has anticipated and met the needs of the reader. The only true test of a document is whether it accomplishes its purpose with its actual readers. One way to evaluate a document before it is published is by reviewing and testing it. This will show you which areas you need to revise.

There are other issues at stake here, as well. Almost everything that technical communicators write is public. A public document has a life of its own: when a document leaves your hand, you no longer have any control over who reads it. However, it still has your name attached to it, so it can affect your career and reputation. Because every document you write reflects your capabilities and judgment, you will want to review and revise as thoroughly as possible. (See chapter 5, "Quality Reviews," for further discussion.)

What Do I Need to Find Out?

There are many questions that can be asked during the review stage: Is it correct? Is it accurate? Is it well organized? Does it conform to the expected standards? Finally, the most important question that can be asked is, is it effective?

How Can I Determine This?

WRITER-BASED REVIEWS

Because writers are responsible for their documents' quality, much of the reviewing is done by writers themselves. This is also a practical matter, since it is easier to arrange for yourself to read something over than to arrange for someone else to. Finally, it is a matter of courtesy: unless you are able to hire someone as a proofreader or fact checker, you should not expect other people to catch and correct all your mechanical errors.

COLLEAGUE-BASED REVIEWS

Because few writers catch all of their own mistakes, good writers look for a second opinion. Often the best people for this purpose are the writer's colleagues, since they usually are somewhat familiar with the project or document.

USER TESTING

An ideal way of checking your writing is to give the document to members of your audience (or to people who are similar to your audience), ask them to read it, and see whether or not it works. Finding out whether or not the document is effective can be the most difficult part of technical writing. It is relatively simple to tell whether instructions have been effective: all you need to do is judge how well readers were able to perform the task. It is much harder to gauge the effectiveness of a proposal or feasibility study through user testing. There are certain tools and strategies for this, however, and while you can never be absolutely certain that a document that tests well will perform well in the real world, almost any information you gather at this stage is useful. User testing is not always possible, but it should be done whenever possible.

Case Study

Susan was aware that her ISN manual would take on a life of its own as soon as she published it. It would be read by not only those who intended to use it, but also by Susan's boss and anyone else interested in the subject or in her career. Therefore, because the document was a reflection of Susan, she wanted it to be of high quality. To assure its quality, Susan decided to use all three methods of testing her document.

First she looked over the document herself, and, except for some surface-level errors (primarily spelling and punctuation), she pronounced it fit for reader consumption.

Then she asked a colleague, another employee at the library, to look it over. Her colleague, who did not have much ego invested in the document, pointed out one or two ambiguous statements and directions.

Finally, Susan asked potential users to review the manual. However, in addition to simply reading the manual for errors, the users tried out Susan's instructions for running interactive programs over the ISN. Although there were only minor problems with the directions themselves, all of the users reported that of the eight programs available, they could run only six. This was surprising, since Susan had had no problem using the programs in the course of preparing the manual. When she checked with administrative computing, they reminded her that the two inoperable programs contained classified information available only to authorized personnel, and none of the people testing the manual had the appropriate authorization. Therefore, Susan indicated in the manual that because those programs contained classified information, the user had first to obtain permission in order to run the programs.

Publishing

The last stage in the writing process is to reproduce and distribute the document. It is your responsibility to make sure the document is legible and complete. There is no excuse for an illegible document. If your report is too light to be read easily, you cannot blame it on a poor printer ribbon. Similarly, there is no excuse for an incomplete document; if a page is lost because you failed to bind the pages adequately, you cannot blame the copy center or a secretary. Decisions on how to reproduce your document and who gets a copy of it are usually based on the audience, the purpose of the document, and budget limitations.

What Would Be Most Suitable?

The decision of whether to mimeograph your document, to run it off on stock paper, or to have it printed on letterhead stationery generally depends on who is to receive it and why it is being sent. The best guidelines to help you with this decision are your own experience and common sense. If you are submitting a formal report to stockholders, you will likely ask for high-quality inking and very good paper. If you are sending a letter to customers announcing a change in procedure, you will likely have the master typed using a laser printer and printed on company letterhead.

How the document will be used will also affect your choice. If you are preparing a document that will require frequent updating, such as a policy manual or a set of instructions, consider a binding that allows easy replacement of pages. If you have prepared a computer manual that users will be following, consider a binding that will allow the manual to lie flat once opened.

What Are the Available Options?

Budget restrictions are usually related to the audience and the number of copies to be distributed. You will have to be sensitive to both of these constraints. If you are writing a one-page letter to the chairman of the board recommending a change in policy, you will have no problem sending it on the best bond paper you have available. However, if you are sending a one-page memo to 5,000 employees announcing a change in policy, it is more economically advisable to find a less expensive paper. Budget concerns are, finally, a matter of common sense and an awareness of the particular corporate environment you are in.

Case Study

Because the manual describing the ISN interactive programs was lengthy, and Susan's budget was limited, and because she had to prepare a large number of copies, she chose to use one of the less expensive papers her university duplicating department offered. Also, because she knew from experience that users of the manual would refer often to the manual while they were performing computer tasks, and because she anticipated that after some use she would have to revise the manual, she decided to use a looseleaf binder that lay flat. That way, users could read from the manual while keyboarding, and revisions could be inserted easily.

By the time she finished the manual, Susan was satisfied that she had done everything she could to provide a first-rate piece of technical writing. After having evaluated the writing situation, Susan knew who her audience was, what her assignment was, and what she had to do to complete the assignment. She then gathered the information necessary to write the manual, determined the best way to present the information, and wrote the manual. Finally, after getting a significant amount of help in reviewing her work, Susan revised and published the manual.

Activities

1. As you examine the steps Susan went through to prepare the ISN manual, you will see that the process is characterized by a number of choices, from how to respond to the assignment to the kind of paper used in publishing the manual. Describe the kinds of choices Susan made as she moved from her immediate response to the assignment to her final act of publishing the document.

2. In a study group, prepare a short report on a technical article of the group's choice. Each person in the group should address one of the following areas, explaining the reasons for the answers given:
   a. What was the writing situation? What is the purpose and audience for this article? Do you think there were any serious time or budget constraints? Was it written by a group?
   b. What kinds of research has the writer done? Is the information well analyzed and well presented?
   c. How would you describe the writing style? What computer software do you think the writer might have been able to use? What design decisions did the writer make?
   d. What review procedures might be used for an article of this kind? Do you think they were used here? Should more reviewing have been done?

e. Where was the article published? (a journal? a magazine? a book? a reference work?) Is the publication appropriate for the journal?

3. As you complete your next writing assignments, keep a writing log, and record the specific steps of the process, from your initial evaluation of the writing situation to the final decision on how to present the document. Compare your process with the one described in this chapter. How similar is it? How different? Is your writing process the same for each piece of writing? If not, describe the similarities and the differences.

4. All writing tasks involve constraints. For each of your writing tasks, both in and out of class, keep a record of the specific constraints (such as time, budget, resources). After four or five pieces of writing, prepare a brief memo to your instructor describing the kinds of constraints you experienced and how those constraints influenced the writing choices you made.

5. Interview three people, such as classmates, friends, or teachers, about their personal composing habits. What process do they use when they write? Are they comfortable with the process? Do they feel it is successful?

Prepare an oral report comparing and contrasting these different methods and offering explanations for why some approaches might work better for some people than others.

6. Using the description of the writing process outlined in this chapter and your own experience, prepare a checklist for the writing process. You might begin with questions such as, "What is the purpose of this document?" "Who is the audience for this document?"

3

Purpose and Audience in Technical Writing

What Is My Purpose?
  Explicit Purposes
  Implicit Purposes

Who Is My Audience?
  Experts
  Technicians
  Executives and Managers
  Novices
  The Mixed Audience

What Are My Audience's Needs and Wants?
  What Is Their Purpose in Reading?
  How Much Do They Know?
  How Receptive Will They Be?

How Can I Adapt My Document to This Audience and Achieve My Purpose?
  Employ Characteristics of All Good Technical Writing
  Vary Content and Presentation According to Audience

Examples
Checklist for Assessing Purpose and Audience
Activities

Stephanie Rosenbaum is president of Tec-Ed, Technical Communication and Graphics Services, Inc., a company that translates the language of technicians into the language of computer users. In an interview with the editors of *Issues in Writing*, Rosenbaum made the point that even if writers are writing for an audience with technical expertise, they should not take anything for granted. Rosenbaum offered a striking anecdote of professional technical writers assuming that one engineer would obviously understand another engineer:

> A few years ago we were writing a manual for a prototyping emulation system. This client had a prototyping hardware system—a microcomputer system—that was being used primarily by engineers who were designing computers with it. We were writing the manual for it and were in our usual interviewing situation talking to the design engineer for this product. We would ask him this and ask him that, and he kept saying, "Any engineer would know the answer to that. Why are you asking me this stuff?" After about an hour of this, we took the information he gave us and we went off and wrote the first chapter. As it happened, his company was in California, so we took the first chapter and we carried it 3,000 miles across the country to one of our consultants who was a senior engineer at a non-competing computer company in the Boston area. And we said, "Would you read this chapter (about five pages) and see if you have any questions?" He starts reading it and his head starts shaking, and he keeps saying, "What does he mean by this? I don't understand this." The consultant, a graduate engineer from MIT, gave us about a page and a half of questions. We took the questions, carried them back across the country, sat down with Engineer A and said, "You know, we asked one of our consulting engineers to look at this because that's your typical audience for this manual, and he had a few questions." We started going over the questions with him and after the first couple of questions he starts saying, "Gee, you know, I guess you could take it that way; yeah, I can see how they might not know that." What we had really demonstrated was that when the first engineer said that any engineer would know the answer, what he really meant was, "Any engineer who had worked for my company for the last five years, and on this development project for the last year and a half, and was working down the hall from me. . . . " It's hard to tell anyone this; it needs to be demonstrated. In other words, you need to take something that the person from Corporate Culture A thinks is complete and give it to someone from Corporate Culture B and start asking questions.

As the anecdote points out, of all the problems and choices technical writers face, none is more critical than understanding the audience and purpose for your writing. To be successful, technical writing must effectively communicate, and communication basically means accomplishing one's *purpose* with a particular *audience*. If a writer doesn't have a good understanding of the purpose and audience of a document, or doesn't use this information effectively, then the document will fail to communicate.

For most technical writing tasks, audience and purpose are the two most important factors shaping both content and presentation. Nothing is superfluous in good technical writing. If an aspect of the document doesn't contribute to communication, if it doesn't reflect a firm understanding of the audience and purpose, then it is wasteful and possibly counterproductive.

This chapter describes the most typical purposes and audiences for technical writers. The task of technical writing is an exercise in problem-solving. This chapter should help you answer these basic questions about your writing: What is my purpose? Who is my audience? How can I adapt my document to this audience in order to give myself the best chance of achieving my purpose?

What Is My Purpose?

Explicit Purposes

Before you begin writing any document, know precisely what it is you are being asked to do. As the following list shows, writers may be asked to

> *report* a solution to a problem
> *describe* services available to customers
> *interpret* lab analysis
> *provide* instructions
> *ask* for an idea
> *gain* authorization
> *initiate* a new procedure
> *order* material
> *change* a course of action
> *submit* a bid
> *define* a position on policy
> *propose* adoption of a plan
> *prepare* a proposal
> *thank* a customer
> *congratulate* an employee
> *announce* a change

Fortunately, these many purposes may be grouped into three main purposes: in general, technical writers write to inform, to help others make a decision, or to instruct. Often there is more than one purpose to a document. For example, in operational audit reports routinely conducted by accountants, one purpose is to inform readers by describing the work done, by listing those areas where there is a potential for improvement, and by making specific recommendations. However, another purpose of the report is to help executives and managers make decisions that will improve the company. Finally, the report often is used to instruct others on specific procedures for improvement.

TO INFORM

The principal *purpose* for the technical writer is the clear and concise transmission of knowledge to the reader. Informative writing can appear anywhere in technical writing. It may be the primary purpose of the document, or it may be one purpose among many in a document. Informative writing often appears in reports, literature reviews, reports of physical research or observations, progress reports, memos, and letters.

Figure 3-1 shows an example of informative writing. It appears in *Positive Feedback*, a publication devoted to issues involving epilepsy. The purpose of the article is to provide information about epilepsy to readers who might know little about epilepsy but who are interested in learning more about it.

This short article demonstrates many of the qualities of good informative writing:

- The first sentence introduces the subject and organization of the article, which follows from one class of epilepsy (partial seizures) to the other (generalized).
- The organization of the article is easy to follow. Each class develops from less serious (simple, nonconvulsive) to more serious (complex, convulsive) symptoms.
- The writer defines terms, such as "partial" and "generalized," that some readers might not be familiar with.
- The use of italicizing reinforces the distinction made in the introduction between partial and generalized seizures.
- The use of details, such as "handwashing, scratching, or teeth grating," helps the reader understand the symptoms.

TO PROVIDE A BASIS FOR DECISIONS

Because the purpose is to help readers decide whether to carry out a course of action, you must focus on the question, "What does the reader need to know?"rather than on, "What do I need to say?"

Figure 3-2 shows a brief report of a new procedure for aligning large cylinders for welding. The purpose of the report is to convince the reader that tooling with brackets and jackscrews is superior to the more traditional method of tooling with mandrels.

This is a good example of writing intended to provide a basis for decision. In this report the writer first introduces the new procedure in very general terms and tells the reader *why* there was a need for a new procedure.

Figure 3-1. Informative Writing

What Are the Different Kinds of Seizures?

There are two major classes of epileptic seizures, each with its own symptoms and causes. The first two types of seizures listed below are called "partial" because they begin in a single, isolated area of the brain. The second two types are called "generalized" because they result from activity deep in the brain, involving more than one part of the brain.

> *Simple partial seizures* are seizures that originate in a small, isolated part of the brain. During a simple, partial seizure, the person remains conscious, and the seizure itself usually lasts only several seconds. The seizure may be a tingling sensation in an extremity, a perception of a bad odor, a vision of flashing lights, or unintelligible speaking.

> *Complex partial seizures* are seizures in which the person acts automatically, as if they were in a trance and performing a set of behaviors that they were programmed to perform. Persons having such a seizure may carry out the motions of handwashing, scratching, or teeth grating, to name a few. Usually, the person appears awake but may lose contact with their surroundings. The seizure may last for less than a minute or as long as an hour.

> *Generalized, nonconvulsive, or absence seizures* usually happen in children. They are just a momentary lapse in which the person loses consciousness for a brief period without passing out. The person will appear to an observer to have just "spaced out" for a few seconds. Since these seizures happen mostly in children, teachers often mistake absence seizures for simple episodes of inattention.

> *Generalized, convulsive seizures* are the seizures most people associate with epilepsy. The person will lose consciousness and fall to the ground. Their muscles may spasm, causing their limbs to twitch, and perhaps the person will thrash about. After the seizure has run its course, the person will return to consciousness, dazed, confused, and exhausted, with no memory of having had a seizure.

Source: What are the different kinds of seizures? (1989, summer). *+Feedback, 1*(2), 2–3. This article has been excerpted from *+Feedback*. Subscriptions may be obtained for $12 per year. 630 Ninth Avenue, Suite 901, New York, NY 10036.

The writer then describes the new procedure. Although there is not a lot of detail in the description, the writer does include an illustration, which becomes an essential part of the argument. The writer ends the report by reinforcing the reasons *why* there was a need for a new procedure.

TO INSTRUCT SOMEONE HOW TO PERFORM A SPECIFIC TASK

The reader's goal here is clear: to learn *how* to do something. Therefore, to be successful, your instructions must allow a reader to take successful action.

Figure 3-3 contains a good set of instructions. Written for technicians

Figure 3-2. Decision-Oriented Writing

Special tooling alines and holds internally stiffened large-diameter cylindrical parts for welding. The tooling replaces large mandrels. The mandrel tooling was expensive and cumbersome and did not mate the parts precisely.

The new tooling consists of brackets that are bolted to the internal stiffening stringers on the cylindrical sections. Push/pull jackscrews on the brackets are adjusted to bring the sections into alinement for welding (see figure). The tooling substantially reduces costs while allowing more precise control and improved quality.

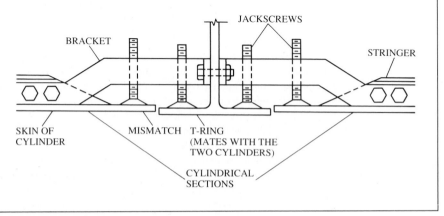

Source: Ehl, J. H. (1986, May). Alining large cylinders for welding. *NTIS Tech Notes*. Washington, DC: United States Department of Commerce, National Technical Information Service.

Figure 3-3. Instructive Writing

Titration for Effective Alkali

- Fill the buret with 0.5167 N HCl to read zero.
- Pipette 5 ml of liquor into a 250 ml Erlenmeyer flask.
- Add it to 50 ml water and 25 ml 10% barium chloride solution.
- Add 3–4 drops of phenolphthalein indicator. At this stage, the solution should have pink color.
- Titrate the solution with 0.5167 N HCl until the pink color disappears. *Do not refill the buret.* The end point is at pH 9.3.

Source: Ring, G. J. F. (1988). *Pulp and paper laboratory methods* (p. 21). Stevens Point, WI: Paper Science Department, University of Wisconsin–Stevens Point.

who want to know how to do a liquor analysis in a Kraft pulp mill, the instructions are from a laboratory manual. This set of technical instructions is clearly written with a specific purpose in mind.

This example uses several techniques that are common for good instructions:

- The writer uses precise terminology to help the reader perform the analysis. The writer assumes the reader will know that a "buret" is a glass tube with a stopcock at the bottom; that an "Erlenmeyer flask" has a narrow neck and a flat, broad bottom; that "0.5167 N" is a measure of the concentration of the solution; and that "HCl" is the chemical formula.
- The writer uses specific verbs to illustrate each step in the process: "fill," "pipette," "add," "titrate."
- Because the readers are familiar with the conventions of their particular industry, the writer is comfortable using the word "pipette" as a verb, rather than as the more traditional noun.
- Even though the readers are technicians, and are familiar both with the materials being used and with similar processes, the writer still offers a warning (and underscores it) at a particular step in the process.
- Because the writer assumes that the readers' purpose in reading is not to learn *why* HCl is used for titrating, or *why* barium chloride is added in the third step, the instructions emphasize *how* to do a liquor analysis.

Implicit Purposes

Whether writing to inform, to provide a basis for a decision, or to instruct, technical writers need to be aware of the broader implicit purposes that inform their writing. Three of the most important implicit purposes are

1. To create credibility
2. To generate trust
3. To persuade

For example, while the explicit or immediate purpose of the article on epileptic seizures is to describe the different kinds of epileptic seizures, the writer must leave the reader with the impression that the information is accurate and that the writer is a credible source of information; otherwise the reader will discount or ignore the information as soon as it is read.

Similarly, while the explicit purpose of the instructions for titration for effective alkali is to provide a set of instructions, the implicit purpose is to generate enough trust in the instructions that the reader is willing to take a chance and try them. Otherwise, the reader probably won't feel confident enough to actually follow the directions, and they will have been written in vain.

And finally, while the explicit purpose of the description of special tooling is to provide a basis for a decision, the implicit purpose is to persuade the reader that the new tooling method is better than the old tooling method. Unless the writer can do this, the decision that is eventually made may be very different from the one desired.

When these implicit purposes are considered, it becomes obvious that every aspect of technical writing can be affected by a document's purpose. When one understands that, at the very least, a technical document is supposed to persuade the reader that the writer is intelligent, well-informed, credible, and thoughtful, then even such small matters as correct spelling, clear diagram labels, and the use of parallel headings can be seen as helping the document accomplish its purpose.

Who Is My Audience?

What you write and how you write it depend on knowing who your readers are. Knowing your audience will allow you to make sound judgments about the nature of information that must be presented in any writing situation. Format, amount of detail, tone, organization, and many other features of the final document will be partially determined by the audience for that particular document. This section describes the most typical audiences for technical writing: experts, technicians, executives and managers, novices, and mixed audiences.

Experts

Who Are They?   Experts are readers with technical expertise in a particular field, such as accountants, engineers, and manufacturing managers.

Why Are They Reading?   Experts are usually skilled readers who read for a variety of reasons. They read to keep up with their field. They read for information on which to base their own research. They read before making recommendations to executives; decision-makers often ask them to analyze and coordinate information presented in the decision-making process. Experts want to know not only *what* an idea is, but *why* it is being proposed and *how* the writer arrived at it. Because they are often key links in the decision-making process, and because they rely so heavily on the writing of others, experts spend much of their time reading the kinds of technical writing that help them formulate their own recommendations: research reports, feasibility studies, progress reports, proposals, and literature reviews.

Technicians

Who Are They?   Technicians are the "builders" and "fixers."

Why Are They Reading?   Technicians are usually interested in *what* something is and *how* it works, rather than why it works. They are often looking for technically detailed instructions. If they want or need to know "why," it is only to the degree that it helps them understand their task or diagnose their problem. Technicians often read technical publications, such as operating manuals, to find out how to work a piece of equipment.

Executives and Managers

Who Are They?   Executives and managers are those in management roles. While many executives and managers have come up through the ranks of an organization, more frequently their backgrounds are in management, which means that generally they do not have a highly technical background.

Why Are They Reading?   While they might read for information or for instruction, executives and managers read to help make a decision: they want to know *why* they should do something. They read to decide on a course of action, especially actions involving the (re)allocation of resources. Because often they are not technically sophisticated, they usually are not interested in the technical details of a proposed project or a solution to a problem. Instead, they want to know the implications of the details being presented: they want to know how much the project or solution will cost, they want to know how it will affect those who are involved. The kinds of technical writing they are likely to consult before making a decision are reports, policy manuals, regulations, proposals, and feasibility studies.

Novices

Who Are They?   Novices are readers with little or no knowledge in the subject area of your writing. Someone who might otherwise be an executive or technician or expert can be a novice in a new field. This is a difficult group to write for since the technical writer knows much more on the subject and must anticipate the needs of the uninformed reader.

Why Are They Reading?   Novices read for a number of reasons. They read progress reports for information. They read manuals and instructions to find out how to do something. They read informative articles to help make a decision. In short, they can have almost any purpose and any interest and can decide to read almost any document. What they do not have is a great deal of background in the field.

The Mixed Audience

Who Are They?   A mixed audience is an audience made up of two or more of the groups described above. A document that is widely distributed will typically have a mixed audience. A report on designing a new computer system in a large corporation, for example, would certainly be distributed to executives and managers, but it also might be distributed to stockholders, many of whom are novices.

Why Are They Reading?   In a mixed audience, each group may have its own reasons for reading the same document. Executives and managers would read the report on designing a new computer system because they have to decide whether to buy it. Experts would read the report because they are the ones who have to implement the decisions made by the executives. Stockholders, although they might not be knowledgeable about sophisticated computer systems, would read the report because they want information about something that might affect their investment.

What Are My Audience's Needs and Wants? ——————

The more you know about your audience, the greater chance you have of preparing an effective and successful piece of technical writing. Knowing more about your audience means understanding why readers need or want to read what you have written, and what their interests and attitudes are.

What Is Their Purpose in Reading?

Readers of technical writing read a particular piece of technical writing for reasons that correspond to writers' reasons for writing. In general, readers of technical communications want either to gain information, to gain information on which to base a decision, or to learn how to perform a task.

READING TO GAIN INFORMATION

Readers who are looking for information as an end in itself want either to increase their understanding of a subject or to locate specific information. These readers want the information they read to be complete, clear, and accurate. The level of detail they're seeking may vary, but they usually want facts, along with some analysis and an overview. They usually don't want a persuasive argument; in fact, they may distrust a piece of writing if it seems "slanted" or trying to sell something.

   Readers seeking information can look in many places. Articles in publications such as popular science magazines, professional journals, and company newsletters provide a wealth of information to readers seeking general knowledge. These articles report solutions to problems, describe services available, explain company policy, or announce a change in procedure. Readers looking for specific information usually turn to reference tools, such as scientific and technical dictionaries and specialized encyclopedias, and to policy manuals, especially those published by various business and industrial organizations and governmental agencies.

READING TO MAKE A DECISION

Readers who are reading as a basis for decision-making are usually not interested in the technical details of a proposed project or a solution to a problem. Instead, they want to know the implications of the details you are presenting. They want to know how much the project or solution will cost, and how it will affect those who are involved.

   Whether they are reading a report on why fax machines are breaking down, a literature review of spreadsheet software, or a memo recommending a new procedure for transporting hazardous waste material, these readers want the writer to interpret the significance of the details that are presented. What can be done to prevent further breakdowns of the fax machines, and how much will it cost? Which spreadsheet software are you recommending, and why? What is the cost, what are the effects on the environment, and in what ways is the new procedure for transporting hazardous waste better than the old method? These readers want documents that analyze and summarize the information with their specific needs in mind, documents that are concise, to the point, and answer all important questions.

READING TO LEARN HOW TO PERFORM A TASK

Readers who want to learn how to perform a task read instructions — often at the same time as performing the task itself. They want to read only the information they actually need to perform the task, and nothing more. They want every step to be crystal clear, absolutely explicit, easy to use, and in its correct place. On the other hand, they don't want to feel like they're being lectured to. Readers in this group tend to be novices, and they generally don't appreciate being made to feel ignorant or dull-witted. They appreciate documents that treat them as intelligent adults.

How Much Do They Know?

The level of knowledge about a subject corresponds roughly to the type of audience: experts know the most about a subject, technicians know less about the theoretical background but have a firm grip on the day-to-day details, executives may only be familiar with the overall concept or process, and novices know little.

   However, remember that the purpose of technical documents is often not only to *inform* the reader about a subject but to allow that reader to *do* something with this information. So, although an executive may not know much about word processing programs or the advantages of various databases, he or she may not need much information in order to make an informed decision about what software the company should be using.

In other words, after you ask yourself, what do they already know?, ask, what do they need to know? Your role as a technical writer is usually not to inform readers about every possible detail and nuance related to a topic but rather to provide readers with whatever information they need but don't yet have. You must fill the gap between what they already know and what they need to know.

How Receptive Will They Be?

Another characteristic of your audience that is important to consider is their general attitude toward the reading situation. Do they enjoy reading? Are they generally interested in this topic? What will their attitude toward this document be? What is their attitude toward you or your organization? The answers to these questions will allow you to judge the appropriate style and tone of your document—how far you must go to create confidence and respect or how far you can take advantage of that confidence and respect once it's been established.

HOW INTERESTED ARE THEY?

If readers are highly interested in what you have to say, they will go to great lengths to make sure they understand what you have written. Such readers might overlook a poor pattern of organization or a superficial error or two. An example of someone with a high interest level is a colleague who has been waiting to hear a report of latest research, perhaps because her own work will be based on it. If, however, readers are not interested in your document or in what you have to say, you will have to work hard to maintain their interest. For these readers, any error will be magnified tenfold. An example of someone with a low interest level is a technician who is required to read a progress report on a project he is not directly involved in.

If you can gauge the interest level of your audience in your topic or document, then you can decide how hard you must work to get their interest. Often it is possible to create interest by pointing out connections that might have been overlooked before: the technician reading a seemingly irrelevant progress report might pay more attention if you point out that this project will serve as the prototype for all future projects—including his—in the coming twelve months.

WHAT ARE THEIR ATTITUDES TOWARD MY DOCUMENT?

If your readers agree with what you have to say or are looking forward to learning what you have to say, they will be likely to accept your information or be persuaded by your arguments. An example of a reader having a positive attitude toward a document is a computer expert who has heard lots of exciting things about hypertext and is finally reading an informative report about the different programs available. If, however, your readers disagree with your position or are anxious about the topic, they will be harder to reach or to convince. An example of a reader with a negative attitude toward a document is a novice confronted with instructions for programming a VCR. Instructions often elicit negative attitudes because many readers have had problems with them in the past and the common VCR is notorious for being difficult to program.

WHAT ARE THEIR ATTITUDES TOWARD ME OR MY ORGANIZATION?

Not every audience will be favorably disposed toward the writer or the writer's organization. Assessing the reader's disposition can be crucial to effective technical writing. For example, writing a progress report for a client who has been disappointed with your organization's performance is very different from writing a progress report for a loyal and satisfied client. By determining whether there is any problem ahead of time, you can make informed choices about the style, the organization, and the level of detail you want to include in your document.

How Can I Adapt My Document to This Audience and Achieve My Purpose?

Employ Characteristics of All Good Technical Writing

Accurate and Thorough Information.   All readers of technical writing want complete information that they can trust. If your facts contain errors or if your argument overlooks a well-known objection, your readers will feel they have wasted their time.

Strong and Apparent Organization.   It is not sufficient to present accurate and thorough information. Successful technical writing must also be clearly organized. Readers must have a clear sense of where they are in a document and where they are going. A confusing description, a wandering argument, or a disorganized set of instructions will be ineffective and ultimately unsuccessful.

Appropriate Type of Document.   The document you present should be a recognizable type of document, that is, a literature review, a research report, a proposal, a feasibility study, and so on. Furthermore, it should correspond to accepted conventions for that type of document. A reader who reads a table of contents and identifies a document as a proposal will be unpleasantly surprised to find it is a feasibility study. Your obligation is to fulfill the reader's expectations of what a proposal should be.

Minimum of Distractions.   In technical writing, everything you do should be purpose-oriented. Most readers of technical writing read for a specific purpose. Anything that distracts from that purpose detracts from the document. Common distractions in technical writing include unnecessary information, flamboyant document designs, and egregious spelling errors. Reviewing for quality is essential for identifying and deleting anything that hinders the flow of information.

Vary Content and Presentation According to Audience

All audiences will appreciate the qualities discussed above. But each of the five audiences discussed above also has its own requirements. When writing a technical document, it is important to identify exactly what your audience needs and wants, and then to supply it. By accommodating your audience and helping them achieve *their* purpose, you will be far more likely to achieve *your* purpose.

Throughout this book, approaches or techniques that are appropriate for one group or another will be mentioned wherever relevant. However, a few generalizations can be pointed out here.

- Because **experts** know their fields intimately, they seldom need background details or definitions. Similarly, it is reasonable to use highly technical terms and abbreviations when writing for this group.
- **Technicians** will likewise be familiar with the tools and processes they use most often, but they will usually need definitions of any terms that might be unfamiliar. Successful documents written for this group also avoid lengthy theoretical discussions. A bit of background is generally all that's needed.
- **Executives** are extremely busy people, so they greatly appreciate anything that makes the reading go faster: summaries, a clear and useful organization, simple language, an economical style. They want a document that has thoroughly analyzed the situation and distilled the information to a handful of important points.
- Effective documents written for **novices** also tend to simplify. Another important strategy for this audience is to present new information in terms of familiar information. Sometimes this is a matter of organization —moving from familiar to less familiar points. Sometimes it is a matter of creating analogies: "A computer's hard disk is like a human being's mind."
- **Mixed audiences** with various levels of expertise and various purposes can be accommodated by segmented reports with clearly identified parts; this way, different people can read different parts.

Examples

Below are three examples of documents adapted to different purposes and different audiences. The first is for a novice audience, the second is for an audience with some technical expertise, and the third is for an expert audience. They all describe the same object—a human spermatozoon—but they vary greatly in tone, organization, level of detail, and level of technicality.

The description in Figure 3-4 appears in a family medical guide. Notice that the technical writer assumed that the readers of the guide are nonspe-

Figure 3-4. Example of Writing for a Novice Audience

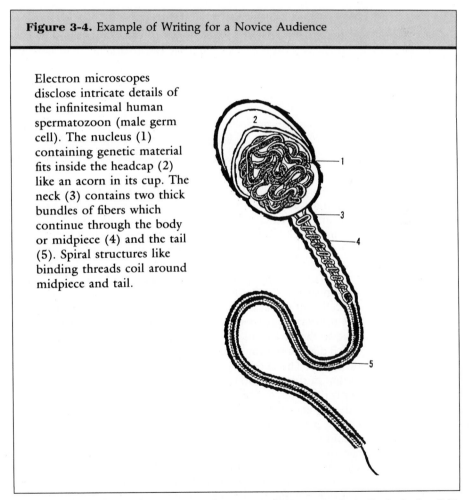

Electron microscopes disclose intricate details of the infinitesimal human spermatozoon (male germ cell). The nucleus (1) containing genetic material fits inside the headcap (2) like an acorn in its cup. The neck (3) contains two thick bundles of fibers which continue through the body or midpiece (4) and the tail (5). Spiral structures like binding threads coil around midpiece and tail.

Source: Greenblatt, R. B., & Weller, C. (1976). The endocrine glands. In D. G. Cooley (Ed.), *Family Medical Guide* (rev. ed., p. 341). New York: Better Homes and Gardens. Used by permission.

cialists interested primarily in general information; thus, the writer included only very basic details. The introduction in this description illustrates the importance of word choice to provide a focus for the description. For example, the reference to the electron microscope in the first sentence focuses on the smallness of the spermatozoon. The smallness is reinforced with the word "infinitesimal" in the same sentence. The part-by-part description is organized spatially, rather than functionally, moving from the head of the spermatozoon to its tail. To reinforce the organization of the description, the writer ties the description clearly to the related graphic. The numbers in parentheses are not a list but refer to the appropriate part of the graphic.

Note the use of figurative language, generally found only in technical descriptions written for less technically sophisticated readers. In this example, the writer uses similes. The first simile ("like an acorn in its cup") helps the nontechnical reader visualize the physical relationship between the nucleus and the genetic material that fits inside it. The second simile ("like

---

Figure 3-5. Example of Writing for a Mid-Level Audience

The mature human sperm cell (Figure 30-9) is enveloped by the cell membrane and consists of three principal regions—the *head, middlepiece,* and *tail.* The head is made up mostly of the haploid nucleus and a small amount of Golgi material. In many animals, including mammals, there is also a structure called the *acrosome,* located beneath the cell membrane of the headpiece, which varies in size and appearance. For example, the acrosome is quite large in the sperm of the guinea pig but relatively small in the human sperm. It is known to be essential for the sperm's penetration of the egg cell, and recent evidence has established that it functions in two ways in this respect. First, the acrosome discharges a filament that joins the sperm head to the surface of the egg; and second, it releases hydrolyzing enzymes which digest a portion of the egg membrane, thus allowing for penetration of the sperm. The middlepiece possesses a pair of centrioles and a rich concentration of mitochondria arranged in a loose spiral around a central core of fibrils that extends the length of the tail. The mitochondria apparently provide the energy responsible for the motility of the sperm. The tail is the longest part of the sperm (about 0.045 mm in length) and accounts for the motility of the gamete. It displays the typical structure of flagella or cilia, consisting of a core of ten pairs of fibrils arranged so that nine pairs constitute a circle about a single pair (see Figure 4-10). A second type of cell also composing the inner lining of the seminiferous tubule is believed to have a role in providing nourishment to the developing sperm.

Source: Nason, A. (1965). *Textbook of modern biology* (pp. 702–703). New York: John Wiley & Sons.

binding threads") helps the nontechnical reader visualize the spiral structures that coil around the midpiece and the tail.

The second description of a spermatozoon (Fig. 3-5) appears in a college biology textbook. The writer of this description has assumed that his audience has more technical knowledge than the general reader and is interested in more than just general information. Because the writer assumes a greater degree of technical sophistication on the part of the reader, he does not feel obligated to tie the description closely to the related illustration.

The last description (Fig. 3-6) is from *Gray's Anatomy*, a standard medical school text. Because it is meant for medical students, who have a strong interest in and knowledge of things biological, this description is highly detailed and unreservedly technical. Note, for example, that while the college biology text tells us that the acrosome "releases hydrolyzing enzymes,"

Figure 3-6. Example of Writing for an Expert Audience

A spermatozoon, or sperm (Fawcett 1961a, 1975, Piko 1969, Rothschild 1957), is a smaller cell than an oocyte, highly specialized to reach the latter and to carry to it its own haploid chromosome complement. Its expanded *caput* or *head* contains little cytoplasm and is connected by a short constricted *cervix* or *neck* to the *cauda* or *tail*. The latter is a flagellum of complex structure, usually divided into *middle, principal* and *end parts* or *pieces.* Volumetrically the tail much exceeds the head, which varies greatly in different species (Rothschild 1957, Phillips 1975), being ovoid or spiriform in man, somewhat flattened at the tip in lateral profile, with a maximum length of about 4 μm and a maximum diameter of 3μm. The tail, about $45-50$ μm in length, displays a greater uniformity between species. . . .

The *head* (2.13A) is an extreme example of chromatin concentration, consisting largely of a dense and visually uniform nucleus, with a distinct bilaminar nuclear membrane and a bilaminar *acrosomal cap* (head cap), the latter covering the terminal two-thirds of the nucleus and partly derived from the spermatid Golgi apparatus. The acrosomal cap is thin in a human spermatozoon but in other species it is often large and more complex in shape. The acrosome has been shown to contain several enzymes including acid phosphatase, hyaluronidase and a protease (*acrosomase*), which are probably involved in penetration of the oocyte. The nucleus and acrosome are enveloped in a continuous plasma membrane without intervening cytoplasm (Fawcett & Burgos 1956, Anberg 1957). The chromatin is stabilized by disulphide bonds, as if to protect its genetic content during the spermatozoon's journey (Fawcett 1975). So densely packed is the chromatin that it appears homogenous even under electron microscopy.

Source: Williams, P. L., Warwick, R., Dyson, M., & Bannister, L. H. (Eds.). (1989). *Gray's anatomy*, (37th ed., pp. 118–119). Edinburgh: Churchill Livingstone.

Gray's Anatomy lists some of the specific enzymes released by the acrosome ("acid phosphatase, hyaluronidase and a protease").

Because it is so lengthy, just the introduction and one section of the description of a human spermatozoon are provided here. Though this introduction is much lengthier than either of the previous introductions, it follows the same basic pattern as the other two introductions: it provides an overview of the parts to be discussed. The part-by-part description begins with the description of the head of the spermatozoon and moves to its tail.

Checklist for Assessing Purpose and Audience ⎯⎯⎯⎯

____ 1. Have I determined my explicit purpose for writing?
____ to inform
____ to provide a basis for decision
____ to instruct

____ 2. Have I identified my readers?
____ primarily experts
____ primarily technicians
____ primarily executives
____ primarily novices
____ a combined audience

____ 3. Have I determined their purpose in reading this document?

____ 4. Have I determined how much my readers know about the subject?

____ 5. Have I determined how much my readers *need* to know about the subject?

____ 6. Have I evaluated their level of interest?

____ 7. Have I considered their attitudes toward my document?

____ 8. Have I considered their attitudes toward me or my organization?

Activities ⎯⎯⎯⎯⎯⎯⎯⎯⎯⎯⎯⎯

1. Working with two or three other students from your class, review the discussion of the three descriptions of a human spermatozoon presented

above. In your library, find at least two or three descriptions of the same object written for different purposes and different audiences. Analyze the descriptions and prepare a brief oral report for the class in which you present the results of the analysis. For your analysis, in addition to the purpose and audience of the descriptions, consider the following questions: Which description is easiest for members of the group to understand? Which is hardest? Why is one description more difficult than another?

2. Find an article in a technical journal. Analyze the purpose and audience for the article. What steps has the writer taken to meet the needs of the readers? What steps has the writer taken to achieve the probable purpose of the article? Are there ways in which the writer could have better addressed the audience and purpose of the article? Place yourself in the position of an editor at the journal, and write a letter to the author explaining these possible improvements.

3. Review the last paper you wrote. Prepare a short analysis of the purpose of your writing and your intended audience. What were the needs of your reader? What steps did you take to address these needs? Were you successful? Why or why not? How could your paper have been improved in this regard?

4. A distant cousin has heard that you are studying [your major] but apparently doesn't know what it involves, since you have just received a letter in which your cousin congratulates you for being able to open a [your major] store after you graduate. Write a letter to your cousin in which you explain exactly what [your major] is and clear up the confusion.

5. Write an evaluation of the purpose and audience of this chapter. Is the purpose of the chapter clear? Is the intended audience clear? Have the authors addressed the needs of the readers?

6. Find a highly technical article in a field you are familiar with. Rewrite a passage from the article so that it is readable and enjoyable for a novice audience.

Technical Style

4

Style Conventions

Specialized Conventions of Technical Style
  Company Style Guides
  Professional Style Manuals

General Conventions of Technical Style
  Principle 1: Help Readers Feel Involved in What They Are Reading
  Principle 2: Select Words Carefully
  Principle 3: Write Clear Sentences
  Principle 4: Create Documents That Are Easy to Navigate
Checklist for Technical Writing Style

Activities

The purpose of writing is to communicate ideas clearly and accurately from the writer's mind to the mind of the reader. Anything that obstructs the communication is undesirable. Therefore, your greatest challenge as a technical writer is to explain clearly and directly so that your intended readers can understand and use what you have written. Although many writing skills combine to make a document easy to read, one of the most important features is style.

Style is the *way* in which we say or write something. Style is not *what* we say but *how* we say it. It is here that writers make the choices that determine whether their ideas will be understood or persuasive. It is here that you choose the words that will move your readers to accept what you have written. It is here that you organize combinations of those words into the sentences and paragraphs that ultimately will inform, persuade, or move your readers.

Style Conventions

Conventions regulating style have been developed to help people communicate more efficiently. Conventions provide writers with a guide for writing and readers with a guide for reading.

There are several levels of conventions you should be aware of. You are probably already familiar with the most important conventions of standard written English: every sentence should have a subject, every sentence should end with a period or its equivalent, and so on. The next level is of conventions about technical writing in general. These concern issues that are more important in technical writing than in other kinds of writing. Finally, the writing within each profession has certain characteristics in common; those characteristics are the conventions of a particular profession. Successful writers understand that in order to be effective they must adhere to the conventions of standard written English, to the general conventions of technical writing, and to the specialized conventions of the profession they are writing for.

It is important to learn and follow the conventions that apply to the sort of technical writing you're doing. If you don't follow conventions, your readers will have a much more difficult time understanding you. Instead of *thinking about* your message, they must first *figure out* your message. Worse, they could misunderstand you; you might not be writing according to conventions, but your readers are probably still reading according to them. Another important reason for following the conventions of a trade, discipline, or profession is because it is one way of establishing yourself as a member of that trade, discipline, or profession. Style is an important factor in establishing and maintaining credibility.

Specialized Conventions of Technical Style ───────

Some aspects of style, in particular mechanical and formalistic details, are governed by specialized conventions within each organization or field. These common traits are the conventions of that particular profession. The best advice we can offer is to find out the conventions of your field. You can do so in a couple of ways: (1) by reading carefully and noting how successful writers in your field write and (2) by checking the style manual appropriate to your field. There are two sources for style manuals that you should be aware of—company style guides and professional style manuals.

Company Style Guides

Most large organizations, and many small companies, have style guides that describe particular stylistic features that that company or organization prefers for those who write for it. If you are going to do any writing at all for the company you work for, and it is likely that you will, one of your first tasks is to find out if there is a company style guide or manual and follow it.

An example of a company style guide is the style guide recommended for use by authors writing for Blair Press, publisher of this book. In a section on abbreviations in the guide, for instance, authors are asked to adhere to the following conventions for punctuating abbreviations.

> In general, omit the period after abbreviations. But note that the abbreviations for *atomic weight* and *inch* are exceptions:
>
> | | | | |
> |---|---|---|---|
> | 50 mph | 6 ml | 22 amp | 6 in. |
> | sin | cos | tan | sec |
> | cot | mm | at. wt. | 12 ft-lb |
>
> *(Prentice-Hall author's guide*, 1978)

If there is no formal company style guide, often there are generally known conventions. The best way of finding out what these are is to pay close attention when you read material written within the organization and to ask colleagues or your supervisor.

Professional Style Manuals

Professional style manuals describe writing conventions that have been deemed appropriate for individuals writing for publication within that profession or for a certain type of document. Listed below are some of the many style manuals available for specific fields:

Council of Biology Editors. (1983). *Council of Biology Editors style manual* (5th ed.). Arlington, VA: Council of Biology Editors.

American Chemical Society. (1978). *Handbook for authors.* Washington, DC: American Chemical Society.

Packaging Machinery Manufacturers Institute. (1973). *Handbook for writing operation and maintenance manuals.* Washington, DC: Packaging Machinery Manufacturers Institute.

American Math Society. (1979). *A manual for authors of mathematical papers.* Providence, RI: American Math Society.

Markus, J. (1978). *Electronics style manual.* New York: McGraw-Hill.

Pollack, G. (1979). *Handbook for ASM editors.* Washington, DC: American Society for Microbiology.

American Psychological Association. (1983). *Publications manual of the American Psychological Association* (3d ed.). Washington, DC: American Psychological Association.

U.S. Government Printing Office. (1967). *U.S. Government Printing Office style manual* (rev. ed.). Washington, DC: U.S. Government Printing Office.

General Conventions of Technical Style

Other aspects of technical style (usually larger concerns) are agreed upon across disciplinary and organizational boundaries. These conventions are called general conventions of technical style. These general conventions aren't arbitrary; on the contrary, they proceed logically from the fundamental objective of technical writing: efficient communication. This fundamental objective can best be met by following the four principles given here. The guidelines within each principle give practical advice on how to apply the principle to your own work.

Principle 1: Help Readers Feel Involved in What They Are Reading

The first rule of effective writing is to help the reader want to understand and be motivated by your document. Whether you are writing a letter, a report, or a manual, you must first understand who the reader is. Then you must decide how best to meet that reader's needs. The guidelines in this section will help you remember that you are writing from one human to another, and you are writing to a human being that you respect and are dealing with honestly.

1. *Address the reader as directly as you can.* People are more likely to read, understand, and be receptive to a document if they feel that it is addressed to them. One way of addressing a reader directly is by name, as in letters and memos. Another way is by using the personal pronoun "you." Manuals, for example, address a wide audience, and direct address of each reader by name is impossible. Thus, use "you" as often as possible. The personal pronoun will give your writing a tone of directness and personal interest. Notice the difference in tone and attitude in the following examples:

DON'T WRITE:

If a company employee suffers an injury of any kind while at work or at any other place while on business for the company, he/she should notify the Personnel Department at once. The injured employee will receive directions concerning medical treatment and reporting procedures. The company carries Workman's Compensation Insurance to cover injuries. Prompt reporting of such incidents protects the employee and helps the company to make the necessary claims.

WRITE:

If you are injured in any way while at work or on business for the company, you should notify the Personnel Department at once. You will receive directions concerning medical treatment and reporting procedures. The company carries Workman's Compensation Insurance to cover injuries. Prompt reporting of injuries protects you and helps the company to make the necessary claims.

2. *Write in the active voice.* Your writing will be more clear and direct if you write in the active voice. In active sentences the emphasis is on the subject, the person or thing performing the action:

Service personnel should unplug the AC power cord before disassembling the cabinet.

In passive sentences the emphasis is on the object, the person or thing being acted upon:

The AC power cord should be unplugged by the service personnel before the cabinet is disassembled.

By writing in the passive voice you create an image of evasiveness, bureaucracy, and pomposity. Sometimes evasiveness is deliberate, especially if the responsible party is unknown ("The policy was established in 1964") or if the emphasis rightly belongs on the object rather than the subject of the sentence ("The driver was taken to the hospital"). But most often the use of the passive is unintentional and unnecessary. When you use the active voice, you give readers information quickly, clearly, and emphatically.

DON'T WRITE:

When reinstating an account with a two-month delinquency, a positive arrangement for the next regular monthly payment should be obtained.

WRITE:

When you reinstate an account with a two-month delinquency, you should get a positive arrangement for the next regular monthly payment.

3. *Avoid sexist language.* Avoid language that could be understood as sexist. Sex-specific language, for example, is acceptable in sex-specific medical usage (e.g., "her pregnancy," "his prostate gland"). However, sex-specific language becomes sexist language when it arbitrarily imposes sex stereotypes ("each nurse should ask her patients . . .") or employs sex-specific pronouns ("she," "he," "her," "his") when the reference applies to both sexes equally. Many common words contain within them the assumption that "man" represents the entire human population, or that men are the only people worth considering. If you use any of the terms in the left-hand column below but are not referring to a specific person, try to use one of the alternatives in the right-hand column.

businessman	business executive
craftsman	artisan
fireman	fire fighter
foreman	supervisor
lineman	line worker
mailman	mail carrier
manmade	synthetic; artificial
manpower	work force; labor force
workmanship	work; artisanry

Exclusive language risks offending readers; at the very least, it will tend to make some of your readers feel left out or ignored. Obviously, if any member of your audience feels offended or ignored in any way, then communication is hindered.

Principle 2: Select Words Carefully

Adjust your vocabulary to match your reader's needs. A basic rule of good writing is be sure that your reader can understand and follow what he or she is reading. Always keep in mind your reader's level of familiarity with the topic, and choose your words with caution. Ask yourself who will read the report, the proposal, the manual. Will different people read different parts? Choose words that are correct and that readers understand. Even expert readers appreciate clear writing.

4. *Avoid inflated language.* Writers who are unsure of themselves or their ideas may attempt to impress their readers by using "big" words. Pompous, stuffy writing is rarely effective because most readers recognize the pretentiousness of inflated vocabulary. Most readers appreciate everyday language. Keep in mind that your task in writing is to inform your readers, not to impress them with your vocabulary. Replace inflated language with more familiar words.

DON'T WRITE:	WRITE:
assistance	help
cease	stop
is contingent upon	depends on
constitutes	makes up
envision	see
endeavor	try
furnish	give
institute	begin
necessitate	require
notification	notice
originate	start
prior to	before
reside	live
retain	keep
terminate	end
utilize	use

DON'T WRITE:

The SecDisk provides a safe, secure place for floppy disks because it *has the capability of being* locked.

WRITE:

The SecDisk provides a safe, secure place for floppy disks because it *can be* locked.

5. *Avoid jargon.* Every field has its special language. Specialized terms are often more precise than general terms. For example, physicians prefer the specific term "myocardial infarction" to the vague term "heart attack." In addition, specialized terms frequently help to condense writing. For example, accountants use the term "depreciation expense" as shorthand for "write-off of original cost of equipment and buildings," and computer scientists use "AI" to designate the study of Artificial Intelligence.

If you use specialized language among members of a trade or profession, you can often communicate more efficiently because all members of the group understand the precise meaning of the term. However, if you use the specialized language of a trade or profession in place of general words that would be more appropriate for an audience having different specialties, you will be using *jargon*.

The disadvantage of using jargon is that the term or terms may be meaningless to people who don't know the jargon, or, worse, may be open to misinterpretation. If, for example, you are describing your work on a computer project, your meaning will certainly be missed if your readership thinks you are discussing members of the order Heteroptera when you suggest that the bugs you encountered delayed your progress in completing the project.

Similarly, to an English teacher the word "diction" means word choice. But, to a speech therapist "diction" refers to the clarity of a person's speech. And, while most readers are quite familiar with the term "liquor," many would be surprised to learn that the term also refers to the solution of chemicals used to break down the cellulose fibers in the paper-making process.

One choice facing technical writers is when to use specialized language. If you are sure that all of your readers will understand your use of the terms "rack rate" and "pot," you may safely use them (although you might be wise to include a glossary of such terms in an appendix or define them in the text). If you are in any doubt, however, replace specialized terms with more familiar words; for "rack rate" you might use "the official posted rate for a hotel room," and for "pot" you might prefer "potentiometer."

6. *Define and explain technical terms when necessary.* If your readers are experts, they know their field intimately and will appreciate the use of specialized vocabulary. However, if you suspect that your audience might not know the technical term you are using, you must define and explain the term or risk losing your audience's attention.

> Our automobile insurance includes a deductible on collision, *that portion of the insurance that repairs damages to Association-owned vehicles.*

> A red square on the radar screen indicates that the return is not squawking, that is, *transmitting any coded identification signals.*

7. *Avoid slang.* Slang is nonstandard language that can be used to establish a humorous or casual tone. However, slang is inappropriate for the relatively formal tone that characterizes most technical writing. Rather than projecting a friendly, casual tone, slang may offend your readers.

DON'T WRITE:

We always attempt to contact customers by phone or mail to advise them to make some sort of transaction (deposit, withdrawal, or update), thereby giving them the chance to avoid *the hassle* of obtaining the funds back from the state.

WRITE:

We always attempt to contact customers by phone or mail to advise them to make some sort of transaction (deposit, withdrawal, or update), thereby giving them the chance to avoid *the bother* of obtaining the funds back from the state.

8. *Make sure you use the correct word.* Words that are inaccurate can be misleading, confusing, and costly. You will be inaccurate if you fail to distinguish between similar words (such as "mallet" and "gavel"), if you fail to be specific ("hammer" instead of "mallet"), or if you simply use the wrong word ("maul" for "mallet"). Find the inaccurate word in the following example, taken from a training manual, and picture the conscientious trainee trying to decide exactly what it is he or she is expected to to.

File safe-deposit box receipts by box number, and thereunder, in chronological order.

9. *Avoid clichés.* Some expressions have the power of originality when they are first coined, but as they are copied and overused they lose their punch and become trite. One good example of this is Harry Truman's famous line, commenting on those in power who find it difficult to make decisions and who can't take criticism: "If you can't stand the heat, get out of the kitchen." Any expression that, while once thought fresh and original, has been used so often that it has become dull and trite is considered a cliché. Here are just a few others. You can probably expand the list.

Busy as a bee.
Dead as a doornail.
Like water off a duck's back.
Quick as a wink.
Selling like hotcakes.
Water over the dam.
Went in one ear and out the other.

Inexperienced writers often resort to using clichés because they are handy. Because readers quickly tire of the predictability of clichés, the ideas as well as the expressions in cliché-loaded writing will seem predictable.

Principle 3: Write Clear Sentences

The challenge in selecting *words* is to be correct and comprehensible. The challenge in writing *sentences* is to convey your overall meaning clearly and without ambiguity. If you follow the next six guidelines, your readers will be able to follow easily even the most complex policies, procedures, and instructions.

10. *Eliminate extra words.* Any word or expression that contributes nothing worthwhile to the sentence is "wasted." It wastes space on the page, it wastes the reader's energy, and it wastes the reader's time by making the sentence difficult to understand. Although you need not squeeze every phrase into the smallest possible number of words, cut out extra words that add length without adding real information.

DON'T WRITE:	WRITE:
a majority of	most
at the conclusion of	after
at the present time	now
at this point in time	now
based on the fact that	because
due to the fact that	because
for the purpose of	to
for the reason that	since
in view of the foregoing	therefore
of the order of magnitude	about
on a daily basis	daily
owing to the fact that	because
prior to	before

DON'T WRITE:

All variations from $.01 and up, over or short, are to be entered on the General Ledger.

WRITE:

Enter all variations on the General Ledger.

11. *Avoid redundancy.* Redundancies are repetitive expressions inserted in the mistaken belief that they will intensify the significance of the message.

DON'T WRITE:	WRITE:
and moreover	moreover
and so as a result	and so
but nevertheless	nevertheless
consensus of opinion	consensus
during the course of	during
few in number	few
never at any time	never
results so far achieved	results

12. *Write short sentences.* Use short sentences (generally under 25 words) when your readers are unfamiliar with your topic, when your readers are unskilled, or when your topic is complex. Short sentences are easier to understand than long sentences.

DON'T WRITE:

Company employees are entitled to two weeks of vacation with pay any time during the calendar year in which their first employment anniversary falls, three weeks of vacation with pay the year in which their fifth anniversary falls (provided that the third week is not taken prior to the anniversary date), and four weeks of vacation with pay the year in which their tenth anniversary falls and thereafter (provided that the fourth week is not taken prior to the anniversary date).

WRITE:

Company employees are entitled to two weeks of paid vacation any time during the calendar year in which their first employment anniversary falls. They are entitled to three weeks of paid vacation the year in which their fifth anniversary falls. (They cannot take the third week of this vacation before the anniversary date.) They are entitled to four weeks of paid vacation the year in which their tenth anniversary falls and thereafter. (They cannot take the fourth week before their tenth anniversary date.)

13. *Make sure relationships between ideas are clear.* The following sentence is from a manual on personal loans. It is intended to offer guidance to managers and loan officers. Notice the potential confusion caused by the failure of the writer to make sure that two different ideas were clearly related.

DON'T WRITE:

This is a convenience loan to a good Second City customer, and we must recognize the impracticality of making this small personal loan (*Standards for Manual Preparation*, 1984).

One purpose of a policy is to help people make decisions. Does the company want to emphasize the convenience to a good customer or the impracticality of making the loan? How is a manager or loan officer to decide? The problem with the sentence above is that the word "and" suggests an even balance between convenience and impracticality. A clear statement would resolve the ambiguity and help the reader make a decision. One suggestion for avoiding difficulties in unrelated sentence parts is to read out loud your own sentences to ensure they make sense.

WRITE:

Although we must recognize the impracticality of making small personal loans, we must remember that this is a convenience loan to a good Second City customer.

OR WRITE:

Though this is a convenience loan to a good Second City customer, we must recognize the impracticality of making this small personal loan.

14. *Rewrite negative sentences as positive sentences when you can.* Negatives are indirect: they imply what we *should* do by telling us what *not* to do. Even well-educated readers have trouble translating negatives into positives. Rather than forcing the reader to perform mental gymnastics, use straightforward positive statements whenever possible. Check your writing to see if you use negatives such as these that can be changed easily into positives.

DON'T WRITE:	WRITE:
not until	only when
not unless	only if
not prevent	permit
not reject	accept
not disagree	agree
not illegal	legal
not fail	succeed

DON'T WRITE:

Although negative constructions are grammatically not unacceptable, they are not easy to follow.

WRITE:

Although negative constructions are grammatically acceptable, they are harder to follow.

15. *Avoid noun strings.* If you use several nouns in a row, your reader may have trouble discovering what you mean. It often takes extra words to rewrite a long string of nouns, but the longer versions are clearer than the shorter ones.

DON'T WRITE:

The following is a collection procedure steps sequence used systematically to control delinquent accounts.

WRITE:

The following is a sequence of steps in the collection process used when an account is delinquent.

OR WRITE:

Use the following series of steps to collect delinquent accounts.

16. *Untangle confusing sentences.* Sentences are sometimes so convoluted that readers may either not bother to try to work out the meaning or may arrive at the wrong meaning. These sentences should be broken into several shorter sentences, each containing only one or two ideas.

DON'T WRITE:

Warning Flag Number 4—Chex Record   Explanation: This flag should be placed on any savings account you open if the customer has a record with Chex inquiries. It should alert others to take the necessary precautions in handling the account, not overriding check holds and the like.

WRITE:

Warning Flag Number 4—Chex Record   Explanation: This flag should be placed on any savings account you open if the customer has a record with Chex inquiries. The flag should alert others to be careful in handling the account. Check holds, for example, should not be overridden.

Principle 4: Create Documents That Are Easy to Navigate

A technical document is usually challenging for one of two reasons: either it presents difficult information, or the reader doesn't have enough time to digest it thoroughly. The following elements of style can help readers get through the document quickly and easily and learn from it as much as they need to know.

17. *Put important information where readers expect to find it.* If readers know ahead of time where they can find important information, they will make their way through a document with confidence and ease; if, on the other hand, they are repeatedly disappointed in their expectations they will become frustrated and even confused. This frustration and confusion is even more trying for the reader who only has time to skim a document. This reader wants to flip through the document quickly, identifying and making a mental note of each important point. Clearly finding important information becomes a much more difficult and tedious exercise if the reader has to search for the main points and then reread the passage to make sure nothing has been missed.

Luckily for both the writer and the reader, some conventions regarding the presentation of important information have been developed. In general, important information is either placed first, placed last, or distinguished in some way from the text around it.

The paragraph is the basic unit of most technical writing, and the topic sentence is the heart of the paragraph. The topic sentence introduces the paragraph or states its most important idea. Not all paragraphs have topic sentences, but most paragraphs in technical writing should.

A topic sentence should usually be either the first element in a paragraph or the last. Topic sentences are most commonly placed near the beginning of the paragraph. This method of organizing the paragraph is clear and straightforward. Topic sentences may be placed at the end of the paragraph as a means of summarizing information, and for variety. The danger of this method of organization is that the reader must hold in mind all the details until the end of the paragraph. This could be difficult for inexperienced readers or readers who do not fully understand the material.

Ultimately, the placement of the topic sentence is determined by considerations of audience and purpose. If the material is highly technical or conceptually difficult, or if the readers might be unfamiliar with the material, place the topic sentence at the beginning of the paragraph.

Following is a paragraph from a report on monitoring hazardous waste in the workplace. The purpose of the paragraph is to describe the limitations of mechanical monitoring of hazardous substances. The topic sentence — the general statement that ties together the information contained in the paragraph — provides a clear sense of the subject of the paragraph and the writer's main point.

> Using meters or analyzing air samples for levels of hazardous substances in the workplace are limited in protecting workers' health. Meters can break; their accuracy is dependent on how well they are calibrated; and because meters are typically set in one fixed location, their measurements may not accurately reflect the varied exposures individuals receive. Moreover, mechanical measurement does not take into account the differences in workers' individual metabolisms. That is, a broad measurement of the overall work environment does not account for the possibility that worker A's metabolism may remove toxic substances from his body more efficiently than worker B's metabolism removes toxic substances from her body. This could theoretically occur even though worker A may be exposed to higher concentrations of the hazardous substance than worker B (Queijo, 1985).

In some types of technical writing paragraphs are not used very often. Instructions, for example, usually consist of a numbered list of steps to be performed, perhaps with a brief explanation at the beginning. How should a particularly important piece of information be presented in this case? Again, there are three main choices: put it first, put it last, or distinguish it from the surrounding text. Because instructions are rarely read before they are used, the second choice — putting it last — is not very practical: a reader could read all the way through the instructions, following each step in turn, before reading an important warning that should have been heeded at the beginning. Here are examples showing how the other two choices could work:

Operating the Immersion Heater

WARNING: Never plug heater in when it is not immersed in water, as this will cause the unit to overheat and damage itself.

1. Fill cup with water.
2. Place unplugged immersion heater in cup up to the red mark.
3. If the heater cannot rest on the bottom of the cup without being immersed above the red line, it may be fastened to the cup with the attached clip or held in place by hand.
4. Plug in the heater.
5. When the water is hot, *first* unplug the heater, *then* remove it from the cup.

Operating the Immersion Heater

1. Fill cup with water.
2. Place unplugged immersion heater in cup up to the red mark.

> **WARNING: Never plug heater in when it is not immersed in water, as this will cause the unit to overheat and damage itself.**

3. If the heater cannot rest on the bottom of the cup without being immersed above the red line, it may be fastened to the cup with the attached clip or held in place by hand.
4. Plug in the heater.
5. When the water is hot, *first* unplug the heater, *then* remove it from the cup.

18. *Use helpful transitions.* Even if every sentence is clear and if every paragraph has an effective topic sentence, readers still may have trouble putting the pieces together. Good technical writing must also be coherent; it must lead readers from one point to the next so that the reader easily comprehends the entire argument or discussion. One effective tool you can use to show the precise relationship between ideas is transitions. Transitional words or phrases signal to the reader where the next phrase, sentence, or paragraph is going. Some words and phrases move the text in the same direction: "in addition," "also." Some move it in the opposite or in another direction: "on the contrary," "however." And some move it toward the end: "and finally," "in conclusion."

In the paragraph above from the report on monitoring hazardous waste in the workplace, the careful use of transitions makes the paragraph easy to follow. The word "moreover" in the third sentence signals the reader that the paragraph is moving in the same direction. The phrase "that is" in the fourth sentence signals the reader that an explanation is coming.

19. *Use lists when you have several items to present.* Technical writing frequently involves lists. Lists are most helpful when each item is brief and the items are parallel in content, and when the relationships among items are clear. Lists are especially useful when you are presenting a sequence of actions or steps and when you are presenting several components that could be confusing. Here is a passage that could be made into a list in order to clarify an awkward and time-consuming set of directions.

DON'T WRITE:

If you suspect a data set or telephone problem and you find it necessary to put in a trouble call to First Data Processing, the condition of the status lights is needed by the representative taking the trouble call. Additional information needed to adequately report your problem includes the FD number, which appears on a silver piece of tape usually located on the top of the data set (e.g., FD 3495-16). We must have this number to process any data set or phone line problem.

WRITE:

If you suspect a data set or telephone problem and you find it necessary to put in a trouble call to First Data Processing, you must be able to give the First Data representative the following information:

- The condition of the status lights.
- The FD number. This number is on a silver piece of tape, usually located on the top of the data set—e.g., FD 3495-16.

When you construct a list, always use parallel structure. Sentences or clauses that are linked in the same way to some major idea should have the same grammatical structure.

DON'T WRITE:

Writing an effective regulation is a process, not a single activity. The process includes at least six steps:

1. Identify the problem
2. Determining the solution
3. Gathering information
4. You should organize the information
5. Evaluate the communication
6. Distributing the communication

WRITE:

Writing an effective regulation is a process, not a single activity. The process includes at least six steps:

1. *Identifying* the problem
2. *Determining* the solution
3. *Gathering* information
4. *Organizing* the information
5. *Evaluating* the communication
6. *Distributing* the communication

20. *Use informative headings.* Use headings (where appropriate) to break up the document into manageable chunks. Headings should give a good idea of what is covered in each section, so that readers have an idea of where they are in the document. Section headings serve three purposes. As documents

Checklist for Technical Writing Style ———————

Specialized Conventions

____ 1. Have I found out about and followed any style guidelines that are specific to my company or organization?

____ 2. Have I found out about and followed any style guidelines that are specific to my profession or field?

General Conventions

____ 1. Have I helped readers to feel involved in what they are reading by

    ____ addressing them as directly as I could?

    ____ writing in the active voice?

    ____ avoiding exclusive language?

____ 2. Have I selected words carefully by

    ____ avoiding inflated language?

    ____ avoiding jargon?

    ____ defining and explaining technical terms when necessary?

    ____ avoiding slang?

    ____ making sure I have used the correct word?

    ____ avoiding clichés?

____ 3. Have I written clear sentences by

    ____ eliminating extra words?

    ____ avoiding redundancy?

    ____ writing short sentences?

    ____ making sure relationships between ideas are clear?

    ____ rewriting negative sentences as positive ones?

    ____ avoiding noun strings?

    ____ untangling confusing sentences?

____ 4. Have I created documents that are easy to navigate by

    ____ putting important information where readers expect it?

    ____ using helpful transitions?

    ____ using lists when I had several items to present?

    ____ using informative headings?

become longer and more complex, readers become subject to fatigue. The sheer mass of information requires some break.

In addition, section headings highlight changes in direction and emphasis, allowing readers to concentrate on the material they find most interesting and important.

Finally, the conventions of certain disciplines require their use. For example, acceptable research reports in the physical and biological sciences must follow a specific pattern: first an abstract, then a statement of the objective, then a literature review, followed by a description of the procedures used, a description of the results, and finally a section discussing the results. In almost every circumstance these required sections should be clearly identified with headings.

Activities

1. Rewrite the following passage, making it less wordy.

Should you wear your seat belt during pregnancy? Yes, you should wear your seat belt during pregnancy. According to medical research, it has been shown that in the event of an automobile accident, the most serious risk to your baby is that you will suffer an injury.

2. Write a brief analysis of the style of an article in a technical journal. Address these questions:
   a. What steps did the writer take to help readers feel involved in what they were reading? What more could have been done?
   b. Did the writer select words carefully? Are there any examples of poor word choice? Are there any examples of particularly good word choice?
   c. Are all the sentences clear? What, in general, would make any unclear sentences more clear?
   d. What steps did the writer take to create a document that is easy to navigate? Can you easily find each major point or section in the report? How did the writer help you to do so? If the document is poorly arranged, what should the author have done to clarify the organization?

3. Below are ten unfinished clichés. How many can you complete? (If you have trouble with any of them, compare your completed clichés with those of your classmates.) Give yourself ten points for each correct answer. What does your score tell you about the predictability of clichés?

At the drop of a . . .
Flat as a . . .
Last but not . . .
Like a needle in a . . .
More fun than a . . .
It's no use crying over . . .
Sly as a . . .
Stubborn as a . . .
My new car will stop on a . . .
The bigger they are, the . . .

4. In small study groups, place yourselves in the position of editors at a large publishing company. Certain sentences and book titles from your publications have been challenged as sexist, and you realize that some changes must be made. You are meeting to consider what action to take. Reach some decision, as a group, on each of the following sentences or titles:

a. Businessmen, who are the backbone of our society, need a stable economy in order to survive.
b. The average working man cannot continue to support his family on a salary that shrinks each year through inflation.
c. *The History of the Black Man in America* (book title).
d. Nancy Moore will be chairperson for the faculty while Alan Young will be chairman of the administration committee.
e. *Man and His World of Science* (book title).
f. The American colonists brought their wives and children with them to the New World.
g. A proud nurse cares about her patients' feelings.

5. Change the following passage from the passive voice to the active voice:

It is well known that advertising is designed to sell merchandise. But it is usually not realized how much planning and work is being put into it. No effort is spared by manufacturers to get our minds used to their products. Once it was considered enough to be better known than the closest competitor. Now advertising slogans are being built into our everyday life.

Many complaints are heard about this. Advertising is being attacked because our press is too dependent on advertising income. It is also said that our radio and television entertainment is being cheapened since practically all of it is sponsored by advertisers. Even the beauty of our country is considered spoiled because our highways are lined by billboards.

These complaints have been answered by a strong counterargument: thanks to advertising, our standard of living has been made the highest in the world. . . .

6. Find three professional style manuals (your library should have many of those listed in this chapter). Prepare an oral report for your class in which you describe the basic differences and similarities among the manuals. Your report should focus on the stylistic features covered in this chapter.

Quality Reviews

5

Determining Quality
  Reasons to Review for Quality
  Common Problems in Determining Quality
  Strategies for Determining Quality
Writer-Based Reviews
  Review for Company Standards
  Review for Accuracy and Organization
  Review for Correctness, Format, and Style
  Guidelines for Writer-Based Reviews
Checklist for Writer-Based Reviews
Colleague-Based Reviews
  Review for Organization and Format
  Review for Effectiveness
  Review for Accuracy
  Guidelines for Colleague-Based Reviews
User Testing
  Comprehension Tests
  Performance Tests
  Evaluation Surveys
  Guidelines for User Testing
Activities

Quality assurance is a major topic in manufacturing these days, with everyone trying to improve product quality. But what makes a quality document? How do you assure quality writing? How do you test for it?

The piece of writing in Figure 5-1 was written as a section of a computer manual. Every word in the excerpt is spelled correctly, all sentences are grammatical, text formatting is used to highlight key terms, and general technical style conventions are followed. But when users tested it, they didn't like it. "The users wanted to know why they would use a command, when they would use the command, how they would use the command, step by step; and how they would know they had gotten the desired results or information" (Rivers, 1991).

There were significant problems. Correct spelling, grammar, formatting, and style were all there, but that did not assure quality. After user testing, the command was rewritten to the form shown in Figure 5-2. That is quite a change, all based on the results of user testing.

"Quality" in this case meant more than correct spelling, grammar, and formatting. Quality assurance meant finding out what users needed, and supplying it.

Figure 5-1. Computer Manual Before Review and Revision

Listing All Users <Users>
<List>

Selecting this item lists all the users on the system. The listing of the users is displayed to the terminal. If the listing is more than one screen long, then press the return or newline key to page down or enter a <-><Return> to page up. Entering a <q> quits the display.

Included in the listing are:

- The user's login name
- The user's group name
- The applications (application users only) to which the user has access

The listing may be sent to the default printer or copied to the specified file. This specified file is indicated by its full path name. For example, to save the list under the name *user_list* in the */usr* directory you would enter */usr/user_list*. the file name must not already exist in that directory.

Source: Rivers, W. E., & Carr, D. R. (1991, January). The NCR–USC document validation laboratory: A special collaboration between industry and academia. *Journal of Business and Technical Communication, 5* (1), 97. Copyright © by Sage Publications, Inc. Reprinted by permission of Sage Publications, Inc.

Figure 5-2. Computer Manual After Review and Revision

Listing All Users <Users>
<List>

Why

- You would use this task:
  - To display all the users on your system.
  - To print a copy of all the users on your system.
  - As a reference for future use.

When

- You should print this list after you have added all the users to the system.

- Every time that you add a new user.

How

- The list of active users is displayed when you complete the menu selection.

- If the listing is more than one screen long, press the RETURN key to page forward or enter a—RETURN to display the previous page. Entering the letter q RETURN key sequence quits the display and redisplays the menu screen.

- Included in the listing are:

  - The user's login name
  - The user's group name
  - The applications (application users only) to which the user has access

Step 1: Indicate if you want to print the user listing.

You . . .	Enter the letter . . .
Want to print the listing	y
Do not want to print the listing	n

Source: Rivers, W. E., & Carr, D. R. (1991, January). The NCR–USC document validation laboratory: A special collaboration between industry and academia. *Journal of Business and Technical Communication, 5* (1), 98–99. Copyright © by Sage Publications, Inc. Used by permission of Sage Publications, Inc.

Figure 5-2. Computer Manual After Review and Revision (Continued)

Step 2: Indicate if you want to save the listing in a separate file.

You . . .	Action . . .
Want to save the listing to a file on the system	Enter the full path name of the file
Do not want to save the listing in a file	Press RETURN without entering any other characters in the field

- For example, to save the list under the name *user_list* in the */usr* directory you would enter */usr/user_list*. The file name must not already exist in that directory.

- If you do not know what a full path name is, refer to the "Describing the File System" chapter in this guide for a description of full path names.

Example listing

NAME	GROUP	APPLICATIONS
root	rootgrp	
shutdown	rootgrp	
startup	rootgrp	
va	rootgrp	
sys	sys	
bln	bln	
adm	adm	
norm	norm	
uucp	uucp	
charlie	special	/appl/special

What is quality, and how can it be assured? First, quality is determined not by the writer, but by the reader. Quality means meeting the needs of the reader for information. It can be measured by the reader's ability to find information and use it, whether that means making a decision, following a procedure, or learning something new.

How do you assure quality? Ultimately you have to test a document to ensure that readers can use it for its intended purpose. This testing can come in many forms and in many phases, but all of the tests have the same end: ensuring that a document does what is needed by readers.

Determining Quality

Reasons to Review for Quality

The reason to review for quality is to see if a document can be used well by readers. But there is some value in breaking this larger goal into four, more immediate goals.

1. *Identifying weak sections of a document.* Any test of a document will find unevenness in quality. Even skilled writers may leave a section or two incomplete, or write an occasional sentence that makes no sense. The longer the document, the greater the number of weak areas there will be. In large measure, one goal of quality assurance is to identify those weak spots.

2. *Learning strategies for improvement.* During the course of testing, writers not only find the weak spots in their documents, but, if the testing is done well, they collect useful ideas for revision. Test readers may suggest new approaches, or even if test subjects say nothing, writers can often discover useful new approaches while watching readers struggle.

3. *Deciding if a document is acceptable by management.* A proposal may be sent out to dozens of potential customers. It may determine how a company is viewed, and if the company is able to get new business. Even a document that stays in-house, such as a report sent between departments, will have an impact on how the home department is viewed throughout the company. Someone and some process must determine if a document is ready.

4. *Deciding if a document is ready for publication.* One common purpose for testing is to make decisions about duplication. The cost of making thousands of copies of an instruction manual, or dozens of copies of a feasibility study, may be quite substantial. Some process has to be used to determine whether a document is ready.

Common Problems in Determining Quality

The example computer manual instructions in Figures 5-1 and 5-2 help explain part of the problem of creating quality technical documents. Even good writers who work hard may fall short of quality, not because they are lazy, but because they face several hurdles in reviewing and revising their documents.

EXPERTISE

A writer's own expertise can create problems. Writers normally write about subjects they understand well to begin with, and as they write they gain even more expertise. Their expertise then shapes their thinking about the subject. They begin to regard some things as obvious. They make assumptions, or leave things unstated because the point seems perfectly clear. They move farther and farther from the position of a novice reader. As a consequence, writers often become terrible judges of the quality of their own writing.

TIME

Time pressures are constantly with us, but they are especially bad for writers of technical documents because documents are often the final phase of a project. A proposal is usually the last thing to come out of a study group. A report is the last work of a committee. An instruction manual is the final piece of a new machine. If these projects were delayed at any point in their proceedings (and they probably were), the amount of time left to write the document was probably reduced. Every day spent writing is one more day before the project can be completed, the action can be taken, or the product can be sold.

The pressures to publish without thorough quality testing are even greater, because quality testing takes time. The writer needs to make serious checks, peer reviews should be attempted, and user testing should often follow. None of this can happen overnight. Too often, the response is to have a colleague give the document a cursory review, dot a few *i*'s, cross a few *t*'s, and send the document to duplicating. The deadline may be met, but seldom with a quality document.

COST

Quality costs money. To begin with, anything that takes a little longer will mean extra personnel costs. If a writer asks a fellow employee to spend two hours rereading a document, that is two hours when other work isn't getting done. If a writer tests a manual with employees, they have to leave their other work. All of that is a cost.

More elaborate review procedures can involve performance testing using additional new materials and trained observers. For instance, a writer might devise a task that a user is to accomplish by following a manual. The writer needs to prepare the materials for the task, time or videotape the test, and then evaluate the results. The equipment and materials needed for testing cost money, the testing takes time, and the test subjects should be paid.

EGO

Possibly the biggest barrier to determining document quality is ego. The more thoroughly writers test, the more likely it is that they will find problems. It is the document that has the problem, but too often the writer will take the problem personally. In an ideal world it might be possible for some writer to produce some document that met all the needs of every reader the first time off the press. In a real world, the best a writer can hope for is enough talent to catch the obvious errors, and enough creativity to design good tests to catch the rest.

Strategies for Determining Quality

Assuring quality is a creative problem-solving task. It requires that you identify exactly what you want to know, create tests that will detect and measure those features, and then implement the tests in such a way that they produce valid results. Document testing essentially involves three basic approaches: a review by the writer, a review by colleagues, and user testing. Which of these review procedures you run will be determined by available time and money and by the importance of the document.

Writer-Based Reviews ———————————————

Most quality reviews are conducted by the author of a document. In part this is a matter of responsibility: you wrote the document and therefore are principally responsible for its quality. In part it is a practical matter: it is easier for you to do your own tests than it is to arrange for someone else to do the tests. Quality assurance for important documents should never be based exclusively on writer-conducted reviews, but with care and practice, these tests should catch many problems. Writers commonly review their own writing for company standards, completeness, accuracy, organization, format, correctness, and style.

Review for Company Standards

Companies are increasingly developing document standards to ensure a common look for all documents and to ensure that documents are complete. Such a review might include such issues as titling conventions, standard organization, and page format. These standards are often put in the form of a checklist like the one in Figure 5-3.

Figure 5-3. Sample Checklist for Company Standards

Engineering Associates
Company Document Standards

____ Titling information is complete.

    ____ Document is numbered appropriately.

    ____ Author and department are included and complete.

    ____ Date is included.

    ____ Version number is included.

    ____ Title is complete and reflects the content.

____ Standard organization is followed.

    ____ Company document outline is followed.

    ____ All sections included.

    ____ Appendixes are attached as needed and labeled.

____ Company format is used.

    ____ Pages are numbered bottom center.

    ____ Chapter titles are continued on top of each page.

    ____ Left margin is 1.5 inches, right is 1 inch.

Review for Accuracy and Organization

Even if there are no stated standards for your company, there are generally accepted standards that all documents are expected to meet. All documents should be well organized and accurate. You should run various checks on your own before asking anyone else to review your work.

REVIEW FOR ACCURACY

So much information from so many different sources is included in long documents that it is possible for some of the contents of such a report to be wrong. The best way to prevent such problems is to approach accuracy from each of several positions.

First, you will want to verify all quotations used. If you are citing a printed source, go back and reread the source to make sure you are getting both the words and the intentions correct. If you are quoting people, it is well worth a phone call to verify that what you are about to print is what they actually said (or remember saying).

Second, double-check any factual information you've presented. Are you absolutely certain that every figure you cite is correct? You may need to check your sources once more, or you may need to recheck your own calculations.

Third, verify your use of all technical terms. This is especially important if you are writing about an area that is new to you. You may have misunderstood a term, or may be using it in a context in which it was not intended. The best way to check is to find the term in other printed matter and check your usage against the other documents. A phone call to someone in the field can also save you embarrassment.

Fourth, be sure that any procedures you describe are accurate. This may involve observing them one more time, or asking someone more familiar with the procedure to examine your description. It only takes one error of fact to destroy your credibility and the value of the document.

Finally, you must rigorously evaluate the quality of your evidence. If you performed a study, will your procedures stand up to scrutiny? Are your sources the best sources available? The evidence you present in your document becomes a logical building block for your conclusions. How firm is your foundation?

REVIEW FOR ORGANIZATION

The organization and structure of a document can greatly affect how easy to use or persuasive it is. For instance, a feasibility study is well-organized if readers find the information they need and if it logically supports the recommendations the writer makes. This aspect of quality can be reviewed by a writer in several ways.

First, write out the outline of the document. By looking at this outline, is it easy to see the original situation? Are recommendations placed where they are most persuasive? How many points of support can you find? Are they ordered for most effectiveness? Such an outline should tell you how logical your presentation is.

Second, examine the document for the amount of evidence it contains. Are you basing your recommendation on only one or two pieces of information? How much support is there?

Third, consider whether your evidence is presented effectively. For instance, if there is a study upon which you are basing recommendations, how well do you cite the study? Do you cite the specific source and quote the most important passages? Do you include examples? Is the evidence presented in the best possible order? How persuasive is your case?

Review for Correctness, Format, and Style

Reviews for correctness, format, and style may be less crucial than reviews for accuracy and organization, but they are still important. Correct grammar and spelling will largely determine if your ideas are taken seriously, while format and style will largely affect how well your document is read. Several common tests for correctness and style are described below. Interestingly, this is the level of review for which most computer revision tools have been developed.

SPELLING

Good writers commonly miss a simple spelling error or two despite proofreading. Unfortunately, it only takes one or two spelling errors for readers to decide that a writer has no credibility at all. There are three procedures you can use to help eliminate these embarrassing mistakes.

First, use a spell checking program. Most word processors now include a spell checking program. These programs take your document, list all the words you have used, then check that list against its own list of words to see if the words match. If the word matches, the program assumes the word is correct. If one of your words can't be found on the program's list, it assumes that the word is probably misspelled.

This method of catching spelling errors is partially successful. The most popular programs have word lists of 80,000 to 110,000 words and do a fair job of catching common spelling errors. But there are two inherent limitations to spelling checker programs.

First, the programs assume that if a word in your document is on their list, it must be correct. The sentence below shows how wrong that can be:

We went to there huse for diner.

"There" should be "their," but the program doesn't catch it because "there" is a word. Similarly, "diner" is wrong. The writer meant to say "dinner." But "diner" is a word, so the program will make no comment. The only error the program will catch will be "huse." As you can see, spelling checker programs will miss many spelling errors.

The other problem with such programs is that they usually have a limited technical vocabulary. A list of 80,000 to 110,000 words may sound like a lot of words, but the vocabulary of a technical area of engineering is not likely to be among them. As a result, the words you may be having the most trouble with are the words least likely to be included. For technical terms and formal names, you are on your own. Many spelling checkers do allow you to set up a "personal dictionary" where you can store any commonly used terms in your field that are not in the program's own dictionary. Be careful, though: if you add "erythormicine" instead of "erythromycin" to the dictionary by mistake, the program will accept the incorrect spelling as correct every time it occurs in your documents.

Because of these limitations, you should use spell checking programs where they are available, but know that they are only doing about half the job. The responsibility for catching errors is still yours. There are two manual techniques you can use for this.

First, print out the document you are writing. People read more effectively on paper than on a computer screen, so you should be able to see errors better on paper. Now that the document is on paper, read it backward. By starting at the bottom and scanning each word in turn, you are less likely to get engrossed in the content and overlook individual words.

Second, have a good dictionary by your side. Often spelling errors come not in the base word, but in adding -*ing* or -*ed* endings. Most people can spell "enter" but get confused over whether the past tense is "entered" or "enterred." A good dictionary will show you the correct spelling of the word in all its major forms. Once you have such a dictionary, use it. When in doubt, look the word up.

GRAMMAR

Good grammar refers to the acceptable structure of sentences in standard written English. Using good grammar means using complete sentences, correct pronouns, and correct verb forms, among others. There are three approaches you can use to help ensure the accuracy of your grammar.

First, keep your sentences under control. The longer and more involved a sentence is, the more likely it is to have an error. Good technical style promotes shorter, more direct sentences—sentences less likely to be ungrammatical.

Second, read your document aloud. For most native speakers of English, the ear is more sensitive to grammatical errors than the mind. If a sentence sounds strange, try to figure out its grammatical structure. You may find some hidden errors.

Third, use a computer for what it can do, and avoid what it can't. There are currently a range of programs available to check grammar. It would be nice if they could. Unfortunately, no program is more than about 30% accurate. Worse, such programs don't just miss 70% of the errors in a paper, much of the time they mark sentences as wrong when they are not. And, ironically, the worse your grammar is, the worse such programs are in correctly identifying your problems. In short, no program will correct your grammar for you.

But the computer can be of some value. There are programs that will list your document one sentence at a time, either alone on the screen or separated on paper. Such programs make it easier for you to proofread your writing. Currently, this is the best use for computers in checking for grammar errors.

FORMAT

A document's format consists of all those elements that make the organization and content clear and easy to comprehend. Most format concerns are related to document design: page layout, text blocking, the appearance of the type, the use of marginal glosses or lists, and the use of distinctive visual features (such as lines, boxes, color, or icons). Other issues are more closely related to the document's organization, such as the use of headings to mark separate sections of the paper.

When reviewing for format, ask yourself two questions. First, will the current format be effective? Second, is the format used consistently? An eye-catching icon or label inconsistently used may be worse than none at all.

In reviewing for format, page quickly through your document. Are the most important headings the most prominent on the page? Can you find at a glance the most important information on the page? Are important technical terms highlighted with boldfacing, underlining, or italics? Are figures and other graphics clearly set apart from the text?

Another way to review for format is to make a list of all the distinctive format features you have in your document and the style you have chosen to use for them. Then go through the document looking for all instances of a single feature: Is every single heading typed in the same manner? Is every single marginal gloss a summary of the paragraph?

STYLE

Style is a matter of presentation. Good technical style generally directs writers to help readers feel involved in what they are reading, to choose words carefully, to write clear sentences, and to create documents that are easy to navigate. (See chapter 4, "Technical Style.") Style can be checked in many ways, but let's start again with the computer.

Common computer programs are available to test reading level, or readability. Readability is measured by a formula created to gauge the relative difficulty of a passage. Such formulas generally produce a grade level for writing, usually 1 to 16. The grade level indicates the amount of schooling the typical reader would need in order to read the passage comfortably. Newspapers, for instance, tend to be written at readability levels of 7 to 8. College textbooks fall in the 11 to 14 range.

There is no ideal reading level or difficulty; text should meet the needs of readers. For instance, if you were writing a report on smoke stack emissions for a local county board, a level of 10 to 12 might be all you could expect. If you were presenting the same report to your engineering board, a level of 15 to 16 might be more appropriate.

Readability formulas are highly controversial. There is some evidence of their accuracy, and they are often specified in military training materials, but there are many weaknesses in their approach, too. Most, for instance, are based on very simple calculations. The Fogg Index is typical. It is calculated by taking the first 100 words of a passage (if the 100th word comes in the middle of a sentence, continue to the end of that sentence), and counting the number of long words (three or more syllables) and sentences in the sample.

Then the sum of the number of long words and the average sentence length is multiplied by 0.4. The formula looks like this:

Grade level = 0.4 (No. of long words + (No. of words/No. of sentences))

Notice that the formula is based entirely on sentence length and word length. It assumes that longer words and sentences are more difficult to understand than short ones. This may generally be true, but there are plenty of one-syllable concepts that are far from simple (like "ohm," "erg," "byte").

As a result, this test should be used as a general guide, or an early indication of how close your document might match the abilities of a reader. If the readability number is too high or too low, you will need to reconsider the needs and abilities of your readers and make a decision about how to respond. But also remember the limitations of readability formulas. They should help you catch obvious problems, but don't assume that just because your formula says a text had a readability of 8, every high school student can read what you just wrote.

Other computer tests are commonly available for aspects of style. Generally, these checks take one of two forms: graphic displays, and string matches.

Graphic displays present your text to you as a series of stars or dashes so you can quickly see certain features such as sentence length or paragraph length. A display might look something like this:

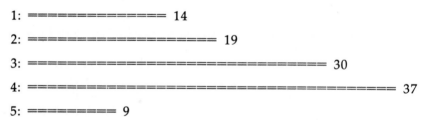

```
1: ==============  14

2: ====================  19

3: ==============================  30

4: =====================================  37

5: =========  9
```

The computer has simply graphed out the lengths of the first five sentences. We can now see they range in length from nine to thirty-seven words. Because sentences of thirty and thirty-seven words are longer than usual for technical prose, you may want to review them to see if they really should be so long.

Computer style analyzers using string matches look for particular strings of characters, such as words or phrases. Some of these strings might be signs of good writing, like the use of transitional phrases — "on the other hand," "furthermore"; others might be signs of problems. Usually string-match routines look for problems. They might look for trite phrases ("in our modern world of today"), clichés ("we'll be writing this report until the cows come home"), or simple style errors such as split infinitives ("we have to quickly go"). Figure 5–4 shows one such program at work.

Figure 5-4. Computer Style Analyzer

```
Your overuse of prepositions may create a clumsy
style.

Total            Total           Ratio:
Prepositions: 73 Words: 520      1 to 7.10

ABOUT            1    AROUND     1
AS               4    AT         1
BETWEEN          1    FOR        13
IN               16   INTO       2
LIKE             6    NEAR       1
OF               7    ON         2
OUT              1    TO         16
WITH             1

Advantages of living in Columbia

There are many places FOR people TO live. The types
OF places where people want TO live depend ON what
they are looking FOR. Some people might LIKE . . .
```

Source: *Writer's Helper*. Conduit, University of Iowa–Oakdale Campus, Iowa City, Iowa. Courtesy of CONDUIT Software.

The power of such computer programs varies dramatically. Some are able to search for dozens of such strings, others for thousands. They all have an inherent limitation: they can only find certain predetermined words or phrases. They will help you avoid silly mistakes, but not major ones. These programs are valuable, but they are no substitute for basic skill as a writer.

Even after you have run such computer programs, there are a number of style checks you should run on your own. (Use the checklist in chapter 4, "Technical Style," as a guide.)

First, read the document from the point of view of your intended audience. If your audience is a group of novices, try to imagine how they would feel reading through it the first time. Encouraged? Confident? Frustrated? Bewildered by the technical jargon? Insulted by the overly casual tone?

Second, read through your document with a pencil in hand and circle any word or phrase that might be reconsidered. Have you allowed pompous, inflated language to slip into your otherwise down-to-earth document? Have you used technical jargon when a simpler term (or a technical term accompanied by a definition) might do? Have you used any slang terms that are inappropriate in a technical document?

Third, read your document aloud. If you stumble from time to time or have trouble understanding what you yourself have written, mark the sentence. Once you've read the entire document, go back to these marked sentences and see if they could be rewritten to be more clear and direct.

Fourth, mark places where the document changes direction, moves to a new topic, or breaks a larger idea into smaller parts. Is there a heading, a label, or a transitional phrase that already signals this change? If not, now is the time to put one in.

Guidelines for Writer-Based Reviews

There are many reviews and tests that writers can make on their own documents to promote quality. With so many to perform, there is some value to thinking about an overall strategy to use in such reviews. Here are several guidelines that seem effective.

1. *Create a plan.* The checklist presented at the end of this section represents one way to plan your revision strategy. Any form of organization or schedule of tests will help, and should be created before you begin. Otherwise, it is easy to become so engrossed in one or two areas of review that you leave out a whole set of tests. You want some way to be sure you test all the major areas.

2. *Wait a day or two.* The more time you can leave between when you finish writing and when you start testing, the easier it will be for you to read as a reader instead of as a writer. It is amazing how much more you will see if you wait two days before starting your reviews.

3. *Divide and conquer.* Rather than try to do all the reviews simultaneously, concentrate on one at a time. Review the entire document for effectiveness first, then for accuracy, and so on. If you try to catch spelling errors at the same time as you are reading for organization, you will neither catch all the spelling errors nor fully examine the organization.

4. *Read as your reader.* Build a mental image of the people most likely to read your document. What is their background? Why are they reading the document? If you can put yourself in their place, you may be able to see how well they will be served by your document.

5. *Read as your employer.* A company has needs and goals, too. How well does your document fit into those general goals? How well does it represent the company? What kind of image of the company does it create?

Checklist for Writer-Based Reviews ———————

____ Completeness
  ____ All sections are included for a document of this kind.
  ____ All quoted materials have source citations.
  ____ All reference materials are cited or included as appendixes.
  ____ All complicated processes are illustrated.
  ____ All illustrations are labeled.
____ Accuracy
  ____ Any quotations have been verified.
  ____ All factual information has been verified.
  ____ All technical terms are used accurately.
  ____ Procedures have been verified.
____ Organization
  ____ Organization is logical.
  ____ Each point has sufficient and sound evidence.
  ____ Evidence is presented persuasively.
____ Technical format
  ____ Sections are labeled.
  ____ Page design is consistent.
  ____ Formatting highlights key terms.
  ____ Points of danger or particular interest are marked.
  ____ Procedures are broken down into single steps.
  ____ Lists are used wherever appropriate.
____ Correct writing conventions
  ____ Spelling has been verified.
  ____ Each sentence has been checked for grammar.
____ Appropriate technical style
  ____ Reading level matches audience.
  ____ Technical terminology is appropriate for audience.
  ____ Abbreviations match conventions.
  ____ Sentences use the active voice.
  ____ Sentences are generally short and direct.

Colleague-Based Reviews

With practice, you should be able to catch many of your own mistakes, but nobody catches all of them. At some point you will need a second opinion. While this is normal, it still creates problems. For one thing, any time fellow employees spend reviewing your work is not spent doing their own. Even if they are interested in helping you, your colleagues may not have the time.

Despite the limitations of colleague reviews, they are still an important step. More experienced writers can give you good ideas about presenting your information and can explain the reasons for the success or failure of past documents. The problem is to use their time wisely and to make sure you make the most of this opportunity.

What kind of reviews can colleagues perform? There are three that are common.

Review for Organization and Format

Someone who has written a particular kind of report many times can quickly look at a document of yours and tell you if it has all the correct parts. Such a review may not require more than a few minutes. The reviewer can look over the table of contents or page through the report and quickly spot missing sections or jumbled organization.

Review for Effectiveness

A reasonable question to ask more experienced colleagues is how well a particular passage meets its goals. Is the source reputable? Is the quotation sufficiently clear? Is there enough evidence to support a point? This draws on your colleagues' expertise as technical writers; you are asking them to share with you some of the skills and knowledge that have made them successful writers.

Review for Accuracy

Someone who has operated a piece of equipment or carried out a process for the last twenty years can probably tell you quickly if your operation description is complete and accurate, whether you are labeling parts correctly, and whether you are using technical terms accurately. The key is *expertise*. The only time you should ask a colleague to review your work for accuracy is when he or she has so much expertise that this is a simple task. It is not appropriate to ask a co-worker who's unfamiliar with the latest budgets to check over your proposed figures for the coming fiscal year.

Guidelines for Colleague-Based Reviews

The reviews described here make best use of a colleague's time and expertise, and treat that person with respect. Here are several guidelines for colleague reviews to help clarify this aspect of colleague reviews.

1. *Choose appropriate tasks.* It is not appropriate to ask a co-worker to verify your spelling. If it is your document, it is your responsibility. You have a dictionary and you probably have a spell checker program with your word processor, so you have no reason to use a colleague for spelling duty. Similarly, you should already know if your document meets the standards of your company, has correct formatting, is grammatical, and is complete. In short, you should not ask a colleague to spend time checking things you can easily check for yourself.

2. *Tell the reviewer the audience and purpose.* In order to give you helpful reviews, your colleagues will need to have some sense of the writing situation; in particular, they will need to know the audience and purpose for your document.

3. *Avoid asking global questions.* Questions such as "Is this report clear?" or "How could this be better?" are overly broad. They ask your reviewer to do all your thinking for you. You are more likely to get the information you need if you ask specific questions, such as "Is the welding description on page 6 accurate?"

These guidelines can often best be followed by putting together a brief question set for the reviewer, one which also provides information about the purpose for the document, the audience for the document, and any other significant information.

Review questions for this report. Note: I was asked to write this report on construction problems in the Highway 18 project. It will be presented to our quality assurance committee on Thursday. If accepted by them, it may be relayed to several of the subcontractors involved in the project.

1. Page 4 contains a technical analysis of the concrete we are using. Should this be moved to an appendix? Is it necessary at all?
2. Page 5 has a diagram of cracks that have appeared in the last week. Is the diagram sufficient, or should I use a photo as well?
3. Page 12 refers to past problems with Warnock Brothers Mixers. I tried to list these problems without much elaboration. Should I have said more? Less?

The point of the questions is that readers know exactly what is expected. They may read the whole document or, if time is short, respond only to the specific pages mentioned. They have been told the role and use of the document so they can judge what changes are best. Most important, their actions are being focused on concerns for them as professionals—what to say, not how to spell it. The questions show a respect for their judgment and time. And you get the answers you need, rather than general responses such as "yes, this looks OK."

User Testing

The third level of review goes to the final link in the communication process —the reader. You give readers your document, let them read it, and find out if what you wrote does the job. However, you may not always be able to do formal reviews with readers. For instance, if you are writing a proposal to another company, you can't ask them to read a draft, you can only wait for their decision. But there are many cases in which you can check with readers directly to see if they understand what you have written. Instruction manuals are one example, as are computer manuals and some kinds of reports.

This opportunity to test your writing for effectiveness allows you to be much more certain about the quality of your work—you can see for yourself if the writing is effective. Such an opportunity should not be wasted. With carefully crafted tests you can dramatically improve your writing.

There are quite a few tests to help you. You are the beneficiary of nearly one hundred years of research on reading. Newspapers, textbook publishers, and educational materials companies have compiled research to improve the effectiveness of written materials. Some of the tests that have evolved are highly technical and involve expensive machinery, but some of the tests are quick and easy to use. Some of the simpler tests are described below.

Comprehension Tests

You are probably used to taking reading comprehension tests of one kind or another. Now put yourself on the other side of the test: how can you be sure that readers understand what you have said? What test would you create to determine which parts of your document are clear and which are confusing? Here are three tests you might try.

CLOZE TESTS

The cloze test dates back to 1953 (Estes & Vaughn, 1978) and is one of the simpler tests to create and administer. You identify a portion of your document that you want to test, and type it with a blank placed where every fifth word was. The reader then writes the words in the blanks while reading.

Which passages should you select? Introductory material is important since it sets the tone for all that follows. There are probably other passages that you already know are either difficult to understand or crucial to the document (like passages of support for recommendations).

To see how cloze testing works, try it for yourself with this example:

Controlling Flare-up. Gas grill cooking is _____ by flare-ups and requires _____ cooking techniques from that _____ with charcoal grills. Because _____ grills must be well-ventilated _____
safety reasons, extinguishing fumes _____ oxygen deprivation is not
_____. Water should not be _____ to extinguish grease fire
_____ it will damage burner. _____ reduce flare-up, trim excess _____ from meat before cooking. _____ sure briquets are
thoroughly _____ prior to cooking to _____ grease remaining
from previous _____. Frequently replace Charcoal Briquets
_____ prevent grease accumulation and _____ flare-ups.

Here are the words that were left out: *characterized, different, used, gas, for, by,
possible, used, as, to, fat, be, heated, carbonize, cookout, to, subsequent.* Cloze
passages are normally scored such that only the exact word is accepted as
correct; no synonyms are allowed. (If you start accepting other words, you
may find yourself making endless judgment calls about what is really close
enough to count.) Count the number of correct words and divide by the total
number of blanks (17, in this case) to get the percentage correct.

What is a good score? A score of 60% or better is usually accepted as a
sign that the reader can understand the material without any help. A score of
40% to 60% usually means that if someone were there to answer the reader's
questions, the reader could work his or her way through the text. A score
below 40% means the reader was frustrated. Because you cannot expect that
your documents will be read with an expert on hand to answer questions,
you should try to produce technical documents that earn scores above 60%.

If your document produces test scores below that level, you have some
rewriting to do. By looking at where the wrong answers congregate you can
probably guess where most of the confusion is originating. A passage that
produces low scores will take more work. It may require an illustration, better
labeling, a redefinition of technical terms, or more straightforward prose. In
any case, the passage is not ready to go out. Rewrite it and then do another
cloze test until you get consistent scores above 60%.

MISCUE ANALYSIS

In the mid-1960s, reading researchers began a new technique to determine
reading comprehension: they had students read aloud, and listened carefully.
They tracked where readers made mistakes while reading, and tried to
categorize the mistakes, or "miscues." The three categories used were *graphophonemic, syntactic,* and *semantic* miscues. The psychological processes
behind each kind of miscue are very elaborate, but at the simplest level, the
researchers listened to determine if the error was at the word level
(*graphophonemic*—confusing "well" for "weld"), sentence level (*syntactic*—

reading two sentences as one), or passage level (*semantic* — replacing the correct word with a word that did not make sense in the passage being read) (Allen and Watson, 1976).

To do a complete miscue analysis requires substantial training, but you can do simple miscue analysis if you just listen carefully while someone reads your passage. To begin with, prepare to hear people read passages aloud that are very different from what you have written. Some adults tend to skip as they read and to guess at words. You shouldn't expect adults to read word for word like first graders — we all learn shortcuts as we gain experience. So you will hear many changes (miscues) in the reading. You want to listen for the *kinds* of changes that are being made. Where do readers seem to slow down or struggle while reading? Which words do they mispronounce? If they substitute a word for the one on the page, is it a synonym, or one with no logical connection to the original?

With a little practice you should be able to tell when readers are taking shortcuts with a document (a good sign) and when they are bogged down (a sign of problems). You should also have a good idea of the level at which problems are occurring. If the problem is at the word level, you may have to define your technical terms better. If readers are getting lost in a long sentence, you may need to rewrite that sentence or break it into several shorter sentences. If they are substituting unrelated words for what is on the page, they have lost the meaning of the passage and you have a major rewriting job on your hands.

READ-ALOUD PROTOCOLS

Read-aloud protocols also involve having a reader read your passage aloud, but in this case you ask readers to describe their thoughts as they read. You might ask them to rephrase the passage, sentence by sentence, in their own words as they read, or to comment about the passage, or tell you how well they understand. You can tape record their comments or follow along on the page and mark places where readers are unclear or seem confused. Later, these passages will be the ones where you can focus your rewriting efforts.

COMPREHENSION QUESTIONS

A more formal evaluation procedure is to ask readers to demonstrate their comprehension when they have finished reading. One simple way to do this is to ask a reader to summarize a passage. You will quickly find out how much they got out of the document. If you are concerned about technical vocabulary, ask the reader to define each of several key terms in your report. If your document involves a process, ask readers to list the steps in the process as well as they remember them. Another approach is to write several different paraphrases of a passage and have readers select the one they feel is closest to the original.

The work in a comprehension test is determining beforehand what your purpose is. It may not be important that the reader understand every term and never miss a single detail. What is important? What are you most concerned that a reader understand? Is it a warning or a particular process that could involve a risk of injury if misunderstood? If you determine in advance what has most value, the comprehension test will follow easily.

Performance Tests

In many cases you will be writing technical documents that must be followed step by step by readers while they perform a task. Instruction manuals are a prime example, but there are places in reports too where the reader must be able to look up information or perform a calculation. If there is a task that has to be performed while the reader is reading, you can set up a performance test. Success is usually gauged in one of two ways, time or completion.

TIME TESTS

In a time test you assume the reader will be able to get the job done, but you want to see how efficient it will be — how fast the reader can work. Presumably a reader will be able to follow clear instructions more quickly than muddy ones. Imagine someone following a manual on how to perform pollution tests. If your instructions are clear, the reader will move from step to step easily and get the job done quickly. If there is a problem, the reader will take long pauses while trying to understand the instructions.

To do a time test, you observe readers as they perform a task. You follow along in the manual and record the amount of time it takes to go through each step. If all goes well, the time it takes this person to perform each step should be similar to the time it would take an experienced person to do the same tasks. If one step seems to take an inordinate amount of time, you know you are confusing readers. When the task is complete, you can talk to the readers about that step, try to identify the problem, and rewrite the instructions. Figure 5-5 shows an example of the notes taken during one such timed performance test.

COMPLETION TESTS

In a completion test, you set up a specific set of tasks you want readers to perform, usually to make sure that a document is complete. For instance, since it is important to be able to access information quickly in a technical document, you may ask a reader to find a certain term, or name an important part in a machine, as a way of finding out how clear your indexes or tables of contents are. Or you may ask a reader to carry out a specific task as a way of seeing if all the necessary information is in the document.

Figure 5-5. Notes Taken During a Timed Performance Test

Timed Performance Test

Task performed: Copy a file.
Subject information: Novice user. Almost no computer ex-
perience, no experience whatsoever with Q-filer.
Testing Conditions: Room B-238 at 3:30. The lab was very
busy and work space was limited. Noise level was very high.
Overall time to complete task: 1 minute 20 seconds.
Completeness of task: Requested file was copied to the
drive exactly as description stated.
Time for each subtask:
Time to find first page of instructions: 17 sec.
Time to start first step: 13 sec.
Time to complete first step: 12 sec.
Time to complete second step: 13 sec.
Time to complete third step: 25 sec.
Results: The task was fully completed in 1 minute 20 sec-
onds. It could have been completed in less time, but the
subject did not notice that a response to a prompt was
needed, and waited a few seconds for something to happen.
This prompt response should probably be included in the
manual as a separate step.

One advantage of a completion test is that you can have readers focus their time on parts of the document that concern you most. Especially in longer reports or manuals, if you have readers follow through the entire document, you and the reader may be tied up for hours, and you might still not have a chance to see the reader perform all possible tasks. Setting specific tasks saves time and leads to more focused testing.

Evaluation Surveys

While we can test how well a reader does on a current version of a document and guess from that what more needs to be done to improve a document, sometimes the simplest thing to do is ask the reader directly. The problem is finding a good way to ask. A question like, "Did you enjoy reading this manual?" is not too valuable. If the reader says "yes," does that mean the document is done? Probably not. If the reader says "no," can you tell how the manual should be improved? Not based on that question. What is needed are direct questions that are simple for the reader to answer and lead to information the writer can use. Here are a few guidelines to consider for asking readers questions.

1. *Avoid global questions.* Questions like "Were the illustrations clear?" lump all illustrations together. As a group they may be fine, but you are also interested in specific illustrations that might be improved. It would be better to ask, "Mark the best illustrations, the worst illustrations, and any place you think should have more illustrations."

2. *Avoid value judgments.* Questions like "Were the instructions clear?" leave everyone wondering what "clear" means. It would be better to ask something less value-laden: "Which instructions seemed hardest to understand? Was the instruction format complete? Mark any instruction that confused you."

3. *Avoid asking the world.* Questions like "What would make this report better?" essentially ask the reader to do your work for you. Try to focus your questions. If you are concerned whether your report is complete, ask, "Are there additional contents that should be in this report?" If clarity is an issue, ask, "Which technical terms could be better defined? How do you suggest that be done?"

The premise of these questions is that a good document is one that is easy to follow and use. That is certainly true in large measure. But there is often a second side to reading. We may be trying to "sell" a product (while consumers are assembling it) or "sell" an idea (while a committee reads a proposal). In this case we not only want readers to understand the document, we want them to appreciate it. We want them to feel good about a product or idea. How do you measure that?

A simple approach is to use what is called a Likert Scale. You set up a range of values and ask readers to mark a place on the scale. Usually this is based on some statement and you ask readers to agree or disagree. Here is one example:

	Agree		Neutral		Disagree	
Library hours should be increased:	1	2	3	4	5	6

The number of points in the scale can range from 4 to 10, but 6 is very common. A 6-point scale also has the advantage of having no exact middle: you force people into either agreeing or disagreeing slightly. A 5-point scale, because it has a neutral 3, makes it easier for readers to simply circle the neutral and register no opinion.

Such a scale is useful in a wide range of surveys. You can ask questions about a product, about a concept, or about a manual itself. In each case, you write an opinion as a fact, and allow the reader to agree or disagree. The example survey below demonstrates this kind of scale

	Agree		Neutral		Disagree	
MirrorCo produces quality products:	1	2	3	4	5	6
Groundwater pollution is a problem:	1	2	3	4	5	6
This manual was adequately illustrated:	1	2	3	4	5	6

Such surveys are especially valuable if you will be collecting a large number of reactions. They can be quickly tabulated and summarized to show the average reactions of readers or customers.

Guidelines for User Testing

1. *Determine what you want to know before you start testing.* If you are concerned with user attitudes, that leads to one set of tests. If you are concerned with performance, that leads to another.

2. *Pick the places in your document most in need of testing.* A few documents are small enough to be tested in their entirety, but this is rarely the case. Usually documents are so long it would be very expensive to test every part of them. Therefore, you need to determine which places in the document are most likely to need rewriting, and test only those.

3. *Choose test subjects from the target audience.* This ensures that you will get accurate responses. If you are writing a manual for a consumer appliance, test your instruction manual on a typical consumer, not on an engineer down the hall who happens to have a few minutes free.

4. *Consider particular qualities of your readers.* Might they be speakers of English as a second language? If so, draw on subjects who have these traits.

5. *Use enough readers to draw valid conclusions.* A performance test done with one subject may say more about the subject than about the document. Use at least three subjects, more if the document is very important.

6. *Test under realistic conditions.* For assembly, it might mean using the tools that are most likely to be present. It might also mean testing under temperature or wind conditions that are common. A manual that seems clear when read under near-library conditions in your office might be much less clear on the factory floor with machines screaming and people shouting and a cold wind rushing through an open door. A true test of your writing should make conditions as realistic as possible. Only then will you know if you achieved what you wanted—a quality document.

Activities

1. Find two similar documents prepared by the same company, such as two reports or two instruction manuals. From examining the documents, draft a standards checklist for the company. Compare your checklist with those of others in the class.

2. The checklist for writer-based reviews described in this chapter is fairly general. What would you add to the checklist if you were doing a review of an instruction manual? A computer manual? A proposal? Create a review checklist for one of these documents.

3. You have probably been asked to review college papers written by friends. In small groups, discuss your experiences as "colleague reviewers." What are you generally asked to do when they give you the paper? How do you respond? How satisfied are they with your comments? How happy are you to verify grammar and spelling? How could they better use your time?

Take a paper that was recently written by someone in the group and collaborate to develop a written request for a review similar to one that would be used in the workplace.

4. What readability level do you think this book uses? Calculate the readability level of one page from this chapter and compare it to your prediction. What level do you think college textbooks should have? How well would a readability formula predict the difficulty level of a physics textbook? A math book? If your college keeps old textbooks, do a readability formula on a book published twenty or thirty years ago. How does it compare to the levels of today? If there is a difference, why do you think this is the case?

5. Create a cloze test for a passage of your choice and have several classmates take the test.
   a. Can you determine why some people did better in the test than others? What kind of backgrounds do they have?
   b. Rewrite your cloze passage so that you think scores will improve, and then give it to additional people. Did scores go up? Which changes seemed to do the most good?

6. To compare the miscues made by a good reader with the miscues made by a novice, take a passage from a physics textbook. Listen to a physics major read the passage; mark the passage as you listen. Now have the passage read by someone who hates physics. How do the miscues made by each reader differ? Why do you think this is so? Identify specific examples to support your reasoning.

7. What style analysis software is available on your campus? Run it on a piece of your writing. What features did it highlight? What errors did it catch? What errors did it miss? As a group, write a proposal to a software company for an ideal style checking program, listing all the features you would like to see.

8. The following list of questions was asked of readers after they had finished reading an instruction manual. In a small group, discuss the strengths and weaknesses of the questions as written, then rewrite them. You may want to consider using a Likert scale for some of the questions.

- Did you like the manual?
- Was the manual easy to follow?
- Were the illustrations good?
- Did you understand the instructions?
- What would make this a better manual?

6

Group Writing

Eight Management Tasks for Group Writing
  Team Building
  Task Analysis
  Task Management
  Time Management
  Document Management
  Style and Format Management
  Technology Management
  Conflict Management
  Case Study
Activities

We often visualize writing as a solitary activity—the novelist working alone on his manuscript, the budget director staying late to crank out her monthly report. This view is reinforced by typical school writing tasks that emphasize individual effort—efforts that are then evaluated so that teachers know how much effort and learning has occurred.

Increasingly, however, writing in the workplace has become a group effort. One recent survey found that 59% of technical employees in large corporations worked with groups on writing projects (Casari & Povlacs, 1988). Other studies have placed the percentage as high as 73% (Forman, 1991). Clearly, writing is not always a solitary activity.

Why is group writing so common? There are actually quite a few reasons why group writing is the only reasonable approach. The size of typical writing tasks is one factor. Although the vast majority of workplace documents are short (one to four pages), on many occasions longer documents are essential. For instance, feasibility studies, proposals, and reports can be quite long. One study of technical documents found that 5% of documents were five to ten pages, and 9% were over ten pages (Casari & Povlacs, 1988). Such longer documents are far more likely to be the work of a group.

Some documents require group writing because of the nature of the task. Reports are good examples: They are often created when one person requests information from lower level personnel or from various offices and compiles all the information into a single document. In the case of special reports, a committee or task force is often asked to investigate a problem or opportunity and produce a report or proposal. Such a report could be the work of the entire committee, or of a subcommittee selected to do the actual writing.

Time constraints are another important reason why writing is done in groups. For example, companies frequently receive requests for proposals (RFPs) asking the company to write a proposal bidding on a particular contract (see chapter 22, "Proposals"). The required proposal may be huge and the time allowed short. If a company wants the work, it must write the proposal quickly. A small group is usually put together to write the proposal.

Eight Management Tasks for Group Writing

Writing in groups is not easy. People have different interests and backgrounds, different ranks within the company, and different attitudes toward writing. And they always have very, very busy schedules. The results may be "group" reports that really reflect the superhuman effort of one member of the group, reports that arrive weeks after they were due, proposals that look like "sandwiches" assembled from different loaves of bread.

These problems can be avoided, but it will require a concerted effort on your part. Whether you are the leader of the writing group or just a member, successful group projects require the work of everyone. Each person in the group must be willing to adjust to the group. Ideally, all members of the group should understand the dynamics of group processes, and understand the importance of their role.

To successfully complete the assignment, the group will have to manage eight areas of the group writing process:

1. Team building
2. Task analysis
3. Task management
4. Time management
5. Document management
6. Style management
7. Technology management
8. Conflict management

Failure in any one of these eight areas will lead to a poor final product and to a poor working experience for everyone involved in the project.

Team Building

With all the work that has to be done on a project, it is easy to start a project without giving proper attention to the team. The people working on the project will be spending many hours together and will be asked to put team goals ahead of their own goals. To make working as a team easier and more productive, members should put conscious effort into learning about each other and developing some camaraderie.

A good way to help the process of team building is to make some effort at the first meeting to learn about each member of the group. People often begin with simple introductions, but that isn't enough. There needs to be more time for people to feel that the others know them. This can sometimes be done by allowing a fair amount of time for general conversation at the start of the meeting. Unfortunately, this only works if people in the group are good listeners and give everyone an equal amount of time to talk. If one or two people dominate the conversation, it will be necessary to use more formal techniques. These might include giving each person five minutes to "interview" another member of the team and then report about that person to the rest of the group, or asking people to introduce themselves individually and then ensure that everyone asks at least one question in reply. Whichever method is used, team building begins at the first meeting.

Another aspect of team building is basic democracy—giving everyone a chance to speak and showing respect for their ideas. Some groups have a rule that no meeting can end and no decision be made until every member of the group has had a chance to talk. Such active participation by all members will help build group cohesion, but it needs to be accompanied by respect for what people say. This comes in two forms. First, it means allowing people to finish what they are saying—no interrupting. Second, it means not immediately attacking what they have said. No one likes to have an idea killed when it has just barely been said. You may want to have a rule that at least one thing good is said about every idea before anyone can criticize it.

A third effort at team building that some companies attempt is building team identity. Computer companies are notorious for having teams design their own T-shirts. Groups members then are easily identifiable to each other and to others outside the group. Even without T-shirts, the identification process can be helped by such things as creating a name for the group.

Another part of team building is leadership. It is often the case that one person has been appointed by management to head the project, or it may be that one person simply has more seniority or rank than the others in the group. If that is the case, then the leader has the job of explaining the leadership style that will be used during the project. Will the leader make all decisions? Ask for recommendations and decide? Respond to a majority? This should be made clear to all so that they understand the procedures the group will use. Even if the leader intends to be a dictator, however, the leader would be wise to follow the other rules of team building, or the leader may find he or she is leading a group of one.

All of these efforts help establish the group as a real entity with goals and procedures and shared experiences. If participation in the group can be made pleasant, more people will be willing to make the sacrifices necessary to get the task done.

Task Analysis

Another important initial step is the careful analyzing of the work to be done. It is amazing how many times people are nominally on the same group, but have a totally different interpretation of what the group should be doing. In the first meeting or two the group should define the major questions required of any writing project. Who is going to read the document? What is the purpose of the document? How does the document fit into the larger context of work?

If each person in the group makes a list of all possible readers, and you then compare lists, the team will begin to come to some consensus about the readers they will be addressing. Similarly, discussions of the various purposes for the document will help create a shared vision for the group. Unless all members of a group share a common sense of purpose and audience, the group will be headed toward a very confused document.

Task Management

Task management consists of breaking the writing task into portions, determining which individuals or teams will work on each portion, and determining how long each activity should take. Consider, for example, writing a feasibility report. The work could probably be divided by two methods: writing an outline of the report, and creating an outline of the writing process. The two outlines might look like this:

    Telephone Switch Feasibility Report
        Recommendations
        Original problem
        Background of the investigation
        Comparison of alternatives
        Conclusion

    Telephone Report Creation Activities
        Plan project
        Collect data
        Review and verify data collected
        Draft main sections
        Collect illustrations and photographs
        Review sections for content
        Proofread and edit for style
        Design and lay out document
        Print, duplicate, and distribute copies

The two outlines make it clear there is much work to do in creating the report. Not only does the report itself contain several sections, but the group will have to edit, and publish the report, as well as write it.

Either approach to the project could be used to divide up the work. If the first outline is used, one subgroup would be assigned to writing recommendations, another to a description of the problem, and so on. The second outline would be used to assign one subgroup to initial project planning, another to collecting data, and so on.

Once the group has itemized the work that needs to be done, three additional tasks are required.

The first task is to identify which activities should be done as a group and which by individuals. One study conducted at the University of California–Los Angeles (Forman, 1991) found that novices tended to do too much as a group. One such group sat down and proofread each line of a report together: "[W]e agonized over crossing *t*'s and dotting *i*'s." Such group efforts waste time and place undue emphasis on trivial matters at the cost of larger issues. Your group should determine early which parts of the report need to be done as a group and which parts are better done by individuals.

The next task is determining equivalency of tasks. Everyone should have an equal amount of work. This is not only fair, it helps complete the job most efficiently. If one person has too much work, the project will be delayed. So you review tasks as a group, trying to determine how long each will take. Clearly, each activity will vary in its difficulty. Drafting the recommendations may take an hour or two, while drafting the point-by-point comparison of alternatives could take days. Proofreading the text might take an afternoon, while desktop publishing the document could take more than a day. Here experience will be a guide, but be prepared for surprises. It may initially appear that proofreading will be simple, but if your group contains a new person with weak writing skills, proofreading may grow into a demanding task. An honest group assessment of task complexity may be needed after a task has been started.

The last task is giving the right people the right assignments. Some people like proofreading, others hate it. Some people are good at writing succinct recommendations, others never get to the point. The group's task is to match up talent with each job. If you have been successful in team building, this matching process should be easier, both because group members will know more about each other and because group members will feel more comfortable about stating their weaknesses as well as their strengths.

If you do a good job of breaking out tasks and assigning the right people to each, you will have a major aspect of management accomplished. But there are many others that you must master as well.

Time Management

Time pressures are a part of life. The clock is always ticking. In a group writing project, these time pressures affect the task in two ways. First, there is the matter of finding common time for your group to get together. Second, there is the matter of getting each portion of the task accomplished in a reasonable amount of time.

Finding a meeting time convenient for all will be the more difficult as the size of the group increases. There are two solutions to this problem. First, don't call meetings unless they are essential. Most tasks should be done by individuals, not by the group as a whole. When you outline your tasks initially, select the few points where the group must work *as a group*. Otherwise, trust individuals to do their work on their own. Second, use updating memos or electronic mail in the place of meetings.

E-mail allows people to post their ideas or voice their concerns just as they would in a meeting. Essentially, the communication that would occur in a meeting is distributed over the course of the day. Group members can participate from their own desks as they have time. (See chapter 27, "Electronic Documents," for more information about e-mail.)

The other aspect of time management is making sure each writing task is being done on schedule. A commonly used management tool for this pur-

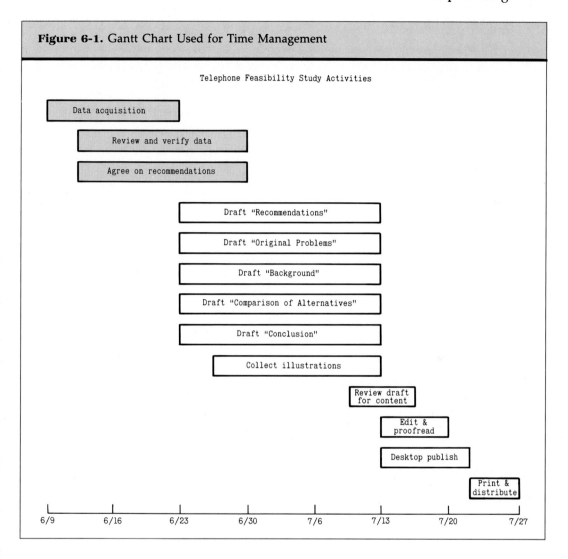

Figure 6-1. Gantt Chart Used for Time Management

Telephone Feasibility Study Activities

pose is the Gantt chart. The chart in Figure 6-1 might represent the principal steps in drafting the feasibility report described earlier.

Dates along the bottom of the chart make it clear when each task is to occur. The chart forces users to make an estimate of how long each task will take, a good exercise in itself. As tasks are finished they can be shaded in, so that anyone looking at the chart can see how well a project is going. Gantt charts are easy to make and easy to understand, and so are a common tool for time management.

Document Management

With larger writing projects, keeping track of the document itself can become a problem. Sections of it may be spread throughout a company as people work on their portions. The other problem comes later in the project when the group begins making revisions. It is easy to go through six or eight or ten revisions. With so many new versions of the report around, how do you know you have the latest one?

This is such a common problem that a number of electronic solutions are evolving. Some computer networks now include word processing software, "groupware," that handles distributed writing projects. A central data store in the network keeps copies of the portions of text being written. This data is electronically accessible from anywhere on the network, so that one person working on the project can see all the portions being written anywhere else in the building. Such software also keeps track of versions, using a clock to mark the time when changes were made.

If such a network is not available to you, you might set a simple format standard to each person working on the project. For instance, you could have people mark each page with the day's date. You might also create a special header for the project with both version numbers and date and time information. Another technique is to use "crossout" characters to show what you are deleting and capital letters to show what you are adding. That way everyone can see your suggested changes. After everyone has had a chance to review and approve the changes, the new version is reprinted in regular type. In either case, keeping track of a large and dynamic manuscript will require effort on your part.

Style and Format Management

Chapter 4, "Technical Style," explained technical style in great detail, and chapter 9, "Document Design," will explain issues of format. At issue is how ideas will be expressed, and how they will be visually presented. A group must agree on these issues very early in the course of the project. If they don't, it can become a significant problem. For instance, it would appear odd if one portion of an instruction manual relied heavily on illustrations and another portion had no illustrations. Or the group could end up with a process description in which one portion had each step defined on its own line, while another portion put steps together in paragraph form. The problem of consistent formatting has led to the development of structured document processors, essentially writing guides that impose a standard format. (See chapter 9, "Document Design," for a complete description.)

Other matters of style are more subtle, but still noticeable to readers. For instance, how much abbreviation will you use? Which technical terms will you use? Which will you define and which will you assume your readers can understand? How will you refer to reader and writer? If all writers in your group adhere to good technical writing style, there should be fewer such conflicts. Also, companies increasingly have predetermined document design standards and style guides for their employees to use. However, in most cases there is still plenty left open to debate.

One way to manage style and format is to create a group guide of your own. Figure 6-2 shows a style guide that was developed by one writing group.

All members of the group should follow a common style guide when preparing material. The alternative is to leave matters of uniformity to the proofreader, an approach that is time-consuming and will make people upset when they see the way their text has been changed.

Figure 6-2. Sample Style Guide Developed by a Writing Group

```
Phrases
  he or she (not he/she or s/he)

Abbreviations
  in (inch)
  ft (foot)
  psi (pound per square inch)
  min (minute)
  sec (second)

Type face
  Geneva 10 point
  Bold face for emphasis
  Underline for titles, not italic

Format
  left margin 1.5 inches
  right margin .75 inches
  top margin 1 inch
  bottom margin 1 inch
  tables and figures on separate pages
  format for figure numbers: Fig 7-3

  Headings
    Section titles all caps, centered
    Subsection titles initial caps, flush left

  Page numbers: centered, bottom, format: 2-13
```

Technology Management

A new problem for group writing is the management of technology. Using word processors and desktop publishing software can save work, but only if the group does some planning. For instance, will you all use the same word processing software? This would simplify the exchange of files, but it also assumes that all know and own the same software. If someone in the group is determined to stay with an uncommon word processing program, you may end up with incompatible files. This may make it impossible to merge files into a single document and smooth out the rough edges; it may totally stop desktop publishing. Some agreement on common writing tools is essential.

Some companies have begun establishing software standards for such common tools as word processors. This ensures that documents will be more compatible. Other companies provide file conversion programs to move documents from one word processing format to another. Computer vendors are also improving technology so that computers can share information more easily. Nevertheless, for group writing tasks it is simplest to find word processing software that most people like, and use it for all writing. This may put a few people at a disadvantage if they need to learn a new word processor, but it will ultimately cause them less pain than if they write large portions of the document only to find that their files are incompatible with everyone else's.

This need for technology consensus extends beyond word processing to such things as electronic mail. There is no sense beginning a discussion of a document on e-mail if two or three members of your group cannot access e-mail easily or refuse to use it. Your group will only function as a group if you all have and use the same writing tools.

Conflict Management

Writing is a sensitive issue. For a few people there is nothing worse than writing; every word is agony. Others seem disengaged from writing and unconvinced that it has any value. Even if everyone in your group is a competent writer who values written expression, time constraints and daily pressures are likely to cause a few short tempers. Conflict during a group writing project is guaranteed.

Social psychologists identify two sources of conflict within groups: emotions and ideas (Likert, 1984). If emotional conflicts exist within a group, they are sure to cause disruptions throughout the writing process. Silence is not golden. A strong group faces such conflicts and responds to them directly. Another sign of a strong group is its ability to deal with conflict of ideas. Novices often fixate on a single idea or approach. They reach closure without examining other possibilities, and they miss good ideas.

How do you make sure differences are aired? If there seems to be an emotional conflict between two members of your group, other members should take some initiative as soon as they are aware of the friction. What is the source of the problem? Does one feel overworked? Is credit for work done being fairly apportioned? Groups can't solve all problems, but if members know what the problem is they have a chance at reaching an acceptable resolution.

Meeting management is essential for staying open to ideas. Often one or two people will dominate. Others will never have a chance to talk, and eventually they will stop trying. A simple technique to get more involvement is to poll all members periodically. Ask each person in turn what he or she thinks of an idea. Give them time to talk. Show you value their opinion. If you take time to listen, people will take time to talk. The result will be more discussion, more ideas, and better solutions. This is an essential component of team building, and a crucial component of conflict resolution.

American work practices and educational practices have been criticized for lack of training in group processes. Those practices are now changing as more and more projects become group efforts. The eight group management tasks discussed above should, if applied by all members of the group, help you negotiate your way through a group writing project.

Case Study

The following case study illustrates how the eight management tasks described in this chapter might be employed in a project. This case study is based on a real writing task, the writing of a feasibility study.

Group Creation.   A large corporation with several branch offices needed to plan for emerging technologies. Branch managers were told to assign two people each to a task force that would meet to plan new phone and computer links between the branches. The group was to make recommendations that the corporation would follow over the next several years.

Task Analysis.   The group spent its first three weekly meetings discussing the range of technologies that might affect their company. They finally focused on communications technology and agreed they would create a report that would give the general background of developments in that field, explain the first impact this might have on the company, and make a series of recommendations for purchases.

Team Building.   One of the group members was appointed chair by central management. All other members had equal rank within the group. Because few of the team members had worked together before, special efforts were made to get to know each other. The group always met for at least half a day at a time, so that everyone had time to talk. The group always had lunch together on days when they met, and twice were entertained at the homes of team members. During discussion, each person was polled for an opinion.

Task Management.   The group identified information gathering as the largest single task and assigned everyone to that task. As discussion of various technologies continued over several weeks, several group members appeared to be expert in important fields. It was decided that each of them would write up a background of that field, that the group would draft recommendations at one of their meetings, and that two group members would serve as editors of the document.

Time Management.   The group had been given a deadline of July 1 for their report—four months from the inception of the group. Despite the tight deadline, the group allocated half of the available time to information gathering and discussion. Their final schedule looked like this:

> March: Goal setting and initial discussion.
> April: Oral reports and recommended readings on each of the major technologies.
> May: Recommendations from the group and first drafts of background chapters.
> June: Edited report presented to whole group for comment and revision. Final report approved.

Document Management.   One of the editors was given the job of keeping all drafts of the report. As section writers finished their work, they sent the material to the editor, who made copies and circulated them to all members of the group.

Style Management.   Before the group began writing its report, it spent an entire meeting deciding on how the report should be formatted. One of the group members brought in another company report that had been done recently, and the group agreed to follow the conventions in that report.

Technology Management.   All members of the group had access to electronic mail. The company had selected IBM-PC microcomputers as the companywide standard, with WordPerfect as the word processing software. For that reason, merging sections of the report together was fairly simple.

Conflict Management.   One member of the group frequently missed meetings and appeared confused about the general goals of the group. Several other members of the group spoke to him about his actions and explained their interest in the group. All of the members agreed to contact him by electronic mail before each meeting to help encourage his attendance. It worked. In the last weeks of the project his attendance was 100%.

Outcome.   The group produced a sixty-page document with a series of recommendations. Senior management reviewed the document and asked each branch manager to comment. After two months of review, the document was accepted as company policy and used to guide technology purchases.

Activities

1. The case study above shows successful group writing. In a small study group, identify the four most important actions the group took to make the project successful. List any improvements you would make in their process. Does everyone in your group agree that such a process could work?

2. A student working on a group project wrote this evaluation of his group:

> Everyone in our group did a great job. We put in lots of effort to get this job done in the past few weeks. However, there were problems. From my point of view, here were the major causes.
>
> *Lack of strong leadership*: There was nobody who could direct others to keep our project tasks on time.
>
> *Lack of communication*: This was really vital when merging texts. Because there was no naming conventions for files, it was really difficult to figure out which belonged to which.
>
> *Failure to recognize the scope of the problem.*
>
> *Some hardware problems*: Incompatibility between the fonts in my machine and the machine in the duplicating room.
>
> *Domino effects*: Paul and Pete finished writing text files on November 5 after noon, and I found that some of them were partially completed. In addition, the table of contents and index were not there at all.
>
> *Reliability of material*: Some texts done by Pete and Paul were incorrect. Therefore, we had to correct and reformat every one of them. This was very time-consuming.

Working in groups, decide how you would have responded to each of these problems if they had occurred in your group. What could you have done at your first group meeting to avoid some of these problems?

3. Create a Gantt chart for your last writing assignment. If the assignment was done by a group of four, what actions could have been subdivided? What would the new Gantt chart look like?

4. Survey the students in your class. What word processors and desktop publishing programs do they have access to and know how to use? Are there common programs? Is there any software conflict? How would you feel if your class had a policy that no one could register unless they knew how to use a certain word processor? What other means could you use to resolve this problem?

5. Form a group to write a large document such as a repair manual for a machine, maintenance procedures for new equipment, test guidelines for a medical examination, or a software manual.

    a. What can you do to build a team?

    b. Outline the tasks you will have to accomplish. What divisions seem reasonable?

    c. If you have four weeks to complete the document, how will you allot your time?

    d. How could you fit group meetings into your schedules?

    e. Create a Gantt chart for the project.

    f. What computer software will you use?

    g. Who will keep the manuscript as it evolves?

    h. Create a group style and format guide.

    i. Suppose one member of the group routinely misses group meetings. He says he is coming along fine with his part of the project, but he has missed two deadlines and refuses to show others what he has written. How does your group resolve this problem?

6. The head of your task force is a dynamic leader who takes charge and delegates all tasks. She works out a complete schedule and keeps everything running smoothly. She calls very few group meetings and keeps those meetings short and direct. Are there any possible problems with this approach? Discuss the pros and cons of this leadership style with others in a small group.

7

Writing with a Computer

Computer Tools
   Information-Access Tools
   Spreadsheets
   Graphics Programs
   Word Processors
   Desktop Publishing Programs
   Group Writing Software

Computer Writing Techniques
   Reasons to Use Shortcuts
   Scavenging
   Templates
   Structured Document Processors
   Boilerplate
   Macros
   Search and Replace

Activities

One of the important trends described in chapter 1 was the growing complexity of the technical communication process. Oral presentations now normally include high-quality graphics. Written documents are increasingly prepared with desktop publishing software that gives the writer control over such things as text size, graphics, and page design. At the same time that both oral and written materials are improving in quality and complexity, companies are cutting back on secretarial support and graphic design personnel, leaving most such work to be done by authors themselves. The reason why companies can expect higher quality technical communication with less personnel support is simple: they assume that technical communicators will use computers.

Word processors add flexibility and legibility to tasks previously performed with a typewriter or pen. In addition to word processing itself, a range of other computer tools is available to today's writer. Spelling checkers do not find all spelling errors, but they find enough to ease the effort of writers. Outline programs provide simple ways to jot down initial ideas and rearrange them while an author struggles to put ideas into an effective order. Style and grammar checking programs can give an initial assessment of a document's "readability," point to possible places where a reader might be confused, and catch such common errors as confusions of "affect" and "effect," and "there" and "they're."

As the computer becomes established as the principal writing tool of our day, other kinds of writing software are being developed, each of them useful to anyone doing technical communication. This chapter describes some of the common computer tools available, and then explores the techniques being developed to take best advantage of these new tools.

Computer Tools

All of the tools described in this section are routinely available or will be available to technical writers in the electronic age. What won't be available are hordes of ancillary personnel, secretaries, designers, and illustrators waiting to do your bidding. Budgets are too tight and timelines too short. You and your computer tools will be on your own.

Information-Access Tools

Many library collections are now accessible by computer catalogs, and most popular indexes can be searched either on-line or by using CD-ROM. On-line information consists of bibliographic information or complete articles that are stored electronically and can be retrieved electronically. Such connections are available through virtually all college libraries. As someone communicating technical information, such on-line information will be especially valuable to you. The information you need to find will seldom be commonly available. By its nature, technical writing tends to address specialized subjects not commonly discussed by the general population. To access information on specialized topics, you will need to search much more thoroughly. Because on-line searches process large amounts of information quickly, you will be able to find sources of information about even the most technical subjects. A further benefit of on-line searches is that they produce the most current information.

CD-ROM (Compact Disk–Read-Only Memory) technology provides a growing resource for writers. Each disk can hold the records of hundreds of books and thousands of published articles. Because CD-ROM disk drives cost about the same as other computer disk drives, it is relatively cheap to set up computers with CD-ROM access capability. CD-ROM technology also has the advantage of being local: the computer and disks are in the library, so that no telephone or database access charges are necessary. The disadvantage is that the disks must be replaced periodically (usually four times a year) at substantial expense, and the information on them may not be as current as that available on-line. Nevertheless, this technology is growing rapidly in popularity because of its convenience. Figure 7-1 shows what a typical CD-ROM screen might look like.

Figure 7-1. CD-ROM Screen

```
File  Edit  Search  Options            Help = F1

     Title: Table manners. (Home & Garden) 30059/1484
  Magazine: Regardie's Mag, Oct. 1990 v11 p225 (3) n2
    Author: Weissman, Eric
    Topics: Etiquette-History * Manners and cus-
            toms-Analysis * Nineteenth century-
            Social life and customs * Washington,
            D.C.-Social life and customs
COPYRIGHT Real Estate Washington Inc. 1990


                      -FULL TEXT-
Table Manners
IN THE EARLY YEARS OF THE 19TH CENTURY, FEW WASHING-
TON residents had the financial means to throw a
party of

TO SEARCH                 TO VIEW:
1) Press SPACE BAR        1) TAB to select article
2) Type query,            2) ENTER to see text
   press ENTER
```

Spreadsheets

Spreadsheets are a common tool in the business world. They look much like the balance sheets used in accounting. But they also have an important role any time numerical information has to be analyzed or presented. The spreadsheet can be used to enter and sort information and do simple calculations. Equally important, it improves the presentation of numbers through its ability to sort in any of several ways and present results in a table. Figure 7-2 shows how a spreadsheet might sort and present a table of pH levels. The same ability to sort and organize information also makes the spreadsheet a wonderful analysis tool (see chapter 17, "Analyzing Information").

Graphics Programs

Any information that can be placed in a spreadsheet can be automatically converted to a bar chart, pie chart, or any other common form of business chart. Graphics are now easy to create, and are easily merged into word processing documents by most word processing software. Graphics are not only an excellent presentation tool, they are also a helpful analysis tool: a picture is easier to assimilate than raw data.

Figure 7-2. Table Generated by Spreadsheet

Acid Rain Observations

Lake	Date	pH Level
6	MAY 24	4.23
4	MAY 24	4.60
2	JUNE 4	4.78
2	MAY 24	5.12
1	AUG 12	5.20
2	JUNE 3	5.40
5	JUNE 5	5.58
1	JUNE 3	5.80
2	JULY 8	6.10
1	JULY 12	6.50

Word Processors

At a minimum, word processors allow users to create, save, print, insert, and delete text. But their capabilities keep growing. Some have multiple type sizes and fonts and are becoming more and more like desktop publishing programs. Others either contain or give ready access to such writing tools as spelling checkers, outliners, style checkers, and more. As such, word processors and related writing software represent a major effort to automate the process of writing, remove some of the drudgery, and improve the appearance of the resulting document.

Desktop Publishing Programs

Basic desktop publishing programs contain facilities for creating and importing graphics, and tools for changing type sizes and styles. As such, they replicate much of the work that used to be done by conventional typesetters. As Figure 7-3 shows, they provide a great array of tools for writers. But they

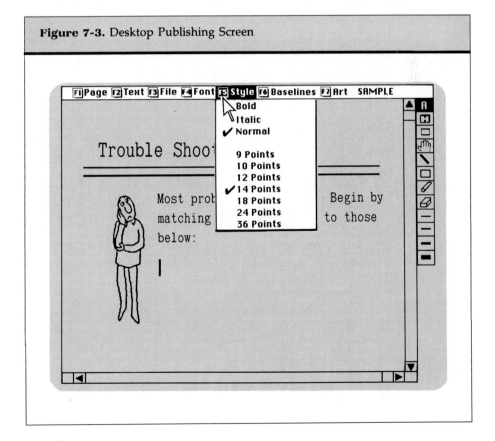

Figure 7-3. Desktop Publishing Screen

also give writers more responsibility. They leave writers with such questions as when to illustrate, which illustration to select, and how best to format text — questions formerly left to graphic designers. As such, they create both opportunities and challenges.

Group Writing Software

One of the newest writing tools in development is designed to simplify group writing tasks. "Groupware" helps keep track of various comments being made on a paper and updates revisions as they are made. The latest versions of such programs allow people at different sites — Dallas, Los Angeles, and Milwaukee, for example — to work on a single report simultaneously, as if all the writers were working in the same office at the same time.

Computer Writing Techniques ━━━━━━━━━━━━

Computer writing tools give writers more flexibility, more control over the appearance of their documents, and more responsibility. When used well, they also improve quality and efficiency. Consider this example.

A community in North Dakota wants your company to propose a solution to their solid waste disposal problems. You are assigned to write the proposal. You can either sit down at your word processor and begin at the beginning, writing steadily for the next 14 hours, or you can load in the proposal you wrote to a city in Louisiana, make appropriate changes and get the document in the mail within an hour.

If you choose the first option, you not only waste a lot of company time (and possibly your own free time), but it is likely that the results will actually be worse. Why? Because by starting from scratch you may miss a portion of the report, or stumble with your spelling, or make a whole host of mistakes that you presumably already caught in the old proposal.

Welcome to computer writing shortcuts. Shortcuts are possible using any professional-quality word processor. If done well, they save time and help you write better reports.

Reasons to Use Shortcuts

There are five reasons — other than saving time — why you should consider using shortcuts.

THE LAW

Much of what you put in writing has legal implications. A simple letter regarding employment might include your company's hiring policies. A proposal might include your credit policies. Reports might include legal descriptions of products or processes. Legal language must be reproduced *exactly* in every document. The best approach is to have someone in the company write formal documents once, have them thoroughly checked by the company's attorney, and then save the documents on a computer disk. Computer shortcuts allow exactly the same text to be used over and over by people throughout the company. Because the text is unchanged, writers can be confident that it is correct each time.

SAFETY

Some processes are inherently dangerous. They involve materials or operations that must be handled with exacting care. In describing these materials or processes to readers, you want to be complete and accurate. The best way is to write the description once, verify it carefully with others, and then use the same description in all future documents.

COMPLETENESS

It is easy to focus on one part of a document to the point of omitting something somewhere else. This is most likely to happen when you start from scratch, and least likely to happen when you have a pattern to build on. The pattern becomes a step-by-step guide for writing your document; it reminds you of information that should be included, and shows you where it should go. Such a pattern can be created once, saved on a computer disk, and used over and over for each new document.

COMMONALITY

Companies are increasingly striving for a common look to their documents. They expect the same outline for all reports, the same order for items in a memo, the same format for proposals. Technical documents must be as easy to use as possible. If everyone in your company knows where to find information in a report quickly, they all save time. Computer shortcuts can help writers accomplish this by providing them with outlines for documents. This both simplifies the work of the writer by helping organize information, and, more important, helps the company by providing standardization and improved information access.

ACCURACY

Every time you type somebody's name you have a chance to misspell it. Every address is an opportunity to get two digits reversed. Every sentence is a chance from which to grammar error make. So don't type them. Shortcuts can provide text automatically. Once the information is right, you can use it over and over with no fear of error. You can focus your proofreading efforts on other sections of your document.

In sum, shortcuts don't just save time. They are an opportunity to reuse text that has been verified for accuracy—legal accuracy, safety of instructions, completeness of information. For these reasons, shortcuts produce better documents faster.

One point should be made here. A few people seem to think the best shortcut they can make with computers is to load in someone else's work, make a few changes, and turn it in as their own. That's not a shortcut, it is plagiarism. Plagiarism in a job situation can get you fired.

In college it is never acceptable to use someone else's work. Every paper or report you turn in should be yours—or your group's—from beginning to end. The only exception is for accurate quotations or paraphrases of another person's work, which must then be clearly and accurately documented. (See chapter 19, "Using and Citing Technical Sources.")

On the job, use the work of others only in two situations. The first situation arises when the company has prepared standard documents and expects all employees to use them. In this case it is well known who the author is. In the second situation you may use parts of someone else's document if you have permission to use it and clearly acknowledge the author in the document. To take credit for someone else's work is always offensive and will eventually result in punishment.

Scavenging

Scavenging entails taking an existing document and modifying it for reuse. Sometimes a writer is able to reuse almost all of an old document, sometimes only small pieces of it. A typical document that could use this kind of shortcut is the proposal. If you have just submitted a bid to one customer for a routine computer system, sending out a bid for a similar system to another customer would require little more than loading the original document into your word processor, changing names and dates, verifying that charges such as shipping and service would be the same, and sending the proposal out.

Even if you have substantial changes between documents, there are often pieces you can still use. For example, if you were writing an environmental impact statement and needed to describe the test results on certain pesticides, background information about those pesticides might already have been included in other impact statements. You could copy those paragraphs out of the old statements and insert them into the new.

One caution with scavenging is that careless proofreading can lead to embarrassment. You might not notice the name of an old customer in a block of text you are including, or you might leave in a section that has no relevance to the new proposal. You will have to reread carefully to ensure that the modified document does what you want it to do. One tool you can use is the word processor's search-and-replace function. For instance, if you were using a document that included the name of a customer, you could search for the name of the old customer and replace it with the name of the new customer. This will solve much, but not all, of the problem. Careful rereading will still be necessary to avoid embarrassment.

There are substantial benefits to scavenging, besides saving time and work. For new employees, a good way to learn company writing standards is to take the work of experienced writers and use it. As they modify the original document they will see how it was organized, what writing style was selected, how often and in what way points were illustrated. The model also forces new employees to be complete: it works as a kind of checklist, showing which sections of a report are necessary, how long they should be, and what information must be present. Scavenging from experts can teach novices much about the writing of a professional community. This assumes, of course, that the scavenging is being done with permission and the resulting document fully acknowledges the authors of all text used.

Templates

Templates are outlines. The most commonly used is the memo template:

TO:
FROM:
DATE:
RE:

By following this template you are reminded to list your readers, identify yourself, say when the memo was created, and state in a single line the content of your message. It is an effective outline that has endured over the years because of its usefulness.

Other common templates are the major sections of reports or proposals. A feasibility study, for instance, will usually take this form:

Recommendations
Original Problem
Background of the Investigation

Comparison of the Alternatives
Conclusion

If everyone in a company agrees to use this form, then the writer can begin writing without wondering about how best to present the information, and readers will know exactly where to look for the information they want.

To help remember to follow this outline, you can simply type it into your word processor and save it as a file. Then every time you begin writing, you can load in the outline (or template) file for each kind of report, and begin with this framework in front of you. You are saved the bother of retyping the headings for your report, and you are reminded to follow the standard format for your document.

Structured Document Processors

The need for standard formats is even more acute in projects requiring the efforts of a group. For example, feasibility studies often include information about numerous products that might be selected for use. To finish the report in a reasonable amount of time, the product descriptions might be split among three people, each of whom is asked to do ten descriptions. This should speed up the work, but it also can cause trouble if the three writers vary too much in the style of their descriptions. The results might look like this:

1. Compu-Tally comes from a small company in Colorado and was originally designed to be used by road construction companies that were required to keep records for state auditors. It costs $3,500.
2. Product: Audit Trail
   Company: Geoworks, Miami, Florida
   Cost: $3,200
   Description: A simple database that records arriving materials, inspection dates and results, test results.
3. Product: EarthAudit
   Cost: $1,000 site license, $300 per year maintenance.
   Description: Jay and I worked every module in this program and were really impressed. It has all we need to manage all the paperwork the state sends our way, and even generates much of the monthly report they require. It works on the hardware we have in house, even those old machines we have out on the job site. I think this will do it for us.

Each writer supplied a product description, but the descriptions are so different in format they cannot be merged into a single report. A reader would be confused about what information was included and how it was organized.

The solution is an emerging writing tool that Rose Norman at the University of Alabama has labeled the structured document processor. With it a writer takes templates one step further and not only creates an outline of the information to be included but puts limits on the size of each section. For instance, the section on "Company" could contain subsections like name, address, and telephone number, so all that information would be included and would be located where a reader could find it. The "Description" could be limited to three lines of type. As a result, every description would contain the same information in the same order and to the same level of detail, no matter who the author was. When the descriptions were merged, they would match perfectly. Figure 7-4 shows an input entry screen for such a structured document processor.

A structured document processor could be created by writing a special computer program to control the order and amount of text being entered, or it could be created by using the data entry control capabilities of simple database programs, or you could get many of the advantages of this structure by setting up a template for your writing in a regular word processor and setting length standards for each of the sections. Any of these solutions respond to the problem posed by group authorship — how to produce a final document that appears to be written by a person rather than by a committee.

Such a program would also be useful to an individual writing a report, since it would ensure commonality of descriptions.

Figure 7-4. Input Entry Screen for a Structured Document Processor

```
                        Part Descriptions


    Part Name:    [        ]    Inventory Control #  [        ]


    Description:  [                                          ]


    Vendor:       [                                  ]
```

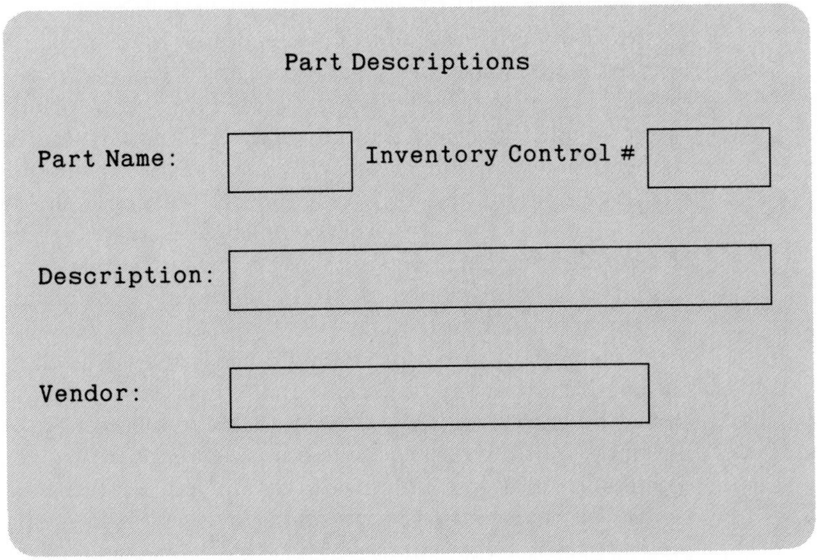

Boilerplate

Boilerplate is any block of text that is used repeatedly without changes. Common examples are hiring policies, position descriptions, and legal definitions. Using boilerplate for these texts is ideal because they are complicated, contain a great amount of detail, have legal ramifications, and seldom change.

Attorneys probably use boilerplate better than most other professionals because they are used to writing contracts that contain references to statutes, or that contain common clauses. Rather than taking the time to reenter the same clauses in contract after contract (and running the risk of typing errors), they read in the clause from their computer system.

Having seen how well attorneys have used boilerplate, companies now commonly use boilerplate for such things as position descriptions, product descriptions, and even company histories that might be made part of a proposal. A common example is the affirmative action statement that nearly all public institutions now publish as part of their documents. Here is an example of such text from a university policy statement:

> We at UW–Stevens Point affirm your rights to equal opportunities in education. We make all our campus programs and activities equally available to all students. In our educational programs, activities and employment policies we don't discriminate on the basis of age, race, color, religion, sex, national origin, ancestry, marital status, sexual orientation, disability, political affiliation, arrest record or conviction record, membership in the National Guard, state defense force or any other reserve component of the military forces of the United States or this state, or other protected class status (University of Wisconsin–Stevens Point, 1987).

Most universities and many businesses have similar statements. Notice that the statement is fairly long, contains a great amount of detail, and has probably come to be worded exactly as is after much discussion and approval of the highest levels of the university. This text cannot and should not be changed at the whim of an individual writer. It should be used without change. It should also be used carefully. A simple slip while typing could leave out one of the clauses of the policy and create a great deal of confusion and possible embarrassment. The best way for this text to be included in campus communications is to be downloaded unchanged through the campus computer system.

Boilerplate is important enough that it should be assembled in a central file by your company, especially if the text has legal ramifications. The file should be protected: it should be possible to read blocks of text from the file but not to edit the boilerplate without clearance from a central authority. As commonly used text blocks are considered for inclusion in the file they should be reviewed by appropriate levels of management, with relevant pieces checked by an attorney. These blocks become official company policy, a common voice always presented to the public.

Macros

Macros are a more personal shortcut. Built into most common word processing programs, macros let you define a special key so that every time it is pressed, the computer automatically executes a series of keystrokes. If there is a name or address or short description you use over and over, rather than type it in each time, you could just press a key and it would appear. Common uses are for addresses, names of departments, complete titles of senior management, or names of common products. Here are a few examples:

Alt-W: Department of Water Quality

Alt-H: Joan Huntly, Senior Manager, Provo Proving Grounds

Alt-R: Department of Natural ResourcesCr1267 4th
St. CrCenterville, MI 46687Cr

F2: TO:CrFROM:Dolores ReyesCrDATE:CrCrRE:UpUpUpUp

In the first example, every time the W key is pressed while the ALT key is held down, the name of the department would be added to the document. In the second example the name and title of a senior administrator would be added. In the address example, notice the letters "Cr." They are an abbreviation for "Carriage Return" and mean that the address would be printed over three lines. The last example sets up a memo on four lines when the keyboard's second function key is pressed. The "From" line includes the name of the author, two lines are skipped between the date and the line that shows what the memo regards, and after printing out all four lines, the cursor is moved up to the top line so the author can enter the name of the person to receive the memo. Other word processing programs would use other commands or keys, but the principle would be the same.

Creating a macro is usually very easy. In general, a writer identifies which key combination to use, and then types in the text to appear every time that key combination is pressed. The computer stores the combination in memory, so that the writer can use it permanently from there on.

As in other computer shortcuts, there are two advantages to macros. It saves time, and it improves accuracy: the writer is assured there are no typing mistakes in the text the computer brings in.

Search and Replace

Any word processing software has a search-and-replace capability. The program scans for a particular word or phrase and replaces it with another word or phrase of the author's choice. This capability is often used by experienced writers in three different ways.

The first is to scan automatically for commonly abused words or phrases. These can be trite phrases, frequently misspelled words, or overly used words. For instance, a writer in the habit of using "really" as an adverb might use search-and-replace to find and delete the word. Another word to search for might be "a lot" spelled as one word ("alot"). In either case, writers who are aware of their common errors can use the search-and-replace function to eliminate such errors.

A second common use is to allow the writer to abbreviate long or complicated names or titles while writing the document. For instance, rather than try to spell "Fyodor Dostoyevsky" correctly each time it is used in a report, the writer might abbreviate it as *FD*. Then when the report is complete, the writer can use the search-and-replace function to find *FD* and replace it with the complete name. The writer saves time and has a better chance of getting the name spelled right every time.

More-creative writers use search-and-replace to help with revision. One technique is to have the word processor ease the burden of revision by breaking the document into a series of sentences separated by several lines. This can be done by having the word processor search for a period and replace it with a period followed by several carriage returns. If this technique were used on this paragraph, the result would like this:

> More-creative writers use search-and-replace to help with revision.
>
> One technique is to have the word processor ease the burden of revision by breaking the document into a series of sentences separated by several lines.
>
> This can be done by having the word processor search for a period and replace it with a period followed by several carriage returns.
>
> If this technique were used on this paragraph, the result would like this:

Each sentence is now easier to edit, and there is more space around it if the writer wishes to print out the page and write in changes between the lines.

Most writers find that as they gain experience as writers and as they become more comfortable with a particular word processor, they are able to use the search-and-replace function in many effective ways, few of which were imagined by the original programmers of the software.

This should be the case with all computer shortcuts. In any professional endeavor, work should get easier as you become more familiar with operations. Computer shortcuts are a quick way to gain some of the advantages that go to the experienced. By taking advantage of these shortcuts you will produce much better work, and have a few spare moments as well.

Activities

1. What CD-ROM indexes are available in your library? Do a search on a subject via CD. Print the resulting bibliography. Look up the same subject in a bound index. Compare the numbers of articles you find there.

2. Interview someone in your field of interest who has been working for several years. What computer shortcuts does that person use when writing? How long did it take to learn the shortcuts? Using a word processor of your own, create a boilerplate, template, or macro similar to one you saw being used.

3. What is the difference between scavenging a paper and plagiarizing a paper? Do companies and colleges follow different rules? Why? How would you feel if other employees wrote proposals by making minor changes to a proposal you had worked on for days? How should they acknowledge your contribution? Compare your ideas with those of others in the class. Is there one method that most people prefer?

4. *TV Guide* was one of the first major publications to begin using a structured document processor when creating its movie and television show descriptions. Find a copy of the magazine. What are the standards for descriptions? Now examine a reference manual for a piece of computer software. What are the standards in the descriptions in that work? Besides coordinating group work, what are some advantages of standardized descriptions? Create a template for standard descriptions for a product.

5. Your college catalog is essentially a contract between you and your university. Identify sections that could be boilerplate, that is, standard text used throughout campus documents, or used in the catalog year after year. Compare your list with those of others in the class.

6. Boilerplate is a shortcut, but it is also restrictive: it allows no creativity in descriptions. How would you justify the use of boilerplate to a new employee who feels that he can do a better job than the current descriptions you have stored in your boilerplate file? Write a memo explaining why the new employee should use the standard company boilerplate.

7. Write a piece of boilerplate text for a product description, company history, credit policy, product label warning, or hiring policy. After you have written the boilerplate, what procedure would you use to verify the text for accuracy? What changes might you have to make in the text to ensure its usefulness in all company publications?

8. Create a template for a technical document of your choice. Do it as a group project and make it as detailed as possible. Once it is completed, see if you can apply it when writing a report. Compare the resulting reports. Were you able to stick to the template? Did the reports end up very similar?

II

VISUAL
PRESENTATION

Graphics

8

When Should You Use Graphics?
   To Condense the Text
   To Emphasize Patterns
   To Clarify Relationships

Common Graphics
   Tables
   Graphs
   Illustrations
   Diagrams

Producing Graphics
   Graphics from Sources
   Manual Methods
   Word Processors
   Spreadsheets
   Computer Scanners
   Computer-Aided Design Programs

Using Graphics
   Integrating Graphics and Text
   Adding Titles and Labels
   Placing the Graphic in the Paper
   The Danger of Novelty

Graphics and Ethics
   Guidelines for Ethical Graphics

Activities

There have been few trends this century as dramatic as the move from verbal to visual communication. A brief communication history of this century might look like this.

books & letters ⟶ telephone & radio ⟶ movies & videotape

This does not mean that we have abandoned books or have stopped writing letters, but at the start of this century, they were the principal way people reached an audience at a distance. Now they are just one of many ways. And of course we still use the telephone and listen to the radio, but they no longer dominate the way they did in the twenties and thirties. All of these communication avenues now have to share with the growing presence of the new visual media — movies and videotape.

The influence of visual communication has become so great that it is now affecting older communication media like books. They carry far more illustrations than they used to. To see this influence most dramatically, compare common encyclopedias. Encyclopedias of the 1930s and 1940s were all words, with the exception of an occasional line drawing and intermittent clusters of photographs. By the 1970s encyclopedias routinely used full-color illustrations. There were far more illustrations, all of better quality than earlier efforts. By 1990, some encyclopedias had left paper entirely and moved to CD-ROM, where entries are not only illustrated but animated. The words are still there, of course, but far more of the communication is being carried by illustrations.

The technical writer should expect to use graphics in technical documents, and should become conversant with new methods of graphic design as they become available. There will always be new media (like CD-ROM) to learn about; learning about bar charts and line drawings will not be enough to last a career. More of the graphics design and production process will be in the hands of the writer, particularly in smaller organizations. Illustration used to be somebody else's business. Companies often had an art staff or used publishing studios for illustrations. With computer software now uniformly available for such work, it is assumed that writers will handle the work themselves. For all these reasons, technical writers will have to know far more about visual information presentation than they did in the past.

When Should You Use Graphics?

Graphic aids supplement the written word, they do not replace it. In general, graphic aids should be incorporated in your document whenever their use would enable you to make a point more concisely and effectively. If the subject involves numerical, relational, or visual concepts, graphics may be an effective means of clarifying information for readers. Consider using a graphic device when you want to do one of the following.

To Condense the Text

By assigning the presentation of data to a graphic, you can often avoid the dense and complex passages of prose that would otherwise be required to convey the same information. The example below presents survey results first as text and second in the form of a table (Fig. 8-1). Both contain the same information, but the table is far more efficient and makes information access easier.

An in-home survey of adults in Louisville, Kentucky revealed public perception of the relative influence of twelve community groups. Television and radio stations were identified as having the greatest influence by 29.7 percent of those responding and as having the least influence by 1.7 percent. Daily newspapers were credited with the greatest influence by 20.8 percent and with the least influence by 3.6 percent. Utilities had greatest influence to 5.3 percent and least influence to 4.6 percent. Industries and manufacturers were identified as the greatest influence by 6.3 percent and as the least by 6.1 percent. Banking and financial institutions were seen as the greatest influence by 7.2 percent, with 8.3 percent seeing them as the least influential. While 9.4 percent saw churches and

---

Figure 8-1. Table Showing Condensed Presentation of Information

Perceived Influence of Community Groups
N = 430

Ranking	Group	% Greatest Nominations	% Least Nominations
1	Television/Radio	29.7%	1.7%
2	Daily Newspapers	20.8	3.6
3	Utilities	5.3	4.6
4	Industries/Manufacturers	6.3	6.1
5	Banking/Financial Institutions	7.2	8.3
6	Churches/Synagogues	9.4	11.0

Source: Smith, K. A. (1984). Perceived influence of media on what goes on in a community. *Journalism Quarterly, 61,* 262. Used by permission.

synagogues as the greatest influence, 11.0 percent saw these institutions as having the least influence . . . (Wresch, Pattow, & Gifford, 1988).

To Emphasize Patterns

Graphics are especially useful in showing changes and patterns in those changes. A line graph displaying the grams of a salt that dissolve in water as the temperature rises shows the reader much more effectively that solubility increases with temperature than would a textual narration of the weight of salt dissolving at each temperature. A pie graph of the national budget stresses the large chunk that military spending takes out of each tax dollar. The example in Figure 8-2 shows the advantage of a bar chart over a table of information. The bar chart helps make the similarities and differences among different computer models more clear than a simple recitation of the statistics would have. A brief glance tells the reader that although the computers tested were almost identical in terms of processors, memory, and disks, the corresponding computer monitors varied greatly in quality.

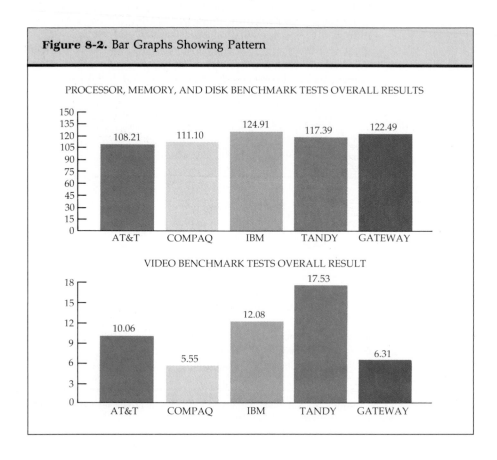

Figure 8-2. Bar Graphs Showing Pattern

PROCESSOR, MEMORY, AND DISK BENCHMARK TESTS OVERALL RESULTS

VIDEO BENCHMARK TESTS OVERALL RESULT

To Clarify Relationships

Presenting a clear impression of the spatial or organizational relationships between all but the simplest subjects is often difficult using text alone. The appearance of a complex mechanism, the relationships between various departments in an organization, and the flow of information through a computer program are all likely candidates for visual aids. The organizational chart in Figure 8-3 clearly shows the connections of departments in the business.

Figure 8-3. Diagram Showing Clear Relationships

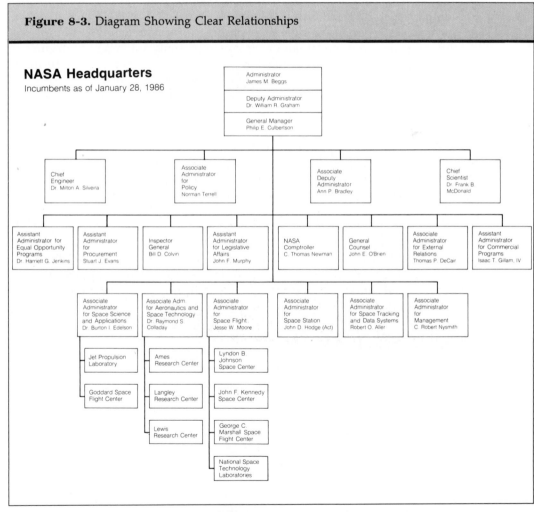

Source: Presidential Commission. (1986). *Report to the President on the space shuttle Challenger accident.* (Vol. 1, p. 226). Washington, DC: U.S. Government Printing Office.

Common Graphics ———————————————————

Tables

Tables systematically present information in horizontal rows and vertical columns. This information can consist of either words or numbers. Decision tables are a good example of a table composed of words. The example in Figure 8-4 shows how such a table can be used in a troubleshooting guide. Because one column groups and lists problems and the other column lists the corresponding solutions, readers can quickly find the information they need.

Tables are used even more often with numbers. The objective is to put quantities of data in a form that is accessible. Items are grouped so that comparisons and relationships can be determined. The table in Figure 8-5, for instance, presents comparative product information. Each column is a different computer model, and each row represents a different performance criterion. The information could have been presented in a paragraph form, but the table not only makes it easier for readers to find specific pieces of information (such as microprocessor speed), but aids in comparison since the information for each computer model is next to all the others.

GUIDELINES FOR TABLES

Since tables often present a sizable body of specific facts, they must be carefully constructed so that they are as clear as possible.

1. *Label columns and rows carefully.* Graphics always need careful labeling, but since tables can contain more information than other types of graphics, special care must be taken. Headings should make the logic and plan of the table clear. Each column should be clearly labeled. Column heads should be of appropriate size for the column. All abbreviations used in the table should be explained in a footnote to the table.

2. *Use rules to separate rows and columns if the table is large.* If there are more than two or three rows and columns, rules will help keep the categories distinct; they guide the reader's eye and make it easier to find a particular piece of information (see Fig. 8-5). For smaller tables with only a couple of rows or columns, these rules are unnecessary (see Fig. 8-1).

3. *Use the body of the text to point out relationships you wish to emphasize.* A table with five columns and eight rows has forty data items for readers to view. Many of those data items are of comparatively little importance, yet they are given equal position in a table with the most important information. Therefore, you will have to direct your reader's attention to those data items that are most important.

4. *Number tables separately from other figures.* Conventionally, tables are numbered in their own sequence; numbers are assigned to them independent

Figure 8-4. Table with Words

Before Requesting Service _____23

Check the following points once again if you are having
some trouble with your VCR.

Power	Correction
No power ...	• Check that the Power plug is completely connected to an AC outlet. • Check that the Power button is set to on.

Recording & Playback	Correction
TV program can't be recorded ...	• Check the connections between the VCR, the external antenna, and your TV. • Check that the band selected matches your antenna or cable TV system. (P.10) • Make sure that the record tab on the back of the cassette is still intact. (P.22)
Timer recording can't be performed ...	• Set the recording start/stop time correctly. (P.16-19) • Make sure that the Power button is set to off. (P.17) • If there is a power interruption before or during a Timer Recording, the Timer may lose its preset time memory and the Timer Recording will not be performed.
Standby OTR can't be performed ...	• Make sure that the Power button is set to off. (P.15)
There is no playback picture, or the playback picture is noisy or contains streaks ...	• Select the same channel on your TV as that of the Channel 3/4 switch. (P.3) • Set the VCR/TV selector to "VCR". (P.3) • Check that channel 3 or 4 of the TV is properly adjusted. • Adjust the Tracking control slowly in either direction. (P.11)
During special effects playback, the picture may contain some vertical jitter when using a TV which has an Automatic Vertical Hold Control ...	• Set the TV's Vertical Hold Control to "MANUAL" and then adjust it.
If the top of your playback picture waves back and forth excessively ...	• Because the VCR playback signal is not as stable as an off-the-air TV signal, the top of your TV screen may be bent or unstable during playback. To solve this problem, slowly turn the horizontal hold control on your TV to correct the wavy picture. If your TV does not have the horizontal hold control or adjusting the control does not help, please contact your local TV service center.

Remote Control & Bar Code Scanner (optional)	Correction
VCR can't be remote controlled ...	• Aim the Remote Control or Bar Code Scanner (optional) directly at the remote sensor on the VCR. • Avoid all obstacles which may interfere with the signal path. • Inspect or replace the batteries (back cover).
Bar Code Scanner (optional) cannot read bar code ...	• Trace across the middle part of the bar code quickly from the starting point completely to right side. (P.19)
Bar Code Transmit or Clear operation does not function ...	• Be sure to aim the Bar Code Scanner (optional) at the remote sensor on the VCR, while holding the Transmit or Clear button down, to allow time to transmit or clear all program information.

Source: *Quasar VHS owner's manual: VH6200/VH6300/VH6400* (p. 23). Elk Grove Village, IL: Quasar Company. Courtesy of Quasar.

of the numbers assigned to figures. If your document has separate numbered chapters or sections, use a double numbering system. The first number is the chapter number, the second number shows that table's place in the sequence of tables in the chapter. So the first table in chapter 6 would be Table 6-1, the second would be Table 6-2, and so on.

Figure 8-5. Tables with Numbers

BENCHMARK TESTS

PROCESSOR, MEMORY & DISK

	AT&T	COMPAQ	IBM	TANDY	GATEWAY
80386 Instruction Mix	4.57	4.48	4.46	4.38	4.12
Floating Point Calculation	8.45	9.67	8.46	8.02	7.19
Conventional Memory	0.83	0.83	0.83	0.74	0.55
DOS File Access (Small)	67.20	64.04	76.59	68.22	72.18
DOS File Access (Large)	5.67	7.31	7.79	9.21	8.75
BIOS Disk Seek	21.49	24.77	26.78	26.82	29.70

VIDEO

	AT&T	COMPAQ	IBM	TANDY	GATEWAY
Direct to Screen	5.17	2.03	6.59	4.29	2.86
Video BIOS Routine without Scrolling	1.65	2.20	1.37	5.44	1.15
Video BIOS Routine with Scrolling	3.24	1.32	4.12	7.80	2.30

*NOTE: The values in the preceding tables represent execution time. Therefore, the smaller the value in the table, the better the computer scored in that test. Please consider this when making a comparison between computers.

Graphs

Graphs, also called charts, display relationships that can be expressed numerically. The change in volume of a solid as its temperature decreases, the populations of several species in a sample environment, the speeds of various computer components — all of these are examples of information that can be effectively conveyed to readers through graphs.

BAR GRAPHS

Use a bar graph to display comparisons between a relatively small number of discrete data items. Bar graphs are especially effective in stressing relative quantities. Properly labeled bar graphs allow the reader to see the trends and the precise value of specific points in a graph. For example, annual totals over a span of years are effectively displayed in bar graph form.

There are four important varieties of bar graphs: the simple bar graph, the multiple bar graph, the stacked bar graph, and the horizontal bar graph.

Simple Bar Graphs. The quickest form of bar graph to interpret is the simple vertical bar graph. Some range of comparison is put on the *y*-axis (such as time, cost, or temperature), and items being compared are put on the *x*-axis. Such graphs are easy to read: the higher the bar, the greater the amount of time, the higher the cost, or the higher the temperature. Figure 8-6

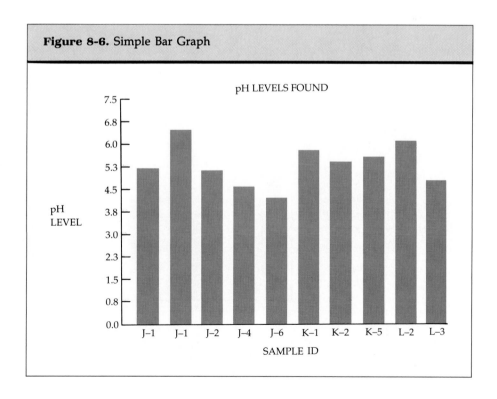

Figure 8-6. Simple Bar Graph

shows a simple bar graph. The graph's format makes it easy to compare the pH levels of different rainwater samples.

Multiple Bar Graphs.   It is sometimes necessary to group sets of information within a single graph. This is easily done with multiple bar graphs. Multiple bar graphs essentially allow you to illustrate two different variables on the x-axis at the same time. The information on the multiple bar graph in Figure 8-7, for example, could have been presented as a series of five graphs, the first one showing the sales in the northeast region by quarter, the second showing the sales in the central region by quarter, and so on. Or, instead of organizing the graphs by region, the information could have been put into four graphs organized by quarter: the first one, for example, would have shown the performance of all regions in the first quarter. Multiple bar graphs allow the writer to collapse these series into a single graph.

A multiple bar graph is more complicated than the simple bar graph. Had there been more than five subgroups or four quarters in Figure 8-7, the graph would have become very difficult to understand. Part of this is the difficulty in labeling such graphs, and part of this is the fact that each bar has to be made smaller in order to fit onto the page. As a result, multiple bar graphs should be used with restraint.

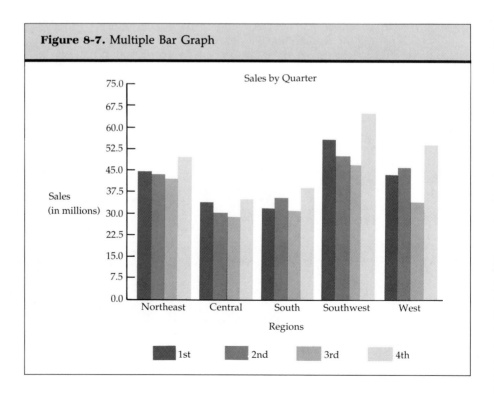

Figure 8-7. Multiple Bar Graph

Stacked Bar Graphs.   A stacked bar graph combines the information for subgroups into a single bar representing the group as a whole. This can only be done if the information can be logically added together. The graph in Figure 8-8 presents the same information as the one in Figure 8-7, but in a stacked bar format. The overall annual performance of each region is represented by the complete bar; performance by quarter is represented by differently patterned segments in the bar.

There are now fewer total items on the graph, so the graph initially seems simpler than a multiple graph. Unfortunately, there are major problems with this approach. First, labeling each segment of the complete bar is not easy. The segments must be differentiated with color or shading, and even then a reader must use a key to understand which color reflects which item. More important, comparisons between subgroups become almost impossible. Consider, for instance, the comparison between fourth-quarter sales in the northeast and in the south. From the multiple bar graph, it is easy to see that the northeast outperformed the south in the fourth quarter. This determination is much more difficult to make with the stacked bar graph, and even if the reader can see that sales in the northeast surpassed those in the south, it's hard to tell by how much. Stacked bar graphs are most

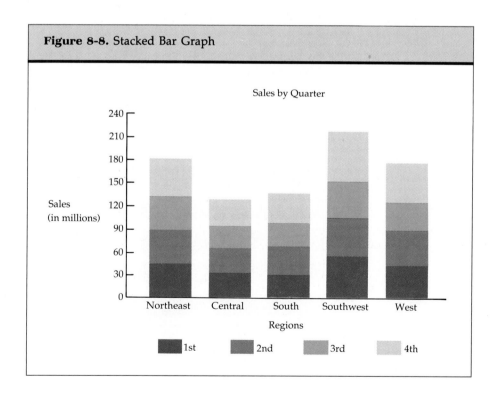

Figure 8-8. Stacked Bar Graph

effective when the sum of the individual segments — the bar as a whole — is the most important information being conveyed.

Horizontal Bar Graphs.   While bars are normally drawn vertically, they are occasionally drawn horizontally when the horizontal dimension seems more analogous to the comparison being made. When talking of time, for example, we usually refer to it as being "long" or "short" (rather than the "high" and "low" used for prices). For that reason, some writers occasionally prefer to express time along the x-axis and use the y-axis for the items to be compared. The same rationale could also be used for distance. The graph in Figure 8-9 demonstrates this orientation in its comparison of tire life. Since some tires run "longer" than others, both in terms of time and distance, the horizontal orientation seems consistent with that measure.

LINE GRAPHS

Line graphs are similar to bar graphs in that they plot some relative quantities on the y-axis, while items being compared are placed on the x-axis.

There are two distinct differences between bar and line graphs. First, the line eases the burden of the reader because it focuses attention on the most important point: comparisons between items. The incline or decline, and the angle of incline or decline, provide quick information about trends.

The second difference is the implied continuity of a line graph. A line

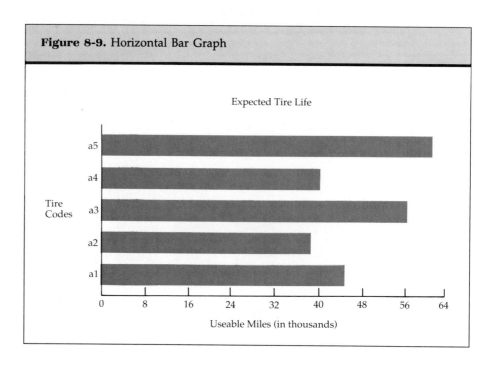

Figure 8-9. Horizontal Bar Graph

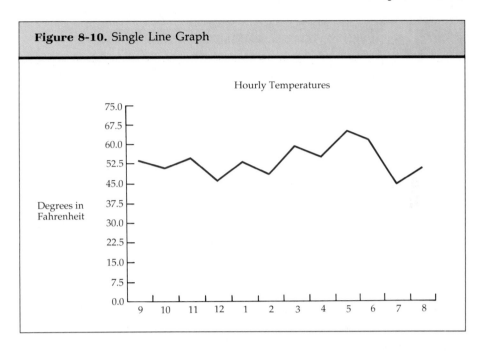

Figure 8-10. Single Line Graph

Hourly Temperatures

Degrees in Fahrenheit

connecting each data point implies regular movement from one point to the next. This is seldom the case. The graph in Figure 8-10 showing temperatures by the hour seems to imply a regular change in temperature from one hour to the next, i.e., if the temperature fell four degrees in one hour, then it fell one degree in the first fifteen minutes, another degree in the next fifteen minutes, and so on. This may not have been the case at all. The temperature may actually have risen for fifty-five minutes (when no temperature was being measured), and only have fallen in the last five minutes. All the reader can actually be sure of is the temperature on the hour, when the readings were taken, even though the line graph implies we knew about the temperature between the hours, too.

There are three important varieties of line graphs: single line graphs, multiple line graphs, and layer graphs.

Single Line Graphs. In a single line graph, information about a single thing is represented. Very often the *x*-axis represents different points in time and the figure as a whole shows the development over time of the event being graphed. In the line graph in Figure 8-10, this technique is used to represent a series of air temperature readings.

Multiple Line Graphs. A significant advantage of line graphs is their ability to plot multiple trends. One multiple line graph, for instance, might plot production of four different products over a series of years, with each prod-

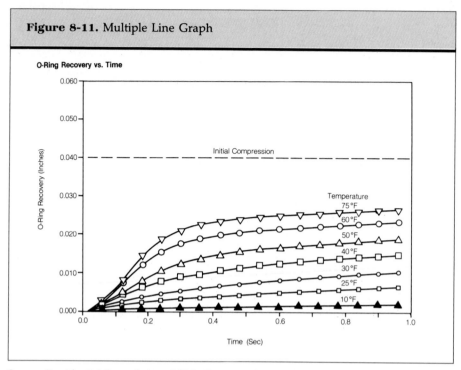

Figure 8-11. Multiple Line Graph

O-Ring Recovery vs. Time

Source: Presidential Commission. (1986). *Report to the President on the space shuttle Challenger accident* (Vol. 1, p. 65). Washington, DC: U.S. Government Printing Office.

uct represented by a different line. From such a graph, a reader could quickly see the relative performance of the four products. Multiple line graphs are often used for such comparisons. The multiple line graph in Figure 8-11 is from the investigation of the *Challenger* accident. This graph depicts the rate of O-ring recovery after compression under seven different conditions. Each condition uses different line markers to make the distinction between lines clear. Notice also that there is only one point of convergence of the lines, so that there is little difficulty in following each line. The effect of temperature is easy to see.

Layer Graphs.   Occasionally writers use line graphs as if they were stacked bars and create layer graphs. This can be done only if the information being plotted is additive. A good example might be a graph showing total manufacturing expenses, with one line for supplies, one for labor, one for capital, and so on (Fig. 8-12). Each item follows its own path, but the sum of the lines determines the overall manufacturing cost, which is identical with the uppermost line on the graph.

Figure 8-12. Layer Graph

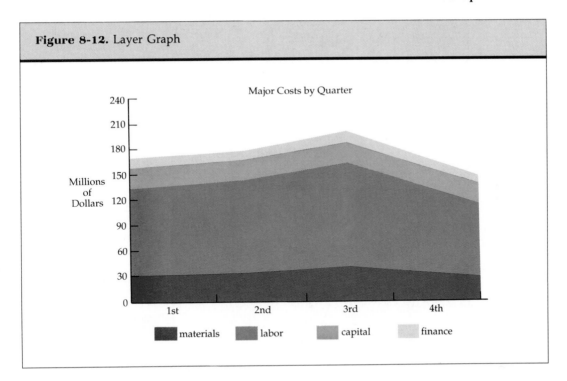

Major Costs by Quarter

PIE GRAPHS

The pie graph (sometimes called a pie chart or a circle graph) stresses the relative proportions of several items that combine to make a whole. You have probably seen this graph used to display how a budget is broken down, with the relative portion of each tax dollar spent on defense, social welfare, agricultural support, and so on. It can just as effectively display the portion of a day spent in various activities, or the relative importance of various food sources in the diet of the black bear (Fig. 8-13). The only restriction is that the items graphed must logically combine to form a whole. Thus, the number of students receiving A's in biology, history, and English literature classes would not constitute a reasonable pie graph since these do not combine to represent all the A's awarded on campus.

Like a stacked bar graph, pie graphs do a poor job of comparing items. For instance, it would be difficult to show a comparison of a given pesticide in water by creating two pie graphs showing total contaminants in two water samples. A bar chart is a more appropriate tool for comparison. Also, if an item of interest to you is only a tiny slice in the pie, this may not be the best way to display that item.

Effective pie graphs follow several conventions. In general, the segments

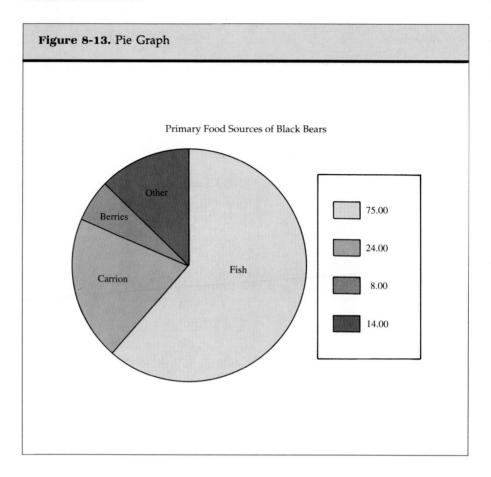

Figure 8-13. Pie Graph

Primary Food Sources of Black Bears

should be ordered from large to small and moving clockwise from noon. An exception to this rule is that a "catchall" segment (often labeled "Other") is usually the last item in the clockwise order. Remember to shade the segments to make each easy to identify. It is best to shade the segments from light to dark, with the smallest segment receiving the darkest shading.

GUIDELINES FOR GRAPHS

1. *Limit bar graphs to a few items.* Bar graphs are best when they represent a few items that can be directly compared. They can be quite dramatic in their impact, but they cannot convey large quantities of information. A simple bar chart with more than five or six bars is probably better presented as a line graph. A multiple or stacked bar graph with more than a dozen items is probably better presented either as a series of simpler bar graphs or as a table.

2. *Limit the number of lines in multiple line graphs.* Five or six lines is usually all that can be neatly displayed.

3. *Clearly distinguish lines on multiple line graphs.* Use different line types (e.g., solid, dotted, dashed) and varying symbols for the plotted data points to avoid confusion between lines. Although your computer may allow the use of color to distinguish lines, remember that your printer very likely won't, so depend on other means to make such distinctions. When possible, avoid graphing lines with multiple intersections.

4. *Be aware of the implications in line graphs.* Multiple lines will be compared even if such comparison is illogical. A famous example is the one of ice cream consumption and crime. The line graph in Figure 8-14 shows how the two seem to go together. A reader might believe that ice cream consumption causes crime (the actual connection is that both ice cream consumption and crime rise in the summer). Line graphs are a powerful presentation tool that should be used with care.

5. *Limit the number of sections in pie graphs.* Pie graphs seldom display more than five or six slices clearly. It is best to graph the largest items and put the rest into a slice labeled "Other."

6. *Be sure each bar, line, or segment is easily identified.* Part of this is a matter of distinguishing among items, when necessary, using different colors

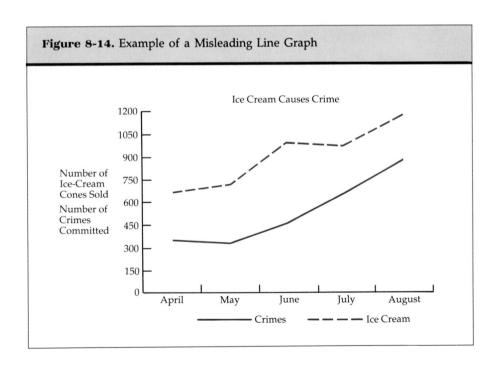

Figure 8-14. Example of a Misleading Line Graph

or patterns. But also make sure that each bar, line, or graph is clearly labeled, using a key or legend if necessary.

Illustrations

The most familiar graphic presentations are those that depict the actual appearance of their subjects. Photographs and drawings show subjects as we expect them to appear in life. Each of these principal forms of illustration has a significant role in technical graphics.

PHOTOGRAPHS

Photographs are the most realistic of all illustrations. They contain every detail that readers might be able to observe if they observed the subject in person. By specialized techniques such as x-ray photography, photomicroscopy, and infrared technology, photographs can capture images not visible to the unaided human eye.

In fact, photographs can be so information-rich they can confuse. Of all the information in the picture, what is it we want the reader to see? What is most important? Figure 8-15 illustrates the problem.

Labeling solves some of the problems of photographs, but not all of them. For one thing, labels look out of place on a photograph: it is rather jarring to have objects of one size and texture dropped on top of photographs with very different features. The lines and labels seem out of place. They can also obscure parts of the photograph.

More important, it is difficult to take good photographs. This is especially true in technical situations, where subjects might be moving, light might be uneven, and much of the subject might be hidden from view. For that reason, photographs are used less often than other forms of illustration in technical documents.

This does not mean that photographs should never be used. They are valuable for establishing realism, and if done well they can be an attractive addition to any technical document. But they should be used with care.

DRAWINGS

Drawings offer almost the same level of realism as photographs; they also allow you to stress aspects of your subject by eliminating irrelevant and distracting details and effects of lighting. For example, a drawing of the tail assembly of an airplane that shows the location and design of the elevator, horizontal stabilizer, and rudder will be clearer than a photograph containing the distracting details of hundreds of rivets, identification markings, and the patterns of light and shade that would be part of a photograph.

Drawings not only can be labeled, they can be easily compared. In Figure

Figure 8-15. Photograph with Labels

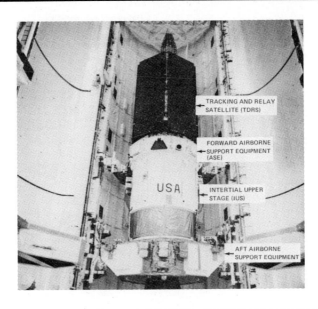

TRACKING AND RELAY
SATELLITE (TDRS)

FORWARD AIRBORNE
SUPPORT EQUIPMENT
(ASE)

INTERTIAL UPPER
STAGE (IUS)

USA

AFT AIRBORNE
SUPPORT EQUIPMENT

Source: Presidential Commission. (1986). *Report to the President on the space shuttle Challenger accident* (Vol. 2, p. L-29). Washington, DC: U.S. Government Printing Office.

8-16, two versions of the same drawing demonstrate the effect of pressure on the joint. The contrast is so clear that the author did not find it necessary to label the point where the joint was separating. As this pair of drawings illustrates, a pair or series of drawings can be a particularly effective way to show a procedure, a process, the effects of time, or the effects of other variables (such as pressure).

Because they are easily understood, simply created and reproduced, and easily integrated with text, drawings are the most popular choice for technical illustration. There are two special types of drawings you should know about: cutaway drawings and exploded drawings.

Cutaway Drawings.   One specialized form of drawing "cuts away" layers of surface in order to show interior information. Such drawings can be particularly effective since the portion that is cut away provides a context for the part that is shown. That way the reader not only sees the detail underneath, but understands how the detail fits into the larger context of the subject. Figure 8-17 shows such a cutaway drawing.

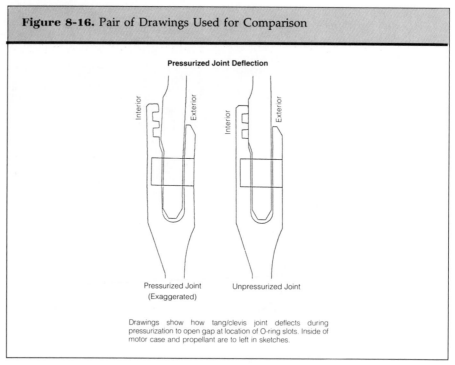

Figure 8-16. Pair of Drawings Used for Comparison

Pressurized Joint Deflection

Interior / Exterior

Interior / Exterior

Pressurized Joint
(Exaggerated)

Unpressurized Joint

Drawings show how tang/clevis joint deflects during
pressurization to open gap at location of O-ring slots. Inside of
motor case and propellant are to left in sketches.

Source: Presidential Commission. (1986). *Report to the President on the space shuttle Challenger accident* (Vol. 1, p. 60). Washington, DC: U.S. Government Printing Office.

Exploded Drawings.   An effective way to show subassemblies within larger assemblies is the exploded drawing (Fig. 8-18). Such drawings essentially show large assemblies several times: first whole, then disassembled to the first level, then further disassembled, until individual parts are visible and labeled. Because exploded drawings provide a context for subassemblies, repair manuals commonly use this technique.

GUIDELINES FOR ILLUSTRATIONS

1. *Use photographs only when necessary.* Not only are they difficult and costly to produce, they are rarely the best method for conveying technical information.

2. *Keep drawings simple.* Draw enough of the object so that the object is clear, but do not include extra information. A small part that isn't pertinent, details of color that don't affect your readers, and adjacent areas of an object all will distract and should not be included. Line drawings are easier to interpret if kept simple.

3. *Label drawings carefully.* There are points you may want to name or features you want to call attention to, but remember that a line that points to an area in your drawing may be indistinguishable from a line that is part of

Figure 8-17. Cutaway Drawing

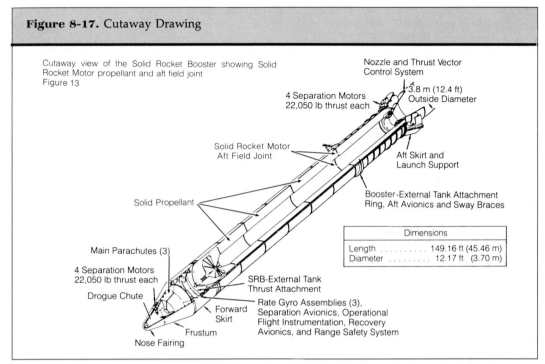

Cutaway view of the Solid Rocket Booster showing Solid Rocket Motor propellant and aft field joint
Figure 13

Nozzle and Thrust Vector Control System

4 Separation Motors
22,050 lb thrust each

3.8 m (12.4 ft)
Outside Diameter

Solid Rocket Motor
Aft Field Joint

Aft Skirt and
Launch Support

Solid Propellant

Booster-External Tank Attachment
Ring, Aft Avionics and Sway Braces

Main Parachutes (3)

4 Separation Motors
22,050 lb thrust each

Drogue Chute

SRB-External Tank
Thrust Attachment

Rate Gyro Assemblies (3),
Separation Avionics, Operational
Flight Instrumentation, Recovery
Avionics, and Range Safety System

Forward
Skirt

Frustum

Nose Fairing

Dimensions	
Length	149.16 ft (45.46 m)
Diameter	12.17 ft (3.70 m)

Source: Presidential Commission. (1986). *Report to the President on the space shuttle Challenger accident.* (Vol. 1, p. 56). Washington, DC: U.S. Government Printing Office.

the drawing. Label only where it is essential, and place your labels and arrows where they are least likely to be misinterpreted. The drawing in Figure 8-19 illustrates the need for careful labeling. At several points it is impossible to determine which line is part of the structure and which line is being used as a label.

Diagrams

Diagrams are also an effective form of technical illustration. In general, a diagram presents conceptual information in concrete form. Typical examples of diagrams are company organizational charts, work-flow diagrams, computer logic flow charts, and electronic schematics.

The organization chart in Figure 8-20, for example, explains the standard reporting and decision-making lines within NASA. It clearly delineates responsibilities. Anyone trained in hierarchy charts could quickly distill the organizational structure from this chart. The diagram in Figure 8-21 could be presented as a narrative, of course, but most readers would find the chart clearer and faster to use.

Figure 8-18. Exploded Drawing

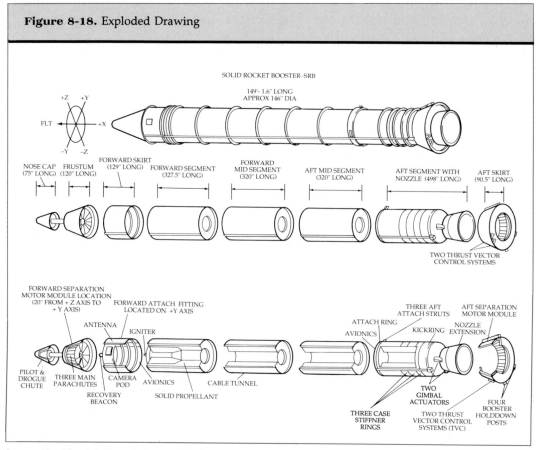

Source: Presidential Commission. (1986). *Report to the President on the space shuttle Challenger accident* (Vol. 2, p. L-34). Washington, DC: U.S. Government Printing Office.

This does not mean that diagrams are always easy to understand. Every diagram represents a set of conventions that must be learned. Take the PERT chart, for example. For those schooled in project management, the chart is a valuable tool in understanding work flow. For those new to the technique, the chart is practically indecipherable. Each profession has its own specialized diagrams. The responsibility of the writer is to follow the conventions generally observed. The writer also must be aware of which diagrams are generally known within the profession, and use those.

GUIDELINES FOR DIAGRAMS

1. *Keep diagrams simple.* Too many boxes or lines on a diagram will confuse most readers. The diagram in Figure 8-21 shows the scheduling of

Figure 8-19. Poorly Labeled Drawing

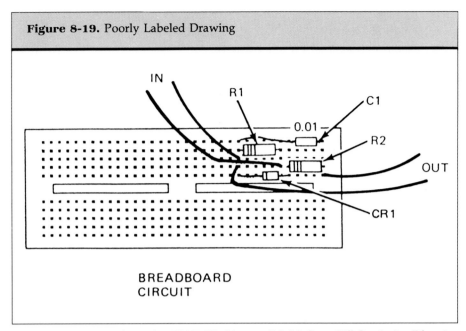

Source: *Contemporary electronics.* (1983). Washington, DC: McGraw-Hill Continuing Education Center. p. 39. Courtesy of McGraw-Hill Continuing Education Center.

various preparatory milestones in the months preceding the launch of the *Challenger* space shuttle; each type of task or activity is shown in its own horizontal row. Notice that the task "Flight Design Cycle 1" is a simple box. In fact, a design cycle encompasses many tasks, all of which could have been shown on the diagram. But the author correctly summarized these separate activities and distilled them into a single point on the diagram.

2. *Limit the type of information.* A chart that shows activities is one thing; a chart that shows activities plus time comparisons, plus interrelationships, plus personnel, and so on mixes too many kinds of information in a single chart. The *Challenger* diagram in Figure 8-21 runs dangerously close to the limit in the amount of information presented. It includes activities, plus dates, plus a count of months, plus dependencies between activities. There is a great deal on the diagram, possibly too much.

3. *Keep conventions consistent.* If each activity is placed in a box, use boxes every time. If a single line shows sequential dependency, don't use a single line one time, a double line another, and a dashed line the third. In the case of the *Challenger* diagram in Figure 8-21, all activities are placed in boxes, except for one placed in an oval. What is special about that activity? The

Figure 8-20. Diagram

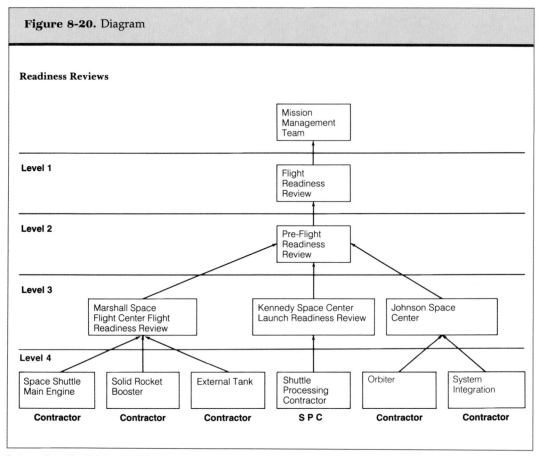

Source: Presidential Commission. (1986). *Report to the President on the space shuttle Challenger accident* (Vol. 1, p. 83). Washington, DC: U. S. Government Printing Office.

reader is left to wonder. If that activity *is* unique, it should be described at length in the accompanying text. If it's just one more activity, it should be presented in the same form as the other activities.

Producing Graphics

Once you have determined that some form of graphic aid can assist you in conveying your message and have selected the most appropriate form, you must now decide whether you will create your own illustration or select one from your research sources.

Figure 8-21. Simplified Diagram

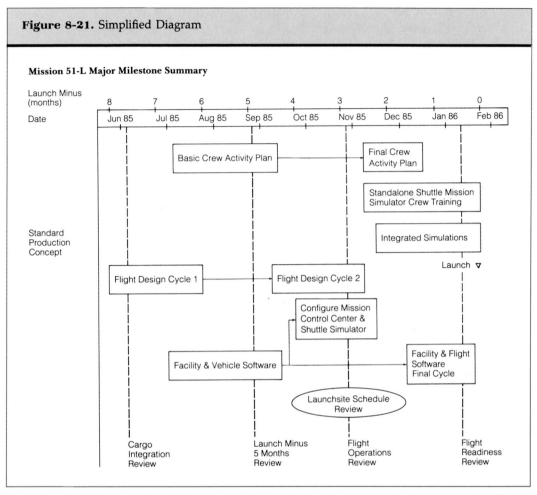

Mission 51-L Major Milestone Summary

Source: Presidential Commission. (1986). *Report to the President on the space shuttle Challenger accident* (Vol. 1, p. 12). Washington, DC: U.S. Government Printing Office.

Graphics from Sources

If you don't have access to a computer graphing program and don't feel confident in your own drawing skills, you can often use photocopies of source graphics. Successfully duplicated (i.e., sharp and with contrast properly set) source copies can often be inserted into your paper with good effect. In general, simple drawings, graphs, diagrams, and tables produce the best photocopies. It is much more difficult to get a good copy of a photograph or of a multiple graph that uses different colors.

Remember, however, that a graphic used in a source may not be suitable for your purposes. A published photograph of a timber wolf (*Canis lupus*) in its natural habitat and containing details such as shadows and underbrush would be an excellent choice if you want to show how the animal fits into its natural surroundings. However, these same details would be irrelevant and confusing if you were using the photograph to illustrate the animal's size and shape. Similarly, a table from a source may contain more information than your presentation requires.

If a suitable photograph, graph, or table is found in one of your sources, be certain to credit that source when you use the graphic. The citation should contain all necessary information for readers to locate the original, should they wish to examine the source. (See chapter 19, "Using and Citing Technical Sources.")

Manual Methods

You may be able to create your own graphics using the standard tools of manual preparation: pencil and pen, ruler, compass, and paper. You need not have artistic talent to prepare acceptable graphics; careful work and attention to detail will see you through the job. For a more professional look, you can use transfer letters and shading patterns. If you select this last option, submit a photocopy of your graphic, as transfer letters break and move out of position with frequent handling of the paper.

Photographs pose special problems. If you wish to create your own photographs, you may want to talk to the people who will duplicate your document before you begin. It is often easier for them if you take black-and-white photographs. These are easier to scan into a computer or to make into halftones (a version of the photograph composed of black dots), which will be necessary if your document is to be reproduced more than a few times.

The content of the photograph should be considered with care. Why do you want a photograph in the first place? — To show the general context? To display the relationship between parts? Determine why you are using a photograph rather than a drawing or diagram, and then take the picture for that purpose only. If what you need is detail, do not substitute photographs for drawings.

Word Processors

Most professional quality word processors have built-in capabilities for drawing lines and making boxes. For this reason the word processor can be an effective tool for making simple diagrams or creating tables.

Even without simple drawing tools, word processors ease the task of creating tables (Fig. 8-22). All you need do is set up the tabs for each column.

Figure 8-22. Table Created with a Word Processor

Composition of Student Population
at Albertville Elementary

	1970	1975	1980	1985	1990
Caucasian	217	220	182	165	112
Black	34	57	97	104	111
Asian	21	18	29	42	65
Native American	12	8	10	6	9
TOTAL	284	303	318	317	297

Remember that all the figures in a column should be aligned consistently, whether this is on the right-hand side, on the left-hand side, centered, or on the decimal point (if there is one):

2,367	282.04
215	992.4
8	6.9987
1,134	1,121.9

Many word processing programs have a feature that allows you to align columns automatically. All you have to do is set the decimal tab at the place where the decimal point should always fall, or the right-align tab where the last digit should be.

For word processing programs without this feature, set the tab for each column at the position to be taken by the leftmost digit in the largest number to be entered. When you tab to a column, use the space bar to properly align the first digit of smaller numbers. An entire row can be moved or deleted, if necessary, without disrupting the rest of the table. One word of caution: remember, partial deletions of a line will affect all characters in that line to the right of the deletion, so insert tabs or spaces as needed to realign your columns.

Spreadsheets

A spreadsheet is a computer program generally used for bookkeeping. It presents a series of rows and columns on the screen much as a ledger sheet might. But there are big differences: this ledger pad is electronic.

To begin with, this makes it ideal for creating tables. It will automatically justify either right or left or center columns, allow different column widths, and save you the trouble of performing calculations. Unlike word processors, spreadsheets allow rearranging and deleting columns without disrupting the remainder of the table. When the spreadsheet table is completed to your satisfaction, it can be saved in a separate file in a form that can be read by your word processor and thus incorporated into your paper. Remember that even though spreadsheets are designed to work with numbers, they can handle text data for tables as well.

Most spreadsheet programs also include simple programs for creating bar, line, and pie graphs. Computerized graphics have the advantage of easy preparation: the program prompts the user to make necessary decisions and then creates the graph and prints it out.

The spreadsheet is an efficient means of producing graphics. It's relatively easy to produce basic graphics, and, unlike manual preparation methods, the spreadsheet allows you to revise a graphic once you have viewed it. If you have used the spreadsheet as part of the analysis process (see chapter 17, "Analyzing Information"), you will find that it offers additional benefits for the production of your paper's graphics. Most important, the data have already been entered. In fact, you may have begun the preparation of your graphics, at least in rough form, as part of the analysis process. All that remains to be done is simple refinement. Generally, this includes adding titles, legends, and labels; selecting shading and line patterns and type styles; and determining proper axis scales. To finalize these determinations, the writer makes the appropriate selections from the graphing package's menu. The computer screen shown in Figure 8-23 is typical of such programs.

Spreadsheet graphics suffer from certain drawbacks. Unless you use a specialized (and, generally, expensive) software package, your choices for graph type, label placement and style, shading patterns, and so on are limited. More important, quality from the dot matrix printers usually available with personal computers is often less than ideal. The quality issue can often be resolved by photocopying the computer graphic in reduced size. As laser printers with their exceptional reproduction ability become more widely available, this issue will no longer be a consideration.

Computer Scanners

The process of marking photos was complicated in the past but is simpler now that computer scanners are available (Fig. 8-24). The photograph can be

Figure 8-23. Using a Spreadsheet to Create a Graph

```
                  Bar-Line Graph Definition
     Data block          Legend       Type   Pattern   Color
     r4c2:6 _____       1st _____     _        1        _
     r5c2:6 _____       2nd _____     _        2        _
     r6c2:6 _____       3rd _____     _        3        _
     _____          _____     _        _        _
     _____          _____     _        _        _


     X-Axis title block: r2c2:6 _

                X-Axis title          Color   Font
     Regions _____     _       _
                Y-Axis title          Color   Font


     _____      _       _

     F6 Define block   Escape to cancel   F10 Finished
     Worksheet: (none)  Loc: r4c1   FN: 0   Font: Standard
     GRAPHICS-define, generate, print, plot, view, edit
```

scanned into the computer's memory, and then marked or labeled as needed. The image is then merged into a document through desktop publishing software. Scanning can result in some loss of resolution (a less focused image), but the ability to highlight points in the picture more than makes up for this loss.

Computer-Aided Design Programs

Computer-Aided Design (CAD) programs are an excellent means of creating drawings of all kinds, and of making diagrams. Many professions are already using this tool. Drawings created by such programs can usually be merged into other word processing or publishing programs. As a result, the drawings that were created for particular machines, for example, can be loaded in, resized, and used without other modification as part of the technical manuals to accompany the machine. There is no need for redrawing. This saves time and increases accuracy.

Figure 8-24. Computer Scanner

Using Graphics

Integrating Graphics and Text

Any graphic in technical documents is an aid to the communication of the ideas, which are principally developed through words. Text and graphics must work together to make a point. Neither one should simply repeat the other, nor should graphics stand alone as extraneous material.

Too often, writers create a graphic or photocopy one from a source and then insert it into a document without any textual reference. Readers are left wondering why the graphic has been included and what its relationship to the text is. At best this puzzlement disrupts the flow of ideas; at worst it may lead the reader into mistaken assumptions. Always reference a graphic in the text, with some comment as to the reason readers should examine it and the message they should receive from the examination: "Figure 1 shows that increased life expectancy leads to increased medical costs. . . ." Under no circumstances assume that a relationship that you feel is obvious in a chart or table will be as obvious to the reader. Always state such relationships clearly in your textual reference to the graphic.

At the other extreme, many writers unnecessarily repeat verbatim the content of a graphic in words. If a table displays a series of temperatures over a period of time, don't repeat these same figures in your text. Such repetition makes the graphic superfluous and fails to recognize its power to condense information. Instead, use the text to emphasize the relationships or ideas that the graphic displays.

Adding Titles and Labels

For a graphic to be truly effective, readers must be able to find it and interpret it easily: it must be clearly titled and labeled. A main title indicating the basic content of the graphic should appear at the top. Tables and figures should also be numbered sequentially throughout the document, or within each chapter or major division of longer documents.

If possible, labels should be positioned so that they can be read without rotating the page. Label important parts of a diagram or drawing; if necessary, do the same for photographs. Each axis of a graph should also be labeled. Simply numbering an axis is not sufficient unless those numbers are self-explanatory: if the axis is divided into 1970, 1975, 1980, and so on, the label "Years" would be superfluous. You may wish to label specific data points on the graph itself; do so if such labeling will not create a confused and crowded graphic. The graph in Figure 8-25 shows how data points can be labeled effectively.

Another important element of labeling is the acknowledgment of sources. Even though you created your own graph or table, if the information was found in other sources, be sure to credit them.

Placing the Graphic in the Paper

Where in relation to the text do you place a graphic? You have three choices:

1. *In the body of the text.* If the graphic is small (less than half a page) and closely linked to the text, it may be inset directly into the text. Small tables can be included this way, as can small graphs or drawings. Most professionally published documents use this approach. It has been the one we have taken throughout this chapter.

2. *On a separate page.* Larger, more complex graphics should be placed on a separate page following the passage in the text in which they are first cited.

Figure 8-25. Graph Showing Labeled Data Points

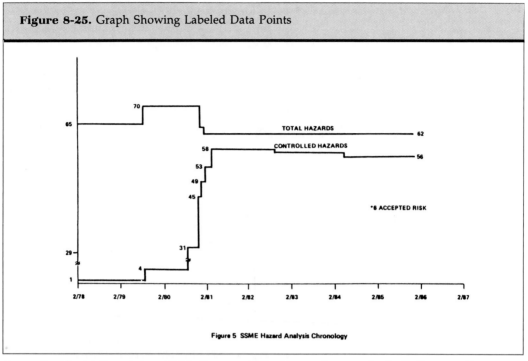

Figure 5 SSME Hazard Analysis Chronology

Source: Presidential Commission. (1986). *Report to the President on the space shuttle Challenger accident* (Vol. 2, p. K-7). Washington, DC: U.S. Government Printing Office.

Because the graphic is separated from the text, it is even more imperative that the graphic be carefully referenced in the text.

3. *In an appendix.* In longer, more formal presentations containing several graphics, the visual aids will often be placed in an appendix. This placement is especially appropriate for complex tables, since these are often included as supporting data that won't always be read with the paper but might be checked only by readers with a special interest in the subject. As always, be certain to provide references to appendix material in your text, and include the appendix in the table of contents.

In terms of location, there is one essential rule: always place the graphic after the text in which it is first cited, explained, or discussed. Because graphics are visually arresting, they attract reader attention. Imagine the consternation of readers who first encounter a graph or table without any idea of how it relates to their reading. Almost certainly they will interrupt their reading to search the text for reference to the graphic. To keep the flow of thought moving, simply be sure that no graphic will appear before the reader has an idea of what it is intended to convey.

The Danger of Novelty

One of the major dangers in integrating visuals in your presentation is the possibility of getting carried away with enhancements. Don't give in to the temptation to show off all the "bells and whistles" in your graphic bag. It is especially tempting to overpower the reader with graphic glitz if you are creating your graphics using a computer program that offers a variety of options. If you give in to these temptations, you will be practicing what Edward Tufte in *The Visual Display of Quantitative Information* calls "chartjunk" (1983).

Showing off every possible type style, all the shading patterns, and three-dimensional enhancements of bar and pie graphs can only lead to confusing presentations that fail to communicate with the reader. When you compound this by adding more data elements than the reader can easily grasp from a single chart, you have assured a failure to communicate. Compare the two graphs shown in Figures 8-26 and 8-27. Which conveys the message most clearly?

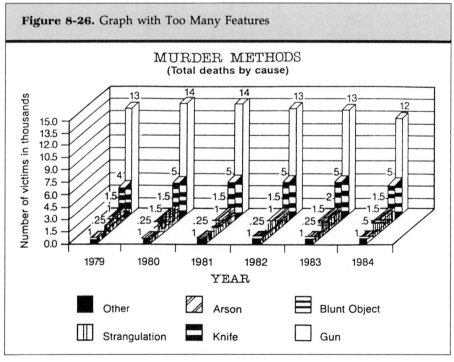

Figure 8-26. Graph with Too Many Features

Source: Wresch, W., Pattow, D., & Gifford, J. (1988). *Writing for the twenty-first century* (p. 227). New York: McGraw-Hill. Courtesy of McGraw-Hill, Inc.

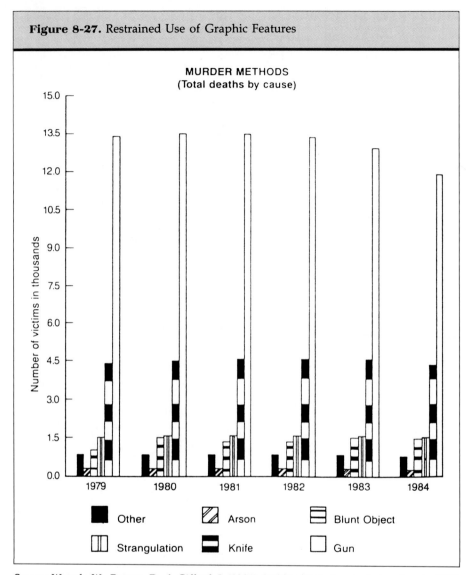

Figure 8-27. Restrained Use of Graphic Features

MURDER METHODS
(Total deaths by cause)

Source: Wresch, W., Pattow, D., & Gifford, J. (1988). *Writing for the twenty-first century* (p. 228). New York: McGraw-Hill. Courtesy of McGraw-Hill, Inc.

Graphics and Ethics

In his book *Doublespeak,* William Lutz describes a number of deceptive tactics used by various political figures to distort graphs. In one of his examples, the Reagan administration created a bar graph that purportedly showed that spending for education was rising dramatically at the same time that high school seniors were scoring much worse on college entrance exams. It is true that spending had been increasing and that scores had been declining, but the graph was rigged to make the changes appear more dramatic than they actually were. Compare the presentation of information in Figure 8-28 with that in Figure 8-29. The first graph shows the information as it was presented by the Department of Education during the Reagan administration. The second follows accepted guidelines. How do each of them make you feel about American schools?

SAT scores on the verbal test range from 200 to 800. The graph in Figure 8-28 begins at 420 and tops out at 470. This has two effects. First, students in 1967 appear to be near the top of possible scores. Actually they were more than 300 points from the top. They were good students, but hardly geniuses. Second, students in 1989 appear to be far worse than the ones in 1967. Their bar is only one-sixth as high, so they appear only one-sixth as smart. The more objective graph shown in Figure 8-29 shows a decline, but shows it as gradual and minor. It shows the full range of scores. It presents a complete picture. It is a much more ethical portrayal of information.

Politicians aren't the only ones who distort graphs. Newspapers and magazines routinely use scales that dramatize very small changes. Sometimes this scale adjustment can be justified as a way of focusing attention on

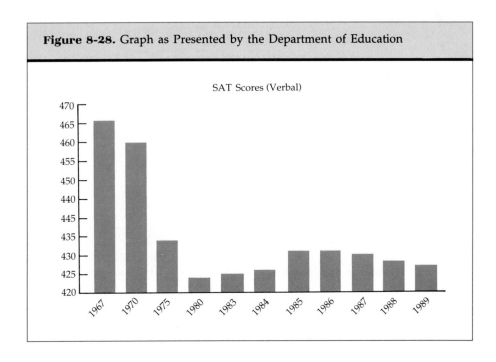

Figure 8-28. Graph as Presented by the Department of Education

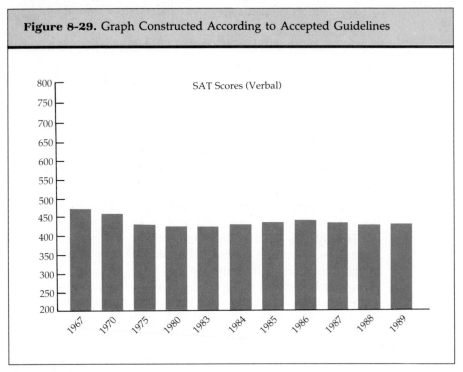

Figure 8-29. Graph Constructed According to Accepted Guidelines

Source: *Statistical abstract of the United States: 1991* (p. 154). Washington, DC: U.S. Bureau of the Census.

a change that might otherwise be missed, but generally scale changes are used to mislead.

Every time you create a graph you face ethical decisions about scaling. Will you show the whole scale, starting at 0, or will you shrink the scale to emphasize differences? Will you let readers make their own conclusions about data, or will you force their interpretation? How ethical will your graphs be?

Guidelines for Ethical Graphics

1. *Start vertical scaling at 0.* This provides a complete picture of the data, even though the resulting graph may be less dramatic. If you must start a graph at a higher point, your text should both state what you have done and justify it. For example, you might include a statement like "scores begin at 0. This graph begins scaling at 500 to emphasize the differences over the years." Such a statement warns your readers about what you have done, but doesn't fully excuse it.

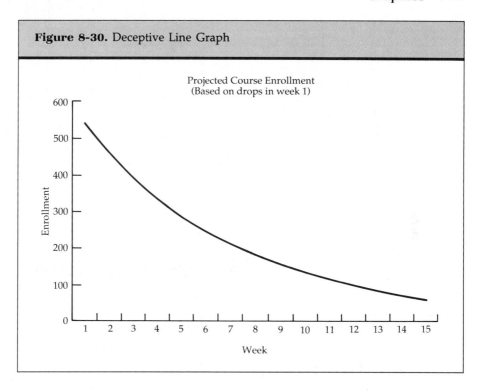

Figure 8-30. Deceptive Line Graph

Projected Course Enrollment
(Based on drops in week 1)

2. *Use an adequate period of time.* The change from one month to the next or from one year to the next might seem very large. But the same change seen over a longer time may be practically unnoticeable. Graphing only a short period of time can be very misleading.

3. *Use more than two points to plot a trend.* Line graphs can be especially misleading in this way. The graph in Figure 8-30, for instance, plots the expected change in engineering enrollment at a university. The change from one week to the next is substantial; it appears that something dramatic is happening. Part of this effect comes from the fact that such a short time period is being examined. Part of the effect comes from the fact that the line continues on into the future as if it were certain to continue in the direction established in weeks 1 and 2, the only weeks for which there are actual data. But in fact, there is no reason to assume that it will. For instance, if we plotted out future heights based on the amount we grew between the time we were twelve and thirteen, we would all expect to end up twelve feet tall by our thirtieth birthday, and fifteen feet tall by the time we were forty.

4. *Use multiple line graphs carefully.* Such graphs lead to comparisons. We have already illustrated the false comparison between ice cream consumption and crime rates. The same might be true for college enrollment and tuition costs, or shoe size and grade point average. Maybe there is a justifi-

able comparison between the two, and maybe there isn't. But real or not, the line graph will make it look real to your readers.

Activities ——————————————————————

1. For each of the following situations, indicate which type of graphic aid would be most appropriate. Assume that any data you need in order to prepare your graphs, photographs, or drawings are available. Be ready to justify your choice of graphic.
   a. You have collected information, using a survey, on students' preferences for new degree requirements for your college. You asked each student surveyed to indicate which of four possible degree plans was best. After two weeks, during which time the student newspaper presented several articles debating the issue, you surveyed the same students again.
   b. You wish to show the effect of protective colorations on the survival of young animals of several species.
   c. You have data showing the total production of radioactive waste during each of the last ten years for each of five major producers of nuclear electric power.
   d. You wish to show a relationship between the length of dress hemlines and economic conditions over the last eighty years.
   e. You wish to display the development of the modern electronic computer from its earliest ancestors to the present day.

2. For each of the situations above, sketch the graphic you would use. Include titles and axis labels, and scales where appropriate. Compare your sketches with those of your classmates. What differences in presentation do you find? Are some approaches clearly more effective than others? In what ways might the different presentations result in different interpretations by readers? Were these differences intentional?

3. The following are hypothetical enrollment statistics for new freshmen at Millard Filmore State University.

Year	Number of Freshmen
1970	1,675
1972	1,678
1974	1,695
1976	1,700
1978	1,704
1980	1,710
1982	1,700
1984	1,654
1986	1,503

As a class, create the following:

a. A bar chart using honest scaling.

b. A bar chart that would make you look good if you had been admissions director from 1970 to 1980.

c. A bar chart that would alarm the college faculty.

4. Find three or four graphs in daily newspapers. What kinds of scales are used? Why? How might newspapers justify using partial scales? How many readers do you think understand what the newspapers are doing? Recreate several of the graphs using complete scales. Would readers interpret your graphs differently from the graphs as printed? What differences would you expect?

5. Find an example of a line graph that seems clear and one that seems confused. How do they differ? In a small group, discuss what could be done to improve the cluttered drawing. Sketch a new graph showing your improvements. You may want to generate a more finished product using available software.

6. Review technical reports that are written in your area of expertise. How often are photographs used? In what situations are they used? Why?

7. What diagrams are common in your discipline? How long did it take you to learn the conventions of these diagrams? Could someone without your training make sense of them? As a class, list the kinds of diagrams you use. Can everyone in the class understand all of the diagrams used by the class as a whole? Find examples to bring in and discuss.

8. Prepare diagrams that would depict the following:

a. The course sequence followed in your major.

b. The daily schedule of a family.

c. The admissions procedures of your university.

d. The hierarchy of your university.

9

Document Design

Principles of Design
 Design and Information Processing
 Design and Visual Coherence

Visual Aspects of Documents
 Page Layout
 Text Blocking
 Type Size, Font, and Style
 Headings
 Marginal Glosses
 Lists and Bullets
 Lines and Boxes
 Color
 Icons
 Graphic Balance

Tools of Design
 Word Processors
 Desktop Publishing Programs

Activities

Technical documents communicate visually as well as verbally. Some of the visual elements are obvious: illustrations, bar charts, photographs. Some visual elements are more subtle: headings, labels, bullets before key points. Others include type sizes and styles, lines and boxes, icons, and placement of text on a page. All of these visual elements are part of *document design*.

As an example, suppose your company has developed a new circuit board for use in personal computers. You are writing an instruction manual on its use, including a page on installation. It is critical that the computer be unplugged before the new board is installed. How do you communicate the importance of unplugging the computer? You will state that the computer should be unplugged, but how do you emphasize the importance of this act? You will probably choose a visual means of emphasis. Here are a few:

1. <u>Underline the command.</u>
2. **Print the command in bold print.**
3. Print the command in larger type.

4. | Put a box around the command. |

 5. Put a special sign in the margin.

6. Separate the command from the other commands so that it stands out.

All of these solutions rely on visual cues to demonstrate the importance of the command. These visual cues complement the verbal message. The two jointly communicate the importance of the command to readers.

In document design the writer undertakes a series of visual decisions. Some are large-scale, such as chapter formats (chapter titles, color coding, etc.) and page designs (margins, headings, page numbering and labeling), while others can be as small as finding a way to ensure that readers read a particular word in a sentence. Document design is concerned with how information is presented on a page; it is concerned with visual communication.

Principles of Design

Document design has been an area for specialists. Companies that produced a lot of printed matter often hired specialists or contracted for their services. Why should a nonspecialist need to know about design principles? There are three reasons. First, in an era of downsizing, many specialists are losing their jobs, and there are fewer ancillary people in many companies. A writer is no longer assured of having a secretary, an illustrator, or a graphic designer readily available. Second, even if a designer is available, some understanding of what designers do is necessary in order to give them directions or to be certain they have done a good job. And third, knowledge of document design helps not just in large projects but in every kind of document, from memo to letter to report.

Design and Information Processing

This principle of design is derived from research into how people process visual information. We know, for instance, how people typically scan a sheet of paper. We also know a good deal about human memory. Insights in both areas can help us make decisions about how we lay out a document.

People in our culture normally begin viewing a page in the upper left quadrant, then scan diagonally down and to the right, then look down and to the left, then look down and to the right. Figure 9-1 shows this process.

Writers have a pretty good idea which parts of a page will be seen first

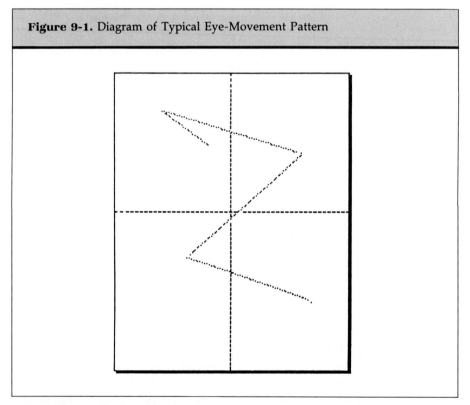

Figure 9-1. Diagram of Typical Eye-Movement Pattern

Source: Killingsworth, M. J., & Sanders, S. (1990). Complimentarity and compensation: Bridging the gap between writing and design. *The Technical Writing Teacher, 17* (3), 216. Courtesy of Association of Teachers of Technical Writing.

and which last. This is important, because memory research shows that we will remember best what we see first and last (known as "primacy" and "recency" effects). We may get the middle muddled, but we will remember the ends well. If you are unconvinced of these observations, try this simple experiment: have a friend read you a list of 40 numbers and then try to repeat the list from memory. The ones you remember most accurately will be those research predicts — those you heard first and last.

What does this have to do with document design? Writers who know how people remember and where they look can use this information to decide where to put the most important information in a document. For instance, if a writer had a warning it was important for readers to see, wouldn't the upper left or lower right areas of the page make most sense? Those are the two most valuable areas of the page, as determined by information processing research, so writers would use them carefully in their page design.

Design and Visual Coherence

The second principle states that the visual impact of items on a page should match their rhetorical impact, or in other words, the appearance of a text should match its content. An important point should look important, a minor point should look less significant. If the warning to unplug the computer before installing the board is really important, the manner in which the information is given should convey that it is important. This can be achieved by the use of larger type, bold print, red color, special icons, and extra space, among other techniques. Most important is knowing which to choose when. One guideline to use when making this choice is to have the visual prominence of information match its rhetorical prominence. Evaluate how important a word or phrase or sentence is, and then choose a visual display that matches that level of importance. This is design coherence.

Design coherence is another way of saying that the appearance of text creates expectations in readers. Everyone is served best if the content of the text satisfies those expectations.

Visual Aspects of Documents ━━━━━━━━━━━━━━━

Page Layout

Page layout is the overall design and arrangement of the typical page; it is usually more or less consistent throughout the document.

Figure 9-2 shows a page from a computer manual. The top line, called a running head, contains the chapter number and title and a page number. A running head is repeated at the top of each page to serve as a quick reminder

Figure 9-2. Computer Manual Page Layout

Formatting Your Document

When you format your Write document, you determine how you want the document to look. You can determine the appearance of characters throughout the document, the spacing and alignment of lines in each paragraph, and the page layout for the entire document. Most of the formats you choose look the same on your screen as in the printed document.

Formatting Characters

The commands on the Character menu control how you format the characters in your document. You can format characters by:

- Using different character styles, such as bold, italic, and underline.
- Creating superscripts and subscripts.
- Using different fonts and font sizes.

Choosing one of the Character-menu commands affects text that is currently selected or text that you are about to type. A new paragraph will have the same character formatting as the previous paragraph.

Write usually displays the same character formats that you will see when you print your document. If your printer cannot print a certain font, point size, or character style, Write attempts to match the selected character format as closely as possible.

Write saves character formats when you save your document. Write also transfers character formats when you move or copy formatted text to other parts of the same document or to another Write document.

Applying Character Styles

Character styles add emphasis to text by making it bold, italic, or underlined. You can apply more than one style to the same characters. For example, you can bold, italicize, and underline the same text.

▶ **To format existing text in bold, italic, or underline:**

1. Select the text you want to format.
2. Choose the style you want to use from the Character menu.
   The selected text changes to the format that you chose.

Source: *Microsoft® Windows™ user's guide.* (1990; p. 237). For the Microsoft Windows graphical environment version 3.0. [Computer Program]. Portions © 1985–1992 Microsoft Corporation. Reprinted with permission from Microsoft Corporation.

to readers of where they are in the book. Just paging through the manual the reader can see what each page is about without needing the table of contents. This header is sometimes enhanced by the use of icons or little pictures corresponding to the contents of the chapter. In either case, a clear, consistent heading simplifies the work of the reader.

Margins are another area of interest. The page shown in Figure 9-2 has a large area to the left of the type area, which the authors used to highlight key points with triangles. The headings also begin in the space to the left of the type area.

Figure 9-3 shows the outer area of the page used for glosses. A vertical line separates the type area from the glosses, which are further distinguished by italic type. The design divides the page into two parallel paths of information.

Figure 9-3. Page Layout with Marginal Glosses

Selecting Color Scanners
Drew Robison

Overview   *June 1991*

Scanning an image is a powerful, easy and fast way to incorporate images into any presentation or application. Selecting the proper color scanner is, however, a difficult task. With a wide variety of cost and capability between individual scanners, and without a means of testing or comparing these scanners, most organizations and academic institutions must rely on a combination of limited product knowledge and blind luck when making a purchase. IAT's technical support group recently researched five color scanners that were rated highly by some noted periodicals.

Implications

It is important to determine the quality of image that is desired. If your organization needs an image for a high-resolution output device rather than a low-resolution output device, then this will certainly affect your decision making when selecting a scanner. These questions and many others must be asked. Finding the scanner that best suits your organization's present needs, yet is flexible enough to support future expansion, requires thorough planning.

Specifying your needs

Some of the questions that should be addressed before making a decision follow:

   1) **Resolution** -- Resolution is usually defined in d.p.i. (Dots Per Inch). This tells you how many dots are in one inch of the scanned image. The minimum resolution should be

With a wide variety of cost and capability between individual scanners, and without a means of testing or comparing these scanners, most organizations and academic institutions must rely on a combination of limited product knowledge and blind luck when making a purchase. IAT's technical support group recently researched five color scanners that were rated highly by some noted periodicals.

1

Source: Robinson, D. (n.d.). Selecting color scanners (p. 1). In *IAT scanner*. Chapel Hill, NC: The Institute for Academic Technology. Courtesy of Institute for Academic Technology.

Figure 9-4. Page Layout with Two-Column Text

Chapter II

Events Leading
Up to the
Challenger Mission

Preparations for the launch of mission 51-L were not unusual, though they were complicated by changes in the launch schedule. The sequence of complex, interrelated steps involved in producing the detailed schedule and supporting logistics necessary for a successful mission always requires intense effort and close coordination.

Flight 51-L of the Challenger was originally scheduled for July, 1985, but by the time the crew was assigned in January, 1985, launch had been postponed to late November to accommodate changes in payloads. The launch was subsequently delayed further and finally rescheduled for late January, 1986.

After the series of payload changes, the Challenger cargo included two satellites in the cargo bay and equipment in the crew compartment for experiments that would be carried out during the mission. The payloads flown on mission 51-L are listed in this table:

Mission 51-L Payloads
  Tracking and Data Relay Satellite-B
  Spartan-Halley Satellite
  Comet Halley Active Monitoring Program
  Fluid Dynamics Experiment
  Phase Partitioning Experiment
  Teacher in Space Project
  Shuttle Student Involvement Program
  Radiation Monitoring Experiment

The primary payloads were the Tracking and Data Relay Satellite (a NASA communications satellite) and the Spartan satellite that would be deployed into orbit carrying special instruments

for the observation of Halley's Comet.

The NASA communications satellite was to have been placed in a geosynchronous orbit with the aid of a booster called the Inertial Upper Stage. The satellite would have supported communications with the Space Shuttle and up to 23 other spacecraft.

The Spartan satellite was to have been deployed into low Earth orbit using the remote manipulator system. The Spartan instruments would have watched Halley's Comet when it was too close to the Sun for other observatories to do so. Subsequently, the satellite would have been retrieved and returned to Earth in the Shuttle payload bay.

Crew Assignments

On January 27, 1985, one year before launch, NASA announced the names of the astronauts assigned to mission 51-L:

Commander	Francis R. Scobee
Pilot	Michael J. Smith
Mission Specialist One	Ellison S. Onizuka
Mission Specialist Two	Judith A. Resnik
Mission Specialist Three	Ronald E. McNair

The mission commander, Francis R. (Dick) Scobee, first flew on the Space Shuttle as the pilot of mission 41-C in April, 1984. Mr. Scobee, a native of Auburn, Washington, received his bachelor's degree in aerospace engineering from the University of Arizona. A former Air Force

10

Source: Presidential Commission. (1986). *Report to the President on the space shuttle Challenger accident* (Vol. 1, p. 10). Washington, DC: U.S. Government Printing Office.

Long lines of text are more difficult to read than shorter lines. For small type on a wide page, the text should be divided into columns. Figure 9-4 shows a page from the Presidential Commission's report on the *Challenger* that uses a two-column format. Note as well the use of lists to simplify the information.

In setting up columns, you must decide the amount of space, or gutter, to leave between them. The gutter should be wide enough to clearly distinguish the columns, yet not so wide as to waste space. In general, it is better to have the space too wide than too narrow. In word processing, use a gutter at least equal to three letter spaces.

Text Blocking

Text has visual impact before it has verbal impact; readers see text before they read it. Where a text is placed on a page affects how readers read. Consider the example page from an instruction manual, shown in Figure 9-5. All the letters have been replaced by x's. (This technique is known as "greeking" since it creates a text that cannot be read—it might as well be in a foreign language, like Greek.)

We note right away that there is boxed information at the top of the page. Boxed text is probably a note of some sort—possibly a warning. The box is in the middle of a numbered list. Since this is an instruction manual, readers can guess that these are steps to follow. The paragraph at the top of the page probably gives general information, and the two columns at the bottom are probably for quick reference.

By comparing the greeked text to the original page shown in Figure 9-6, you can see how much information the writer communicated by blocking and positioning text. Before we read a single word, we knew a good deal about what we would find.

You can create the same effect in your documents. Simply follow three general principles.

1. *Information that serves the same purpose should have the same appearance.* In Figure 9-6 there are two paragraphs of explanation and three paragraphs with action steps. Each group is positioned consistently. From the layout, readers know that the content of each item in a group is similar to the content of other items in that group.

2. *Position determines importance.* In an outline, main ideas are farther left and subordinate ideas are indented under them. The same expectation holds true for text. Readers expect to find main ideas positioned first and to the left.

3. *Empty space emphasizes text.* To draw attention to a paragraph, separate it from surrounding text. The more space around the paragraph, the more important it will appear. But be careful not to overuse this. If every paragraph is surrounded by space, then none of them stand out. Only use space when emphasis is warranted.

Many word processing software programs will automatically greek text as a step prior to printing. This is a good opportunity for you to see what you are communicating with the appearance of your text.

Figure 9-5. Greeked Page Design

Xxxxxxx X Xxxxxxxxxx   XXX

Xxxxxxxx x Xxxxxxx

Xx xxx xxxx xxxxxxxxx x xxxxxxx xxxxxxxxx xx xxx xxxxxxxxx xx Xxxxxxx X, "Xxxxxxx Xxxxx," xxx xxx xxxxx xxxx Xxxxxxxxxx xxxxxxxx. Xxxx Xxxxxxxxxx xxxxxx x xxxxxxx, xx xxxxxxxxx xxx xxxxxxxxxx xx xxx xxxxxxx, xxxxxxxx xxxx xxxxxxx xxxx xx xxxxxxx xxx xxxxxxx xx xxx xxxxxx xxxxxxxxxx xxx xxxxxxxx xx xxx xxxxxxx xxxx.

▶ **Xx xxxxx xxx xxxxxxxxx xxxxxxxxx xxxxxxx:**

X. Xxxxxx Xxxxx xxxx xxx Xxxx xxxx.
  Xxx Xxxxx xxxxxx xxx xxxxxxx, xxxxxxxxxx xxx xxxxxxxxx xxxxx xxxxxxx xxxxxxxxx.

X. Xxxx xxx xxxxxxxxx xxxxxxx xx xxx xxxxxxx.
  Xxx xxx xxxxxxxxx xxxxx xxx xxxxxxxxxxxx xx xxx xxxxxxxxx xxxxxxx.
X. Xxxxxx XX.
  Xx xxx xxxx xx xxxxxx xxxxxxx, xxxxxx Xxxxx xxxx xxx xxxxxx xxx xxxx xxxxxxx.

Xxx xxxxxxxxx xxxxx xxxxxxx xxx xxxxxxxxx xxxxxxx xxxxxxxxx xx xxx Xxxxx xxxxxx xxx

<u>Xxx xxxx xxxxxx</u>	<u>Xx xx xxxx</u>
Xxxxxx xx xxxxxx	Xxxx Xxxxxxxxxx xxx xxxxxx xx xxxxxx xxxx xxx xxxx xx xxxxx.
Xxxxx	Xxxxxxx xx xx-xxxxxxxx xxxx xx xxxx xxxxxxx xxxxx xxx xxxxxxx xxxxx xx xxxx xxxxxxx. X xxx xxxxxxxx xxxxxxx xxxx xxx xxxxxxx xxx xxxxxxxx, xx xxxxx xxxx xxx xxxx xxx xxxxxx xxx xxxxxxxxxx xxxxxxx Xxxxx xxx Xxxxxx.

Figure 9-6. Page Design Showing Text Blocking

Printing a Drawing

If you have installed a printer according to the directions in Chapter 5, "Control Panel," you can print your Paintbrush drawings. When Paintbrush prints a drawing, it preserves the proportions of the drawing, ensuring that objects such as squares and circles do not become rectangles and ellipses on the printed page.

▶ **To print the currently displayed drawing:**

1. Choose Print from the File menu.

   The Print dialog box appears, containing the different print options available.

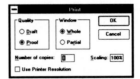

2. Make any necessary changes to the options.

   See the following table for explanations of the different options.

3. Choose OK.

   If you want to cancel printing, choose Cancel from the dialog box that appears.

The following table explains the different options available in the Print dialog box

Use this option	To do this
Number of copies	Tell Paintbrush the number of copies that you want to print.
Draft	Produce an un-enhanced copy of your drawing using the fastest speed of your printer. A few printers support only one setting for printing, in which case you will not notice any difference between Draft and Proof.

Source: *Microsoft® Windows™ user's guide*. (1990; p. 309). For the Microsoft Windows graphical environment version 3.0. [Computer Program]. Portions © 1985–1992 Microsoft Corporation. Reprinted with permission from Microsoft Corporation.

Type Size, Font, and Style

Many word processing and desktop publishing programs allow you to select type sizes. A type size is measured in points, with each point equal to 1/72 of an inch. That may not sound like much, but it can make a difference. In the following examples, the same paragraph is printed in 6-, 8-, 10-, and 12-point type.

Group all specimens into one large plastic bag, which may be sealed with masking tape. Place bag into cardboard box insulated with absorbent cotton or paper toweling to absorb any leakage from the enclosed specimens.

Group all specimens into one large plastic bag, which may be sealed with masking tape. Place bag into cardboard box insulated with absorbent cotton or paper toweling to absorb any leakage from the enclosed specimens.

Group all specimens into one large plastic bag, which may be sealed with masking tape. Place bag into cardboard box insulated with absorbent cotton or paper toweling to absorb any leakage from the enclosed specimens.

Group all specimens into one large plastic bag, which may be sealed with masking tape. Place bag into cardboard box insulated with absorbent cotton or paper toweling to absorb any leakage from the enclosed specimens.

Which size did you find easiest to read? Most research suggests that 9-point to 12-point type is best. Smaller than that, and some people have trouble; larger than that, and it takes many more pages to print the text. So you may want to set most of your text in 10- or 12-point type.

If the text body is typed in 10 point type, you may want larger type for titles and headings. The larger type should not be excessively large and should be used consistently for heads of equal value. In general, find a size in the 12- to 18-point range and stay with it for titles and headings with the same level of importance.

You should also be consistent with type fonts. A font, or type face, is a particular shape or look to a set of letters. Fonts can be quite different from one another. For example, the following text is set first in Times Roman and then in Souvenir:

Group all specimens into one large plastic bag, which may be sealed with masking tape. Place bag into cardboard box insulated with absorbent cotton or paper toweling to absorb any leakage from the enclosed specimens.

Group all specimens into one large plastic bag, which may be sealed with masking tape. Place bag into cardboard box insulated with absorbent cotton or paper toweling to absorb any leakage from the enclosed specimens.

The words are the same, but a different visual impression is achieved in each case. For technical writing, select an unobtrusive font and use it consistently throughout the document. Too often novices try to enliven their writing by using many different fonts within the same document. The result can be garish. There may be very special occasions such as warnings or long quotations that warrant a different font, but such occasions are few and far between. Which font should you choose? If your purpose is professional, your font should be too. Select one that seems quiet, respectable, unobtrusive — unless you or your company is trying to project a creative image.

The situation is similar for type styles. It is easy to have text appear bold, underlined, or in italic. The difficulty is learning when to use a different type style. Consider this warning sentence:

Be sure the computer is unplugged

It could use underlining in any of these ways:

<u>Be sure</u> the computer is unplugged.
Be sure <u>the computer</u> is unplugged.
Be sure the computer is <u>unplugged</u>.
<u>Be sure the computer is unplugged</u>.

Each approach to underlining has a slightly different effect. The visual emphasis is different. The first sentence seems to be a reminder. The second emphasizes the equipment, the third emphasizes the action. The fourth emphasizes the entire command. Visual emphasis works when one word or group of words appears different from the surrounding text. If too many words are marked in the same way, they are no longer unique. Emphasis only works if used sparingly (Kostelnick, 1990).

Emphasis should also be used consistently. If you print key words in boldface one time, you should use it throughout. Switching from boldface to italics confuses everyone.

A useful watchword for all questions of type size, font, and style is restraint. Pick a comfortable 10-point type and only change the size, font, or style of a word if you have a good reason.

Headings

Headings in technical documents mark off sections of material so readers can quickly see the content and purpose of each section. Headings are so common they are often used in two ways: as section heads and as labels. A section head marks a major block of text and is often centered or printed in larger type. A label can mark units of text as small as a paragraph and helps readers jump to particular areas within sections. It is often placed to the side of text it identifies. Both prepare the reader for the contents to follow, and allow readers to jump to text that contains the information they need. Technical documents use section heads and labels frequently. Figure 9-7 illustrates both kinds of headings.

To make your use of headings most effective, here are four guidelines to follow.

Figure 9-7. Page Design with Section Heads and Labels

64   Working with Documents

Working with Documents

This section explains tasks that are common to many different Windows applications: opening and saving documents, and doing basic text editing. See Chapter 3, "Program Manager," and Chapter 4, "File Manager," for information about organizing your applications and documents within the Windows environment.

NOTE   The descriptions in this section apply to the applications you received with the Windows package. Other applications you use might work differently.

Opening Documents and Files

▶ **To open a document:**

1. Choose Open from the application's File menu.
2. Move to the Directories list box.
3. Double-click the directory that contains the file you want to open.
   Or press UP or DOWN ARROW to select the directory and press ENTER.
   Windows displays the names of all files in that directory in the Files list box.
4. Move to the Files list box and select the file you want.
   Some applications provide a check box to specify that the file be *read-only* (which means that changes cannot be made to the file). If you want the file to be read-only, select this check box.
5. Double-click the filename or choose OK.

Saving Documents and Files

With many applications, the File menu contains two different commands for saving files: Save and Save As.

You use the Save command to save changes to an existing file.

You use the Save As command to name and save a new file or to save an existing file under a new name. For example, you might want to make changes to an existing document without modifying the original. With Save As, you can save another copy of the original file by giving it another name.

Source: *Microsoft® Windows™ user's guide.* (1990; p. 64). For the Microsoft Windows graphical environment version 3.0. [Computer Program]. Portions © 1985–1992 Microsoft Corporation. Reprinted with permission from Microsoft Corporation.

1. *Use headings often.* Most documents have natural sections: introductions, recommendations, actions necessary, background information. Each section should be clearly labeled so readers know the purpose of the section. Within a section, there may be subsections that should also be given a heading. Whenever the content of a document changes, a label should say so. These labels simplify the work of readers, who use them to know in advance the contents of each portion of the text.

2. *Keep headings short.* Usually a word or a phrase is sufficient. If the heading is longer than that, the reader must puzzle over the meaning of the heading itself. If you have trouble creating a short heading for a section, it may signal some confusion over the purpose you are trying to achieve. The solution may be a rewrite of the section.

3. *Use a consistent format.* If you start using nouns for section heads, use them throughout. For example, if your first three heads are "Introduction," "Recommendations," and "Background," it would be inconsistent to make the next head "Evaluating Options" or "Possible Action Steps for Accounting." Use short nouns throughout, or verbs, or short phrases. Whatever your choice, once you have made it, stay with it throughout.

4. *Make labels easy to find.* Headings should be easy to find. To accomplish this, writers generally use one of two strategies: they set the heading in larger and darker type, to make it more visible, or they position the heading so that it stands out. In several of the example pages earlier in this chapter the headings are placed to the left of the type area. This helps distinguish them from the rest of the document and makes it easier for someone scanning a document to find them.

Marginal Glosses

Marginal glosses are a running summary of a document. They allow a reader to grasp main points without having to read the main document. Because of their value to busy readers, they are becoming increasingly common.

Glosses pose a design problem. They are not part of the main text but an alternative to it — in a sense, a competitor for the reader's attention. In some cases a reader might become attracted by a gloss and decide to read the fuller explanation next to it, but frequently the gloss is read instead of the document.

How do you signal this fact to your reader? In the page reproduced in Figure 9-3, from a technical report on scanners, the author signaled the function of the glosses in three ways. First, he drew a vertical line between the text and the gloss so that a reader would not mistake the two. Second, he added extra space around the gloss so that it appears distinctive. Third, he used both larger and darker type to show that the gloss is not part of the regular text; it is visually distinct. All three are good practices to follow in similar situations.

Lists and Bullets

One of the simplest strategies writers use to clarify prose is to break paragraphs into lists. Consider this example of greeked text.

xxxxxxxxxxxxxxxxxxxxxxxxxxxxxxxxxxxxxxxxxxxxxxxxxxxxxxxxxxx
xxxxxxxxxxxxxxxxxxxxxxxxxxxxxxxxxxxxxxxxxxxxxxxxxxxxxxxxxxx
xxxxxxxxxxxxxxxxxxxxxxxxxxxxxxxxxxxxxxxxxxxxxxxxxxxxxxxxxxx

   xxxxxxxxxxxxxxxxxxx
   xxxxxxxxxxxxxxxxxxxxxxxxx
   xxxxxxxxxxxxxx

xxxxxxxxxxxxxxxxxxxxxxxxxxxxxxxxxxxxxxxxxxxxxxxxxxxxxxxxxxx
xxxxxxxxxxxxxxxxxxxxxxxxxxxxxxxxxxxxxxxxxxxxxxxxxxxxxxxxxxx
xxxxxxxx

Even without reading the text, the arrangement of words on the page communicates a number of things to the reader. The spatial cues we get from the page tell us

1. There are three points to be made.
2. They must be important, since they are separated out from the rest of the text.
3. They must be roughly equal to each other in importance, since they are all indented the same amount.

That is quite a bit of information gained without reading a single word (Kostelnick, 1990).

Such highlighted lists are common in technical prose, for obvious reasons: they give writers a chance to emphasize important points, and they give readers a quick way to find important information. Such lists can be enhanced when it is important to draw attention to them.

First, they can be numbered. By placing a number in front of each item, the writer reinforces the fact that there are specific key points to remember. Numbering is also useful when there is a sequence to follow, such as steps in an instruction manual.

Second, they can be set off with bullets. Bullets are small circles placed in front of each item in a list. They help set off the item for added emphasis.

There are three guidelines for lists. First, if possible, keep the items in a list short. Lists have more visual impact if each item takes no more than a single line. If an item requires more than one line, indent the second line so that it begins under the first. Second, keep items in a list roughly equal in importance. Since they are visually equal, they should also be rhetorically equal. The example list below shows how comical an unequal list can be:

Reasons to buy a new computer:
Save labor costs
Speed record keeping
Improve information flow
Play computer games

Third, use lists with some discretion. They emphasize points greatly and so create more impact. This is usually good, but there are times when it is inappropriate. For instance, none of us would like to receive a letter beginning, "Reasons why you are being fired."

Lines and Boxes

Many word processing programs and all desktop publishing programs make it easy to create simple lines and boxes around text. Like many new features, these capabilities can enhance documents, but they can also distract. Here are some suggestions for their effective use.

PAGE HEADERS

Our first example of page design (Fig. 9-2) showed a good use of lines. A simple line across the top of each page separated the running head from the rest of the page. By using the same line at the top of each page, the author created a pattern throughout the document that helped unify the document.

SIDE GLOSSES

Another good use of lines occurs when there is a logical disconnection between elements on a page. In the example of page design showing the use of side glosses (Fig. 9-3), the line distinguished the two kinds of material.

TABLES

Simple tables of just a few rows should not be boxed since they can be integrated with the rest of the text. But large, detailed tables may take real effort to understand. Setting them off from the rest of the page helps delimit them, and makes them easier to skip over if readers feel intimidated by their complexity.

IMPORTANT INFORMATION

Lines and boxes are commonly used to set off information that is particularly important, such as warnings or comments on the text itself. Figure 9-8 shows

Figure 9-8. Page Design with Lines around Warning

350 *Ending Your Communications Session*

Ending Your Communications Session

To end a communications session, you save the settings file (if you want to use these settings again), disconnect from the remote computer, and exit Terminal. Descriptions of these steps follow.

Saving Settings Files

You can save Terminal's settings so that you can load them for the next communications session.

▶ **To save the settings file:**

- Choose Save from the File menu.

   If you have not saved the settings before, a dialog box asks you for a filename. Terminal automatically assigns the extension .TRM to the file. You can choose Save As to save the settings under a new filename or to overwrite an existing file.

Disconnecting from the Remote Computer

Disconnecting from the remote computer is an important step in ending your Terminal session.

▶ **To disconnect from the remote computer:**

1. Type the exit command specified by the remote system—typically, *bye*.

   This command hangs up the remote modem.

2. Choose Hangup from the Phone menu to hang up your modem.

CAUTION   It is important to signal the remote computer with an exit command before choosing Hangup. If you choose Hangup first, you will be disconnected from the remote computer, but its phone connection will remain off the hook. Your account might continue to be charged for connect time until the remote system detects that the connection has been dropped.

Source: *Microsoft® Windows™ user's guide.* (1990; p. 350). For the Microsoft Windows graphical environment version 3.0. [Computer Program]. Portions © 1985–1992 Microsoft Corporation. Reprinted with permission from Microsoft Corporation.

a page design that uses lines around a warning. By using lines and boxes to set such information off from the rest of the text, a writer can make it seem particularly important—so important, in fact, that the reader should consider reading these words first. This is exactly the effect intended for warnings, of course, but it can be distracting if it occurs too often, so use lines and boxes sparingly.

Color

Good technical writers have found many ways to signal the content of their text to readers. One practice that has found its way into growing numbers of instruction manuals is the use of color. Each chapter can be printed on a different color paper, for example. Copy machines make the practice easy, yet it is very effective: if the page is yellow, I must still be reading the chapter on maintenance. More formal reports may not be appropriate for multiple colors, but where the practice is possible, it works.

There are two guidelines to consider when using color in this way. First, some colors make text harder to read. Black print on a dark gray paper, for instance, creates problems for everyone. Second, give some thought to color combinations. A purple chapter followed by a pink chapter followed by an orange chapter would be a lurid mess, not an enhancement. Stick to simple combinations of a few primary colors.

Icons

Icons are simple pictures. Most professions have a group of icons that are well-known. Computing is rife with icons, from the page with a corner turned symbolizing a word-processed document, to the waste basket symbolizing "erase" or "delete." Electronics has the symbol for a ground, etc. No profession is without at least a few icons (see Fig. 9-9).

The value of icons in technical documents is their effectiveness in communicating quickly and clearly. One symbol can carry the content of a host of words. Because of this value, icons are often used in three ways in documents.

HEADERS

Icons are increasingly used in headers to signal the intent of some section of the document. By placing the same icon on each page of a section, the author unifies the section. For instance, a chapter on maintenance might have a small wrench on the top of each page. The result might look like this:

Chapter 6—Maintenance

The word "maintenance" is highly visible, too, but the wrench reinforces the function of the chapter and links this page to other pages in the chapter.

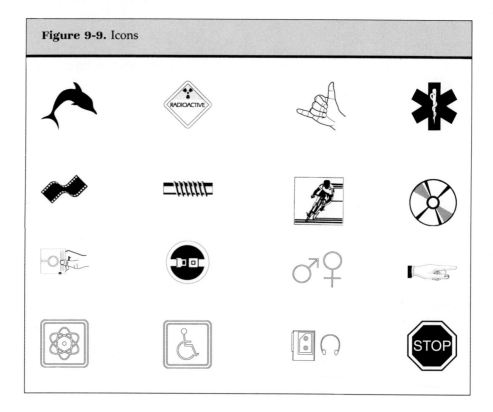

Figure 9-9. Icons

WARNINGS

Increasingly, instruction manuals include a stop sign to warn of danger if a step is missed or a part is malfunctioning. Here is an example:

   Make sure gas is off at LP tank.

The stop sign is universally understood. Its shape and color make it stand out from the rest of the text, and its meaning is clear. It will get a reader's attention far better than words like ''caution'' or ''important.''

HIGHLIGHTS

The pointing hand is most common for this purpose. It gets more attention than a simple arrow would, yet is less dramatic than a stop sign or other danger signal. Here is an example of its use:

If you cannot get ignition, check the following:

 Is there gas in the tank?
Is the burner connected to the valve?
Is the valve open?

While icons are an effective way to draw a reader's attention, they get the most attention when used sparingly. A stop sign in front of each step ceases to have any impact. Decide where you really need to put additional emphasis, and use the icons there alone.

Graphic Balance

Graphics have real impact in documents. Ironically, they may sometimes have too much impact. Normally the burden of creating a coherent communication is left to the words, with graphics adding clarity and depth. Problems arise when a document contains large numbers of graphics. In this case the reader may be tempted to ignore the words altogether and get an explanation totally from the pictures. This may in fact be how many people "read" news magazines like *Time* and *Newsweek*. While this may work at the newsstands, it can lead to errors when reading technical documents.

The example in Figure 9-10 is a case in point. Notice how the two graphics take over the page. There are several problems with this page. It would appear from the commands that each step involves a choice for the user and each screen shows the choice. Unfortunately, while the choice is displayed for step 3, it is not for step 2. Essentially the two graphics show the computer screen before and after step 3; neither has any direct connection to step 2.

Another problem is that these are very "busy" screens. They contain a host of information. Yet the only information that is germane is the box in the lower righthand corner.

The real problem is that the intent of this manual is to guide a reader step by step through an action. The steps are there, but they are separated so far by very busy graphics, that the graphics take over. The manual would have been much improved if a single screen had been referenced by all instructions, or if small portions of the screen had been displayed as needed. In either case, the real job of the writer was to emphasize the steps required of the reader, and these actions should have dominated the document. They don't.

The chief way to prevent graphics from distorting a document is to control their number and size. If they occupy more than one third of the page, they no longer supplement the text, they dominate it. Your text no longer supplies the coherence for your document, the graphics do. Sometimes, as with very large tables, this is unavoidable. But that is a special circumstance. In general, a writer needs to keep graphics under control.

Figure 9-10. Page Design Showing Poor Graphic Balance

---

2. Select **Change Font** from the Set Attributes panel. The File Services panel is displayed.

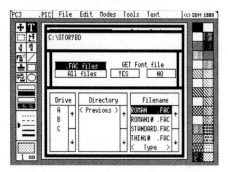

3. Select **THIN10.FAC** in the Filename box of the File Services panel.

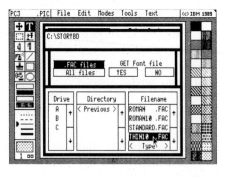

Source: *IBM storyboard plus.* (1990). Armonk, NY: IBM, p. 2-50. Courtesy of IBM.

Tools of Design

The word processor appears on the computer screen much like a blank sheet of paper. The software invites the writer to think of words alone. Nothing on the screen suggests that graphics or design are important issues. Desktop publishing programs, on the other hand, look very different on the screen (Fig. 9-11 is a good example). They display rows of tools available to the writer, with a restricted work space on the screen for the writer's text and graphics. As such, they constantly remind the writer of how many publishing options are available—of all the choices to be made.

Figure 9-11. Desktop Publishing Program

This point is worth considering during your own writing. While it is common to write a document in a word processor first and then use a desktop publishing program to reset the text or bring in graphics, it doesn't have to be that way. Each tool has an impact on your writing. There may be times when the best way to create a document is right in the desktop publishing program. At other times a good word processor may be enough for the entire job. Which you choose should depend on your purpose.

Word Processors

A point often made is that word processing programs are becoming more and more similar to desktop publishing programs. They increasingly can set columns of text, handle multiple sizes and fonts of type, and even merge in graphics. This means that for most reports a writer can begin work in the word processor and have sufficient power in the software to stay with the word processor all the way to the finished document. Word processors also have the advantage of connections to related software such as spelling and style checkers. For that reason alone, many writers prefer word processors.

But there are disadvantages to word processors, too. Many popular word processors still have limited ability to handle graphics, some have limited text sizes and styles, not all can merge graphics with text. They are called "word processors" because they were designed to handle words. The industry in general is moving beyond that, but the movement is incomplete and uneven. For many documents a word processor is insufficient.

Desktop Publishing Programs

The fundamental strength of desktop publishing programs is their ability to take the old jobs of typesetting, paste-up (pasting columns of text and pictures onto a larger sheet of paper), keylining (drawing in lines), and halftone creation (turning photographs into an image that can be duplicated by a printing press) and merge them into one job. Essentially, they reproduce the function of commercial art departments in a personal computer. Figure 9-12 illustrates some of those capabilities.

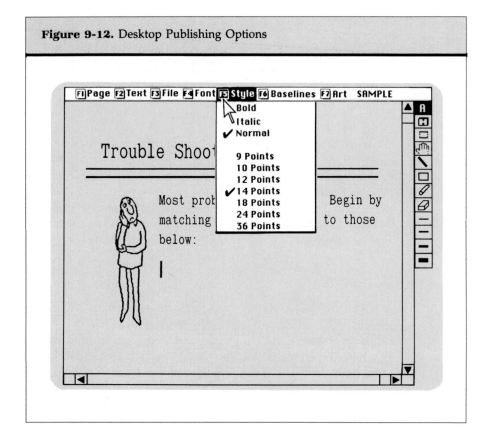

Figure 9-12. Desktop Publishing Options

One menu pick allows the writer to create boxes of any size (and to control the thickness of the box lines). Another pick might be ovals or circles. The program allows the writer to create simple line drawings on the screen. If such drawing appears too difficult, the program comes with "clip art," or a collection of drawings that can be loaded from a floppy disk and used in any document (see Fig. 9-13). Of course, the program will also import, resize, and modify graphics generated elsewhere. Such a program is especially valuable if you are creating a document that will be heavily illustrated.

Despite all these strengths, some writers resist using desktop publishing software. Bad spellers resist using publishing programs because they don't usually include spell checkers (or other ancillary software like grammar and style checkers). Others are troubled by the complexity of the software. With multiple menus the new user is often faced with unknown options and disappearing choices. Then there is the fact that some users are not only new to the software, but to the publishing concepts that underlie it. They not only need to learn computer terms, but publishing terms as well. The result is more resistance to such a powerful tool than you would normally expect.

Figure 9-13. Clip Art

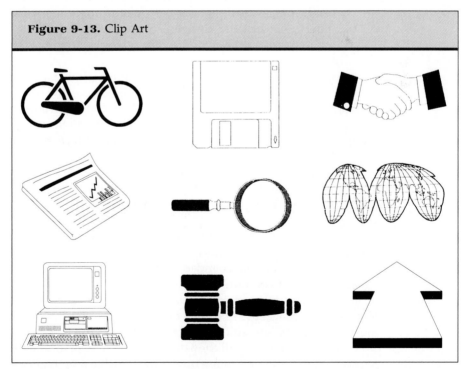

Source: *WordPerfect® for IBM personal computers and PC networks.* (1989; pp. 777–779). Courtesy of WordPerfect Corporation. WordPerfect is a registered trademark of WordPerfect Corporation.

As someone creating technical documents, you will probably have occasion to use both tools. Both give writers far more control over documents than we could have dreamed of just five years ago. They give us all the power we need to design effective documents.

Activities

1. Find three different page designs. What makes them different? In each case, why did the author select that format? Which seems most effective? Which is the most popular among your classmates? Try to identify which features are most attractive.

2. Create a page design for each of these documents: an instruction manual, a feasibility report, a department memo, a letter to a customer.

   a. First try designing the pages by hand. What are the advantages and disadvantages of this approach?

   b. Next try designing the pages using a word processor. See if it has special capacities such as lines, boxes, multiple columns, multiple fonts, different type sizes, and graphics. What are the advantages and disadvantages of designing with word processing software?

   c. Show your favorite designs to your classmates and ask for their opinions. Do they find your designs appropriate? Effective? Appealing?

3. Review the headings used in this chapter. How many are there? Time your ability to go back and find the section on icons. How much help was the heading? Where do you think more headings could have been used? Compare several books on technical subjects. How frequent were the most frequent headings? How frequent were the least? Which book was easiest to read?

4. Find a technical document that includes boxed text. What was the author trying to accomplish? Do you agree that the text should have been boxed? As a group, try to determine other ways the text could have been highlighted.

5. Use your word processor to "greek" one of your documents. Ask others in class to guess about the importance and purpose of various text blocks. Is that what you intended? What changes could you make to match visual design and content?

6. Take any magazine that uses glosses and scan several past issues. When did the magazine first start using glosses? Why? Which magazines do you think will follow suit? Do you think glosses are a permanent part of professional writing, or will they disappear or change to something else? Poll your classmates and determine how others read text with glosses. How many read only the glosses? How many read the main text? How many read both? If they read both, which do they read first?

7. For a professional journal article that does not use glosses, write glosses you think would be appropriate for the article. What kind of glosses did you use, summaries or quotations? Why did you make this choice?

8. Primacy and recency effects can be quickly demonstrated by having subjects recall lists of words, lists of numbers, or pictures of objects. In groups of three or four, try several lists and determine which is stronger, the primacy effect or the recency effect. What implication does that have for information placement on a page?

9. Some complain that desktop publishing programs are complicated and difficult to learn. Does that seem true of the programs you have access to? If so, what makes them difficult? Ask others in class if they are aware of simpler programs.

III

TECHNICAL WRITING TECHNIQUES

Technical Definition

The Role of Technical Definition

When to Use Technical Definition

Types of Technical Definition
Informal Definitions
Formal Definitions
Stipulative Definitions

Writing a Technical Definition
Writing a Single-Word Definition
Writing a Phrase Definition
Writing a Sentence Definition
Writing an Expanded Definition

Guidelines for Technical Definition
1. Use the Least Intrusive Method Possible
2. Place Definitions Carefully

Examples of Technical Definition

Activities

The Role of Technical Definition ───────────────

Technical writing is a process of mutual understanding. The technical writer must be certain that his or her audience understands the means of communication in the same way the writer does. Imagine you are observing a set of signal flags being raised by a ship; obviously a message is being transmitted, but, since you are unfamiliar with the meaning of the individual flags, you are unable to receive their message. Or suppose that you know the traditional meaning of the flags, but, in the interest of security, the "sending" ship has adopted a code; the meaning you receive from the flags will be, at best, garbled beyond understanding, or, worse, misleading. In both cases, successful communication has been precluded by the fact that the means of communication are not equally understood by sender and receiver.

A shared basis of communication is especially important in technical writing since the technical writer is often far removed from the audience; the communication is essentially one-way. Unless the writer has a chance to have a reader assess the piece of technical communication, the writer has no opportunity to observe a reader's reactions, and readers in turn have no way to query the writer if a point is not clear.

Consider the following passage from a request for proposal (RFP) for a telecommunications system:

> A cutover on or before August 1, 1985 is required, and a $300 per day penalty will be assessed for a late cutover (*Request for proposal: Telecommunications system*, 1984).

Some readers may know that the term "cutover" describes an area cleared of trees. But what are readers to make of the term "cutover" in the context of the RFP? Does it mean that some area needs to be cleared of trees by August 1, 1985? Unless readers check the glossary in the back of the RFP, they might never know that "cutover" refers to "the actual time at which the new system is put into service for all users." Because the possibility of failed communication always exists, writers must carefully consider the need for definitions.

A definition gives a precise meaning of a term by identifying the particular qualities of an item and by distinguishing the item from all others. A definition may be of a tangible object, such as a "solvent cement," a process, such as "flame hardening," or a concept, such as "artificial intelligence."

The purpose of definition is to help readers understand a term you are using. Definitions are a way of ensuring communication by establishing shared means of communication.

When to Use Technical Definition ────────────

Sometimes it is difficult for technical writers to know when to define. In general, technical writers define whenever they think readers will not understand the word or term they are using. But this will mean different things in different circumstances.

Define any word or term that might be unfamiliar to many or most of your readers. Why might a physician call a "heart attack" a "myocardial infarction"? Why not say a "germ" caused your illness, rather than a "virus" or a "bacterium"? Why not call *The Last Supper* a "painting"; why use the term "fresco"? Why use an unfamiliar word or term at all? The main reason for using an unfamiliar word or term is that the unfamiliar term provides a level of specific information not available through the use of the everyday word choice.

Whether or not the reader is unfamiliar with the word or term you are using depends on the reader's level of knowledge. For example, you might have to define "fresco" for a reader unfamiliar with Renaissance art, but you would not have to define "fresco" in a book on Renaissance art. If readers are experts or technicians, they often don't need definitions. However, if readers are executives or novices, they often will need definitions. And, if readers are a mixed audience, successful writers will usually assume the lowest level of knowledge.

Familiar words used in an unfamiliar way should be defined. Consider the word "bug." We all know what a bug is (especially those insects that bite). If, however, you are describing your work on a computer project, your meaning will certainly be missed if your readers think you are discussing members of the order Heteroptera when you suggest that the bugs you encountered delayed your progress in completing the project.

Typically, the problem of using familiar terms in unfamiliar ways arises when writers discuss their technical specialty—in other words, when the writer falls into jargon or shoptalk, when ordinary words take on specialized meaning for practitioners of a specialized field. For example, a "Christmas tree" is not only the familiar holiday conifer; in drag-racing parlance it is the vertical arrangement of staging and starting lights; to a naval engineer the term might mean the lighted panel in a submarine which indicates if hatches are opened or closed.

If you think that using a familiar word in an unfamiliar way might lead to reader confusion, either define the term or substitute a word or phrase that is both familiar to your reader and will readily identify the technical concept at hand. For example, a computer program bug might be called an "error in logic." When such substitutions can be made accurately and economically, they have much in the way of increased effectiveness to recommend them. However, there are also times when shoptalk can be justified on the basis of economy, accuracy, or simply tone. In those cases, you must be aware of possible problems of reader comprehension and consider the need for definition.

Some writing tasks are more likely than others to need definition. For example, a feasibility study almost always needs a stipulative definition, but a literature review often does not require one. However, any writing task, whether a literature review, a proposal, or a feasibilty study, may require the use of definition. The determination to use a definition is not related to the kind of writing you are doing but rather to the ability of the audience to understand the words, terms, and phrases you are using.

Types of Technical Definition

Knowing when to define is not enough: successful technical writers must also use definition skillfully. Once the writer has determined that simple, everyday words cannot effectively convey the message, he or she must judge the level of formality and detail necessary to assist the reader.

Informal Definitions

There are many instances when all readers need is a quick operational definition of a term. This is especially useful when the writer does not wish to interrupt the flow of the presentation for a fuller, more formal definition. An informal definition of this sort may be a single word, a phrase, or even a sentence, depending on the level of detail and thoroughness the writer feels readers need.

Informal definitions are, essentially, synonyms: a familiar word, phrase, or term is substituted for an unfamiliar word, phrase, or term. For example, the following list of recommended fats and oils appears in a brochure published by the American Medical Association, *What you should know about high blood cholesterol*. The writer assumed that readers would be familiar with all the oils except "rapeseed." Thus, the writer included in parentheses the synonym for rapeseed.

> Unsaturated vegetable oils: corn, olive, rapeseed (canola), safflower, sesame, soybean, sunflower (American Medical Association, 1988).

Remember that informal definitions are exactly what their name implies: informal and incomplete.

> Most members of the family Gruidae (cranes) are being threatened by man's incursion into their nesting grounds.

"Crane," for example, refers both to any of various large wading birds of the family Gruidae and to similar birds, such as herons; but, for the purposes of the example sentence, which requires a brief and nondistracting definition, the informal definition is sufficiently precise.

Formal Definitions

If precise, detailed understanding of the term in question is critical to your presentation, an informal definition will not be adequate. In such cases you must use a formal definition. A formal definition is a precise, rigidly structured, logical construct. It consists of three parts organized in a specific form: the *term to be defined* is placed in a *general category*, then differentiating characteristics are supplied to distinguish the term from other members of the same general class. Consider the formal definition as a mathematical equation:

Term = General class + Distinguishing characteristics.

Every object has characteristics that make it different from objects that are like it. A formal definition works by narrowing from general to specific. Notice in the following example how the formal definition is a process of gradual tightening and increasing specificity. The term "solvent cements" is placed in the general class of "adhesives," and then placed in the increasingly more specific categories of adhesives that contain "organic solvents other than water."

Solvent cements are adhesives that contain organic solvents other than water (Smith, 1969).

Stipulative Definitions

There are times, of course, when the technical writer recognizes that readers will be aware of a number of possible meanings for a word or phrase. Each definition may be perfectly accurate and acceptable in different circumstances, yet the reader must know exactly which meaning is intended by the writer. One way to ensure such understanding is to state that for the purpose at hand, you are defining the term in a particular way; the writer thus "stipulates" which of the possible definitions he or she intends.

For example, suppose you are writing a proposal for updating voter registration records for an ecumenical group. In the process of preparing the proposal, you administered a questionnaire to determine the religious affiliation of members of the target population. The question might arise as to what it means to say that a person is Roman Catholic. That could be construed as "baptized as a member of the Church" or "born of Roman Catholic parents" or "self-identified as Roman Catholic" or "attending Mass at least twice a month" or any of a number of other possibilities. Failure to recognize the possible interpretations of the identification is certain to lead to questions concerning the validity of your results. This is an ideal opportunity to use the stipulative definition. Simply state:

> This study identifies as "Roman Catholic" all respondents indicating that they attended services at a Roman Catholic church at least six times in the past year.

While some readers might take issue with your definition, and you might well wish to discuss the justification and implications for this use of the term, at least your readership has a very specific awareness of what your terminology means.

Writing a Technical Definition

The basic strategy for writing a technical definition is that the definition should be simpler and easier to understand than the thing being defined. For example, "facilitate (help)" is good, but "help (facilitate)" probably is not. But knowing when and why to define are not enough: you must also apply the methods of definition skillfully. Once you have determined that simple, everyday words cannot effectively convey your message, you must judge the level of formality and detail necessary to assist your reader.

Writing a Single-Word Definition

If the term in question might be unfamiliar to the reader, but an accurate synonym can be found, a single word may suffice. In the following examples, the writers chose to define a term because of their perception of their audience. In the first example, although the document is primarily intended for an informed audience, the parenthetical insertion defines the term "Gruidae" for the nonspecialist reader.

> Most members of the family Gruidae (cranes) are being threatened by man's incursion into their nesting grounds.

In the second example, because the writer anticipated a combined audience that would include many readers who might not know what the word "tack" means in the context of the document, she included a single-word informal definition.

> Generally, dextrins possess good humidity resistance and high initial tack (stickiness) (Smith, 1969).

Writing a Phrase Definition

When a single-word synonym cannot properly perform the definition, a synonymous phrase inserted into the sentence that first introduces the unfamiliar term can often do the job. In the following definitions, the writers used words that many readers might have heard but might not fully understand. The first term appears in a brochure on high blood cholesterol published by the American Medical Association. The second term appears in an Environmental Protection Agency guide to methods of reducing radon.

> Your risk of developing atherosclerosis and coronary heart disease (narrowing down and blockage of the arteries that supply blood to your heart muscle) is directly related to the amount of cholesterol in your blood (American Medical Association, 1988).

> Performing screening and follow-up measurements prior to a decision to mitigate (that is, to reduce radon levels), is strongly encouraged (U.S. Environmental Protection Agency, 1987).

Writing a Sentence Definition

The informal definition may even extend to full sentence length. The following definition of a process was written for manufacturers who use hardenable ferrous alloys and gray or pearlitic malleable cast irons. In this example, the technical term, not the definition, is in parentheses. While longer than previous examples, the definition still is informal because of its tone and relatively imprecise and general level of detail.

> Essentially, flame hardening is a process used to harden steel by heating it above the transformation range (austenitizing) by means of a high temperature flame and then quenching it at a rate that will produce complete hardening (Pavesic, 1957).

Writing an Expanded Definition

Often a simple word or phrase of informal definition, or a sentence-length formal definition, will clarify the meaning of a term enough to allow you to go on with the business of your writing. There will be times, however, when you feel that further amplification is needed. If your readers' understanding of a term is crucial to their understanding of the point you are making, if distinctions need to be made to avoid misinterpretation, you will need to expand your definition. Expanded definitions, which are usually based on a formal sentence definition, add to both the definition and the readers' understanding by employing one of the following techniques.

DEFINITION OF WORDS WITHIN THE DEFINITION

For example, the definition of "solvent cements" above might well be amplified for inexperienced readers with additional definitions of "organic solvents." A concrete instance of the term defined often provides the reader with a "handle" on the term. For example, the following definition of "durable goods" appears in an article on "Protective Packaging Problems."

> Durable goods are manufactured products capable of long utility, such as power driven tools, lawnmowers, stoves, high fidelity reproducers, television sets, washing machines, sewing machines, office machines, furniture, motors, and engines (Burton, 1958).

You should, of course, try to draw your examples from the readers' experience; most readers are unlikely to better understand the term "carnivore" if your examples of carnivores are "the genus *Vulpes* and the genus *Herpestes*." And, in the following definition, the example is ineffective if the reader does not know what a "roller conveyer" is.

> **Skate-wheel conveyor.** This is a fixed or portable unit similar to a roller conveyor for carrying packaged material down an incline (Asimow, Morris, & Bosticco, 1958).

LINGUISTIC DERIVATION

By providing the history of the term or breaking the word into its component parts, you may implant the meaning of the term more firmly in the readers' mind. This technique is often used parenthetically for terms that are acronyms — words formed from the first (or first few) letters of several words:

> radar (radio detection and ranging)

> FORTRAN (formula translation)

GRAPHICS

Very often a graphic helps the reader to understand and remember a definition. Usually the most helpful graphics for technical definitions will be illustrations (photographs and drawings), which can simply and clearly represent the physical appearance of the item being defined.

Sometimes the item being described is one that the reader has probably seen before but might not know (or remember) the term for. In this case a simple representative drawing or photograph will often do. Labeled drawings are quite helpful if the item being defined is part of a larger whole or if it

---

Figure 10-1. Technical Definitions Using Drawings

flange (flanj) *n.* ⟦ < ? ME *flaunch*, a lenticular space on a coat of arms < OFr *flanche*, side, var. of *flanc*: see fol. ⟧ a projecting rim or collar on a wheel, pipe, rail, etc., to hold it in place, give it strength, guide it, or attach it to something else —*vt.* **flanged, flang'ing** to put a flange on

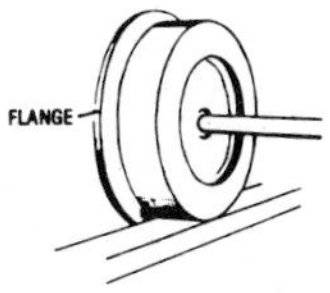

FLANGE

neu·ron (nōo'rän', nyōo'-; -rən; noor'än', nyoor'-; -ən) *n.* ⟦ModL < Gr *neuron*, NERVE⟧ the structural and functional unit of the nervous system, consisting of the nerve cell body and all its processes, including an axon and one or more dendrites: also **neu'rone'** (-rōn') —**neu'ro·nal** (-rə nal) or **neu·ron|ic** (nōo rän'ik, nyōo-) *adj.*

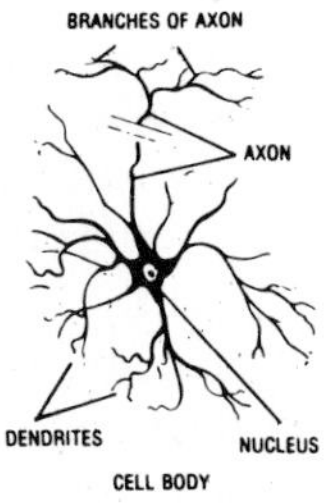

BRANCHES OF AXON

AXON

DENDRITES   NUCLEUS
CELL BODY
NEURON

Source: From the book: *Webster's New World Dictionary* (3rd college edition) © 1988. Used by permission of the publisher: Webster's New World Dictionaries/a division of Simon and Schuster, New York.

contains several parts whose relationships are difficult to describe in words. In this case the drawing can either provide the visual context for the item being defined, or it can visually identify each of the parts that have been named in the description. Figure 10-1 shows two examples of drawings being used in this way.

Guidelines for Technical Definition ⎯⎯⎯⎯⎯⎯

1. Use the Least Intrusive Method Possible

If you need a formal, extended definition, by all means use it. However, if you believe that a parenthetical insertion will suffice, it is preferable to do that.

2. Place Definitions Carefully

You have three choices for *where* to place a definition: in the text, in a footnote, or in a glossary. The decision is usually based on the writer's assessment of the audience for a particular piece of writing.

PLACEMENT IN THE TEXT

The obvious advantage of placing a definition in the text is that the reader need not hunt for the meaning of the term. In-text placement is appropriate for unobtrusive informal definitions, but it is also feasible for lengthy, obtrusive definitions in these cases:

- If it is absolutely essential that the reader have a thorough and precise understanding of the term (as in a legal text or a feasibility study).
- If the writer is quite sure that almost none of the readers will understand the term.
- If the writer is particularly concerned that the readers not *mis*understand the term (e.g., if you write "bug" and you're reasonably sure that most readers won't think to look in a footnote or glossary until it's too late, you might want an in-text definition, even though it might be interruptive).
- If there is no other place it could reasonably go (i.e., memos usually do not have footnotes or glossaries).

The following excerpts, taken from the same piece of writing on bat guano, are examples of in-text informal definition. In the first excerpt, the writers chose to insert the definition of "guano" parenthetically, the least intrusive of the three methods of defining. Clearly, in this instance, the writers did not feel that the readers needed more than an informal definition of "guano."

> Unlike other organic fertilizers, bat guano (manure) is unique in containing three crucial nutrients—nitrogen, available phosphorous, and potash—while most contain only one (Summar, 1984).

In the second excerpt, however, the writers felt that it was important for the readers to know precisely what is meant by the term "leached." Because the concept is essential for their argument, the writers presented a fuller definition in a separate sentence.

> According to organic-farming consultants, bat guano is especially valuable as a fertilizer because it's one of the few manure products which hasn't been "leached." That is, since it has been composted by time inside caves, it hasn't been rained on—which pulls valuable minerals out of manure (Summar, 1984).

PLACEMENT IN FOOTNOTE

If your definition is lengthy and you do not want to include it either in the body of the text or in a glossary, place the definition in a footnote. The advantage of placing a definition in a footnote is that it is readily available to the reader who needs it but does not intrude on the reader who does not need it.

One section of an RFP submitted by the Department of Health and Human Services (1985) for fabrication of cardiovascular devices asks for information about the company replying to the RFP. One of the questions is about the *representation* of socially and economically disadvantaged individuals within the company:

> *Representation.* The Offerer represents that it
> _____is   _____is not
> a small disadvantaged business concern.

The definition of "small disadvantaged business concern" was considered too lengthy to be included in the body of the text; therefore, the writer placed the definition in a footnote:

> "Small disadvantaged business concern," as used in this provision, means a small business concern that (1) is at least 51 percent owned by one or more individuals who are both socially and economically disadvantaged, or a publicly owned business having at least 51 percent of its stock owned by one or more socially and economically disadvantaged individuals and (2) has its management and daily business controlled by one or more such individuals (Department of Health and Human Services, 1985).

PLACEMENT IN GLOSSARY

A glossary is a list of definitions of selected words, phrases, and terms. If the list is short, it often appears at the beginning of a document; if the list is lengthy, it often appears at the end of a document. Glossaries are used for two reasons.

First, if you are writing for a mixed audience, readers who have varying levels of knowledge about the subject you are writing about, and you have a lot of highly technical terms that need defining, place the definitions in a glossary. For example, in the definition of flame hardening above appears the term "austenitizing." Rather than define the term in the text, for those readers who want to know the meaning of the term, the writer chose to include it in a glossary where it would be least intrusive.

> **Austenitizing.** The process of forming austenite by heating a ferrous alloy to temperatures in the transformation range (partial austenitizing) or above the transformation range (complete austenitizing) (Pavesic, 1957).

Second, if you want to offer definitions that are more technical than already appear in your document, use a glossary. Similarly, the writer chose not to include a definition of "austenite" in the definition of "austenitizing." For those readers who are not familiar with the term "austenite," the writer also included the definition in the glossary:

> **Austenite.** A solid solution formed when carbon and certain alloying elements dissolve in gamma iron (Pavesic, 1957).

The advantage of a glossary is that readers who need definitions will have them all together alphabetically in one place. The disadvantage of placing definitions in a glossary is that the reader must continually turn to the back or the front of the document to find out the meaning of unfamiliar terms. Thoughtful writers will be considerate of readers by giving special treatment in the text to words that are included in a glossary. One method for indicating that a word, phrase, or term is glossed is by inserting an asterisk before or after the word, phrase, or term. Another method is by annotating the word, phrase, or term with a parenthetical notation, such as "(see glossary)."

Examples of Technical Definition

Below are two examples of technical definition. Figure 10-2 contains an expanded definition that appears as part of the background section of a proposal for a process of genetic manipulation. The readers of the proposal are a mixed audience of executives, who are educated but not specialists in genetics; managers, many of whom are highly knowledgeable about ge-netics; and technicians, who have varying degrees of knowledge about ge-netics. Because the proposal is highly technical, the writer felt it important to provide this information for readers who needed the background but were not highly knowledgeable about cell reproduction. Note how the writer skillfully uses informal one-word definitions of the two forms of cellular reproduction in sentence 3, and informal phrase definitions of the life cycle of a cell in sentence 6 and of chromosomes in sentence 12. Note also the use of analogy in sentence 12, which refers to the nucleus as the "command center." The writer further expands the readers' understanding by providing an analysis of the process to which the term is applied, and, in the first two sentences, provides justification for the space she will devote to the definition by pointing out the pervasive importance of the terms to be defined.

Figure 10-3 contains a glossary from the *Walnut Acres Catalog.* While most glossaries appear at the end of the document, sometimes a glossary will appear in the front of the document. The customers of this catalog requested definitions for the various terms used in the catalog to describe the quality of their products, and the writers chose to present the glossary of terms *before*

Figure 10-2. Example of a Technical Definition

Cell Reproduction
Mitosis and Meiosis

New cells are produced by existing cells. Plants, animals, even fungi, every living organism must reproduce itself in order to perpetuate its species. The two forms of cellular reproduction are *mitosis*, or asexual reproduction, and *meiosis*, or asexual reproduction. Most one-celled organisms, such as algae and yeasts, reproduce by mitosis. Most multicellular organisms reproduce by meiosis. The life cycle of a cell, the time required for one cell to reproduce itself one time, ranges from 20 minutes to 24 hours.

With few exceptions, most cells have the same construction. The cell is surrounded by a plasma membrane or outer wall. This membrane contains a thick fluid substance called *cytoplasm*. The cytoplasm converts food into energy and synthesizes compounds such as proteins into building blocks for the cell. Cytoplasm contains several structures, but only one, the nucleus, is important in reproduction. The nucleus, the "command center" for cell reproduction, is made up of chromosomes, rod-like bodies containing DNA, the substance encoded with the hereditary information necessary to reproduce a new cell as a copy of the original cell.

Source: Department of Health and Human Services. National Institutes of Health. (1985, October). *Fabrication of cardiovascular devices* (Request for Proposal No. NHLBI-HV-86-01). Washington, DC: U.S. Government Printing Office.

Figure 10-3. Example of a Glossary

Throughout this catalog, you'll see three symbols used with product descriptions:

Walnut Acres-certified organic. Grown at Walnut Acres, where we've been farming and preparing food without synthetic chemicals since 1946; or supplied to us by reputable outside sources who've been farming without synthetic chemicals for at least five years (most go back further), and with whom we've worked over the years.

* *Certified organic by others.* Produced by growers whose land is associated with one of the standard organic food certification programs.

/\ *Transitional organic, or unsprayed.* Products that partially meet the requirements for certification by any group. In all cases, no synthetic poisonous chemicals have been used on the product itself.

Source: *Walnut Acres catalog.* (1989, winter). Penns Creek, PA: Walnut Acres.

describing items in the catalog to make it easier for readers to understand the terms and make their choices. The following glossary of three terms appears on page 3 of the catalog, preceded by a statement indicating to the reader how the products are identified.

Activities

1. With two or three other students, identify as many types of definition and as many ways of writing a technical definition as you can in the example, "Cell Reproduction: Mitosis and Meiosis." When you have finished, confer with other groups to see if your group or the other groups missed anything. Ultimately you should arrive at a complete list.

2. A prospective employer would like your technical writing team to demonstrate its writing skills by revising a piece of writing. With two or three other students, rewrite "Cell Reproduction: Mitosis and Meiosis" for a novice audience of freshman English majors.

3. The term "cutover" appears in a passage at the beginning of this chapter. The term is defined in a glossary. How many different ways are there to present the definition? In a report to your classmates, describe each way of presenting the definition and discuss under what circumstances you might use each definition.

4. Amplify the definition of "solvent cements" to include an additional definition of "organic solvents." How many different ways are there of presenting that amplified definition? Under what circumstances might you use each of the different methods?

5. Find four technical definitions. In a brief oral report to the class, discuss what types of definitions you found (informal, formal, stipulative) and why the definitions are or are not appropriate for the intended audience. As you analyze the definitions, consider such items as the length of the definition (e.g., single-word, phrase, sentence) and the placement of the definition (e.g., in the text, in a footnote, in a glossary).

6. Choose one word, term, or phrase from the expanded definitions of "Cell Reproduction: Mitosis and Meiosis," and write six definitions of that term or phrase. Your audience is your classmates.
   a. Informal (single word)
   b. Informal (phrase)
   c. Informal (full sentence)
   d. Formal
   e. Stipulative
   f. Expanded

Technical Description

The Role of Technical Description

When to Use Technical Description

Types of Technical Description

Writing a Technical Description
Introduction
Part-by-Part Description
Graphics

Guidelines for Technical Description
1. Use an Appropriate Level of Detail
2. Use Figurative Language
3. Indicate Subdivisions Clearly

Examples of Technical Description

Activities

The Role of Technical Description ──────────────

Technical descriptions are generally used to provide basic information as a framework for further discussion. For example, if an automotive supply company wants to direct attention to its new fan belt, it will include a description of the belt in the company catalog. In addition to giving economic reasons for purchasing a new fax machine for the office, a manager may also describe the new machine. An architect presenting a plan for a mall will include a description of the building and the grounds around the building. Because a technical process description is a specialized form with some unique characteristics, it will be discussed in a later chapter.

To describe is to create a mental picture. Technical description is a kind of translation. Whether the object being described is a cold chisel, a jet turbine, or a wetland, technical description is the conversion of one form to another.

Whether the description is part of a larger piece of writing, such as a manual, a research report, or a report of observation, or whether it is the primary goal of the writing, the purpose of technical description is to help the reader accurately visualize the thing being described, whether an object, a mechanism, a place, or a process. Technical writers describe to help readers understand easily and clearly what an object, place, or mechanism looks like.

When To Use Technical Description ──────────────

Although technical description may be an end in itself, more frequently it accompanies a larger piece of writing (such as a proposal, a manual, or a report). As the examples in this book demonstrate, technical descriptions are found in user manuals, environmental impact statements, systems specifications, field reports, research reports, instructions, and even in sales brochures. In fact, almost every technical writing task requires some amount of technical description: research reports include descriptions of the subjects, materials, equipment, and sites of the research; proposals and feasibility studies include descriptions of the original situations and of possible alternatives; progress reports often include descriptions of the equipment or materials used in a project; and most complex instructions include a description of the object or mechanism as a whole so that the reader will know how the steps to be performed relate to the finished product.

Types of Technical Description ──────────────

A technical description usually describes either an object (such as a cactus plant or a photon), a mechanism (such as a turbine engine or a Geiger counter), or a place (such as the nesting sites of herons or the plan of a city).

Technical descriptions of processes will be treated in chapter 12, "Technical Process Descriptions."

Technical descriptions can also be grouped according to their subsequent use. **Instructional descriptions** are appropriate when the reader will use the description to replicate the object, mechanism, or place described. The processes involved in assembling or creating some things are so simple that a step-by-step description of how to perform the process may not be needed. For example, once most readers have read a description of a typical memo format and seen an example, they have all the information they need to create memos of their own. In these cases, technical descriptions are a form of instruction. The writer of an instructional description should be sure to include enough information that the reader can replicate the item described without difficulty.

Informational descriptions are appropriate if the reader is reading primarily for information. That is, the writer assumes that the reader will not actually use or create a rainbow, a lunar buggy, or a deep salt mine. Most descriptions are of this kind. Although these descriptions can be either quite broad and general or quite specific and detailed, as a rule they require less detail than instructional descriptions.

Writing a Technical Description

Descriptions of objects, mechanisms, and places often follow a two-part structure: (1) a brief, general introduction that includes either an overview of the description or a statement of the focus of the description, and (2) a part-by-part description of the thing being described.

Introduction

The introduction provides any information the reader might need in order to understand the description that follows. Most readers will need some or all of the following information:

Definition of the object to be described
Overview of the description
Purpose (of a mechanism)
Location (of a place)
The physical characteristics of the object

Knowledgeable readers will need only an abbreviated form of this information. Readers who are not highly knowledgeable about the object, mechanism, or place, however, will need more complete information.

Part-by-Part Description

The major section of a description is the part-by-part description of the mechanism or place. The first task when writing this section is to choose the method of organization. The most common organizations for the part-by-part description are functional, spatial, and chronological.

A functional organization describes how something works. A functional organization might be used to describe how a fax machine, a jackhammer, or a rotary engine works. If you organize a description functionally, you would likely describe a mechanism according to how it works.

A spatial organization describes an object, a mechanism, or a place from one vantage point to another. If you organize a technical description spatially, you might describe something from left to right (or right to left), from east to west (or west to east), top to bottom (or bottom to top), or inside to outside (or outside to inside).

A chronological organization is used for describing how something is put together. Chronological organizations are used when a writer needs to describe an object or mechanism in the order in which it is assembled. Typical objects and mechanisms that are described chronologically range from swing sets to nuclear reactors.

Once you have chosen the general organization of the description, you will have to partition it, that is, divide the description into its parts. The problem for technical writers is deciding just what is a part. For example, for some audiences and purposes, a general description of the handle and the blade would be sufficient to describe a kitchen knife. However, for an audience that needed more details, the writer would include a description of the butt, the back, the edge, and the point.

Again, depending on audience and purpose, you might have to subdivide further into subparts. If, for example, it were important for the description of the kitchen knife to be even further subdivided, the writer could include a description of the tang and the rivets in the description of the butt.

After you have chosen the parts to be described, identify each part, either with a heading or with introductory language that clearly identifies the part being described. The description of the sheet machine that appears later in this chapter demonstrates the use of introductory language:

A.2.2 *The outer trough* for overflow water is . . .

A.2.3 *The inside of the funnel* is carefully . . .

A.2.4 *The grid plate* consists of a . . .

A.2.5 *The top hinged part*, A, of the apparatus . . .

A.2.6 *The sump* consists of a large cylinder . . .

Graphics

Although graphics are subordinate to the text in technical description, and rarely substitute for the written word, graphics can be powerful aids in descriptive writing. Just as the use of similes helps clarify thought, the use of graphics can make technical description easier to understand by showing relationships more dramatically than words alone, helping to focus on key portions of the description, and giving a great deal of information efficiently.

The most common kinds of graphics that accompany technical description are illustrations (photographs and drawings) and diagrams. Photographs show the reader exactly what a mechanism or place looks like. The reader can get an accurate sense of proportion and see more realistically the relationship between the parts. There are disadvantages to photographs. Generally, the reader sees only the outer surface features of the thing photographed. Even if you use a cropped photograph, the reader still sees more than is necessary, which might lead to confusion. Finally, the reader is limited to a two-dimensional view of the object, mechanism, or place.

An example of an appropriate choice of a photograph for an illustration is the cover of the owner's manual for the exercise machine pictured in Figure 11-1. Here the primary concern is not to show certain parts of the object but to present a realistic illustration of the machine. The machine as a whole is

Figure 11-1. Technical Description Using a Photograph

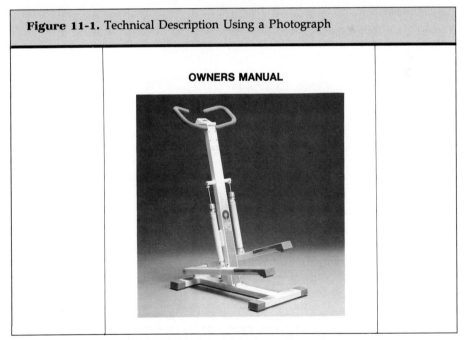

OWNERS MANUAL

Source: Reprinted, with permission of the editors, from *Owners manual.* (1990). Variable resistance climbers (models C401, C405, C406, C440). Bellevue, WA: Tunturi.

described in the manual, but the photograph constitutes a description in its own way. Once readers have looked at it, they have as much information as if they had read a lengthy description of the machine's appearance and construction.

Use a drawing when you want to focus on specific aspects of the mechanism or place you are describing. A drawing has three advantages: (1) it directs the reader's attention to specific parts of the mechanism or place, (2) it offers greater detail than a photograph, and (3) it can present objects, mechanisms, and places in three-dimensional views. A cutaway drawing, for example, can show the internal workings of a mechanism. Because the cutaway drawing in Figure 11-2 provides so much detailed information about the

Figure 11-2. Technical Description Using a Drawing

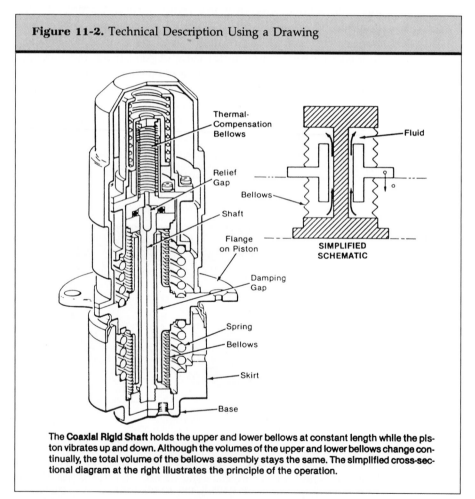

The **Coaxial Rigid Shaft** holds the upper and lower bellows at constant length while the piston vibrates up and down. Although the volumes of the upper and lower bellows change continually, the total volume of the bellows assembly stays the same. The simplified cross-sectional diagram at the right illustrates the principle of the operation.

Source: National Aeronautics and Space Administration. (1989, January). *NASA Tech Briefs.* Washington, DC: U.S. Government Printing Office.

nature and placement of the parts in the mechanism described, the textual description itself (which appears beneath the drawing) can be much briefer and clearer.

The disadvantages of a drawing are that the reader does not see exactly what the mechanism or place actually looks like, may not get an accurate sense of proportion, and does not see a realistic relationship among all the parts.

Use a diagram when you want to emphasize the relationship among things rather than the appearance of things. The advantage of a diagram, such as a flowchart, an electronic schematic, or a structural blueprint, is that it allows the writer to show the abstract relationship among the parts of the mechanism or the place. For example, the schematic diagram in Figure 11-3 accompanies a description of the National Electrical Code for antenna grounding. Although the diagram does not show the size of the mast, the height of the supporting structure, the distance from the antenna to the ground clamps, or what an antenna discharge unit looks like, the diagram does accurately describe the proper grounding of the mast and supporting structure, the proper grounding of the lead-in wire to the antenna discharge

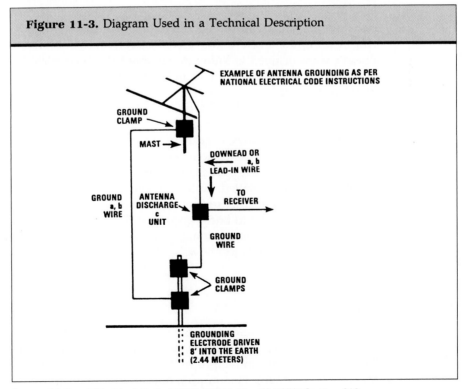

Figure 11-3. Diagram Used in a Technical Description

Source: National Electrical Code (ANSI/NFPA No. 70-1978) Section 810.

unit, and the location for the antenna discharge unit. The disadvantage of a diagram is that because it tends to be abstract, the diagram is both nonrepresentational and nonproportional; thus, the reader does not get a sense of distance, size, or shape.

Guidelines for Technical Description

1. Use an Appropriate Level of Detail

Depending on your audience and purpose, sometimes you will describe something in great detail, sometimes in broad outline. There are basically two audience characteristics that will affect your presentation of the material: the readers' level of expertise and their reason for reading (directional or informational). The amount of detail you need to supply is not necessarily proportional to the expertise of your audience. That is, sometimes novices need quite a bit of detail. An example of this would be a description in a proposal of an inadequate water filtration system; even novice readers (and most of the executives who read the proposal will be novices in the field of water filtration) will need enough information to understand how the system is supposed to work and why this one doesn't. At other times novices will need only broad, general information. A casual reader of a popular science magazine reading an article about recent astronomical developments probably doesn't want or need to know much about the theoretical background and potential consequences of black holes; this reader only wants a general idea of what a black hole is. The same can be true of other audience groups: executives, experts, and technicians. You must not only assess how much your readers do know about the thing you're describing, you also have to consider how much they need or want to know.

You must also consider your readers' general purpose in reading. If your readers are going to use or recreate your mechanism or place, the description should be highly detailed, and you will have to alert the reader to any potential trouble spots. Also, you as author will have to determine exactly what the reader *needs* to know and what the best order is. However, if your description will be informational, your goal will be to determine just how much the reader *wants* to know.

2. Use Figurative Language

In addition to the traditional tools writers use to help readers visualize things (color, shape, size, and texture), successful writers use similes when such language is possible and appropriate. Descriptions are often visual and lend themselves to similes. The purpose of a simile, in which two dissimilar things are compared, is to help clarify a complex or unfamiliar thing by comparing it with something the reader is familiar with; for example, consider the description of DNA's double helix as a "spiral staircase." Though the use of simile is not yet widespread in technical writing, experiment with them. When they are used, they can effectively render complex or unfamiliar objects clear to readers.

3. Indicate Subdivisions Clearly

Though technical writers have adopted a variety of methods for indicating subdivision, the two most common methods are verbal and numerical. If you subdivide verbally, introduce each section of the description with a heading that identifies the part being described: a description of a one-cup filter-cone coffeemaker might have sections labeled "filter," "filter cone," and "mug." If you subdivide numerically, introduce each section of the description with a number. If you do choose to subdivide numerically, consider reinforcing the description of each part with words that introduce the particular part being described:

1. The outer trough
2. The grid plate
3. The sump

Examples of Technical Description ─────────

Figures 11-4, 11-5, and 11-6 are examples of technical descriptions. The first describes an object, a plant. It appeared in an article advocating the use of houseplants to reduce indoor pollution. The article appeared in an environmental journal aimed at readers who are interested in environmental issues

---

Figure 11-4. Example of a Technical Description

The resinous juice in the fleshy leaves of this popular plant is often used as a home remedy for minor burns, and in soaps, lotions, and other cosmetics that abound in health-food stores. The plant's pointy, pale green leaves are armed with hard, whitish-to-reddish spines on the margins, and they grow in compact rosettes that get about 12 inches tall. New rosettes sent up from the base of the plant soon fill a pot with foliage.

Source: Marinelli, J. (1990, March/April). Plants for healthier homes. *Garbage, 2*(2), 36–43.

Figure 11-5. Example of a Technical Description

A.2. Sheet Machine

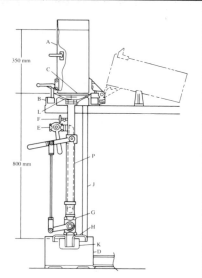

A.2.1 The complete sheet machine is shown in Figure 2. The main parts of the machine consist of a base, B, connected by a drainage pipe, P, and cock, G, to the sump; a wire-covered grid plate, C, on which the sheet is formed, and which rests in the top of the base; and a cylindrical deckle, A, which is provided with a hinge and which fits over the plate and the base.

A.2.2 The outer trough for overflow water is connected to the sump by means of two brass drain pipes, J. Around the top of the funnel of the base is machined a recess on which the grid plate rests.

A.2.3 The inside of the funnel is carefully machined and a removable 4-fin baffle, L, made of 1.6 mm brass sheet, rests in it. The function of the baffle is to prevent any possible swirling. A horizontal disc is fitted in the baffle over the drainage pipe to distribute the suction uniformly over the grid plate.

A.2.4 The grid plate consists of a cylindrical plate with square hole perforations, a backing wire and a surface or forming wire on which the wet sheet is deposited during drainage. An outer ring clamp stretches and holds the forming wire securely in place.

A.2.5 The top hinged part, A, of the apparatus is a cylinder which is stepped out at the bottom to fit over the base of the machine and carries a soft rubber gasket ring, which makes contact around the edge of the grid plate. The cylinder is provided with a hinge and clamp, so that when closed, water-tight joints are made between the cylinder, the upper edge of the top face of the plate and upper edge of the base. The drainage valve, G, is operated by a lever mechanism as shown in Figure 2.

A.2.6 The sump consists of a large cylinder, D, with an outlet hole on either side as shown, or, if required, in the center of the bottom. Three lugs are provided for screwing the sump to the floor. A cross bar, H, carries a circular baffle to prevent splashing. An overflow vessel, K, is bolted with spacers to the underside of the bar. The distance from the surface of this vessel to the top of the wire is adjusted to exactly 800 mm. This is the suction head on the wire. The drainage pipe is connected through a T and elbows to the water supply valve, E.

Source: From Technical Association of the Pulp and Paper Industry. (1981, April). *Forming handsheets for physical tests of pulp* (corrected) (T 205 om-81). Norcross, GA: TAPPI. Reprinted with permission of the editors.

but who are not highly knowledgeable about science or technology. Because one purpose of the article was to persuade readers that plants can help reduce indoor pollution, the *introduction* to the description of the aloe plant stressed the medicinal uses of the juice of the plant. However, because the description is not related to the plant's medicinal or antipollution qualities,

Figure 11-6. Example of a Technical Description

Ferrand and Le Caisne's concept is divided into three parts: a large lawn area, parterres and a "romantic garden." From west to east, there will be first a series of multipurpose grass areas bounded by a quincunx of existing trees at the Sports Palace. This large, adaptable open space is punctuated by groves or rows of old plane trees. The traces of warehouse courtyards and streets further divide the area. New 10-foot-wide, east-west allees will be raised slightly above grade and will be continuous across the site. Turfed stairs will rise from the lawn to a raised terrace; the lawn areas will serve as large public gatherings for outdoor theater. To the east of the lawn areas are the parterres.

The separate garden parterres will be "embroidered" in modern graphics on the base of the old lines, like a colored carpet. Within the network of squares will be medicinal, rose, scent and didactic gardens, and a labyrinth. Bercy's viticultural history will be recalled in trellises and vine-colored pergolas. A neoclassical pavilion will remain in the centermost square, representing the heart of the park. Rows of grapes will be planted along a canal to a lake in the third part of the garden.

The romantic garden retains features of the same orthogonal layout, but in its use of water features, a grotto and the presence of several historic structures, it takes on a "romantic" flavor. The two parts of the garden, now divided by the rue Dijon, will be linked by a canal that passes through a grotto installed in a hypostyle that supports the roadway overpass. After passing under the rue Dijon, the canal will terminate at a circular lake. A renovated historic building will serve as a restaurant sited on a square island in the middle of the lake. Berms will conceal the street from within the park, and two footbridges will pass over it, adding up to a total of five ways to enjoy both sections of the park when the terrace bridge at the Seine is included.

Source: Clemons, M. (1989, January). Gardens of memory. *Landscape Architecture, 79,* 30.

the writer provided only a superficial description of the plant. In this manner, the writer developed the *part-by-part description* spatially, moving from outside ("the plant's pointy, pale green leaves") to inside ("rosettes sent up from the base of the plant").

Figure 11-5 shows a technical description of a mechanism. The sheet machine is one of several pieces of complex equipment used to form test paper handsheets from wood pulp in order to determine the physical properties of the pulp. This description appeared in an Official Test Method document prepared by the technical committee of TAPPI, the Technical Association of the Pulp and Paper Industry.

The *introduction* (section A.2.1) provides an overview of the whole machine. The *part-by-part description* (sections A.2.2 to A.2.6) develops functionally—that is, in addition to describing the mechanism, the description explains how the mechanism works. Because readers will use the object described, the writer has chosen a directional description, which is highly detailed. In fact, in addition to subdividing the description of the sheet machine into individual assemblies, the writer has subdivided each individual assembly into subassemblies.

To add specificity to the description and to help the reader more readily identify the various parts of the assemblies and subassemblies, the writer used locational and positional words ("[a]round the top of," "in the center of the bottom"). Finally, because effective communication is often the result of both verbal and visual description, the author included a drawing of the sheet machine.

Figure 11-6 is a technical description of a place. This description constituted approximately half of an architectural review of an extensive park project in Paris, Parc de Bercy.

Although the description of the park project is lengthy and dense with architectural details, it is easy to read, even for a layperson. In the *introduction*, the writer provided an overview of the organization of the description. In the *part-by-part description*, the writer helped the reader visualize the park by moving in one direction, from west to east. The writer also used directional terms, such as "center," "from," and "to," to help guide the reader from one segment of the description to another. In addition, the writer used metaphor ("'embroidered'") and simile ("like a colored carpet") to help readers visualize the garden parterres.

Activities

1. Prepare a detailed analysis of one of the three examples of technical description that appear at the end of this chapter. Your analysis should include a discussion of the type of technical description used, the introduction, the method of organization, the appropriateness of the level of detail, and any use of figurative language.

2. Find a brief technical description in a trade journal. Identify the method of organization (functional, spatial, or chronological). Rewrite the description using a different method of organization. Prepare an oral report to the class discussing any problems you faced in rewriting the description (such as possibly needing additional information).

3. The sheet machine described in Figure 11-5 is organized according to the parts of the machine. Are there other ways of organizing the description? Under what circumstances might you use a different method of organization?

4. The following description is a brief and very broad description of the storage capacity of the reservoirs in the Wisconsin Valley Improvement Company (WVIC) system of 21 storage reservoirs. The description is in-

tended to provide nontechnical readers with an understanding of just how immense the storage capacity of the reservoirs is. With two or three students, analyze the description. Is the description clear? If so, what makes it clear? If not, how could it be made clearer?

> The reservoirs range in size from the smallest natural-lake reservoir, which is 312 acres, to the largest man-made reservoir, which is 7,657 acres. Total surface area of all reservoirs is 66,600 acres. They are surrounded by 650 miles of shoreline. Storage capacities range from 22 to 4,457 million cubic feet. To help you visualize it, total storage capacity is 17.4 billion cubic feet, which is enough water to fill a mile-square 625-foot tall tank. Expressed another way, total storage would also flood 400,367 acres one foot deep, which would be the same as flooding 625 square miles or three quarters of Vilas County (Two reservoir types, 1990).

5. The following description appears in a lengthy article in a popular science magazine. The article discusses various plans for constructing and operating a barrier at different sites in the Thames estuary. This particular description, written by one of the engineers for Tidal Engineering, Hydraulics Research, is of the barrier eventually built.

> The structure spans 520 m of the river in Woolwich Reach, 11 km downstream of London Bridge. It consists of four main rising sector gates housed in sills in the bed, each closing a 61-km span of river between piers. The four main gates are flanked by two small rising-sector gates and four small falling gates 31.5 m wide. In all there are 10 gates, 9 piers and 2 abutments (Kendrick, 1988).

The audience for this description is an educated but nontechnical audience. Is the description appropriate for that audience? If so, what specific techniques has the writer used to ensure that the description is appropriate? If not, rewrite the description for its intended audience.

6. Your instructor's office has been widely touted as representative of a typical professor's office. Your writing team, consisting of you and one or two other students, has been asked by a technical writing journal to prepare a technical description of your instructor's office. Your group will have to consider, among other things, the most appropriate type of organization and whether to use a graphic.

Technical Process Description

The Role of Technical Process Description

When to Use Technical Process Description

Types of Technical Process Description
Process Descriptions That Inform
Process Descriptions That Instruct

Writing a Technical Process Description
Accurate Information
Strong Organization
Appropriate Level of Detail
Description of the Purpose of the Process
Description of Observation Methods
Description of Equipment
Graphics

Guidelines for Technical Process Description
1. Use an Appropriate Level of Detail
2. Describe Time Spans as Needed

Example of a Technical Process Description

Activities

The Role of Technical Process Description ————

In a process description a writer reports on a series of actions. This is the sort of thing we do every day as we describe the last plays in a basketball game, how we decided which gift to buy, how snow looked as it melted. Technical process descriptions are similar to such common descriptions, but have one significant difference: they require a level of specificity far beyond that of daily speech.

Mark Twain comments on this difference in *Life on the Mississippi* as he describes his training as a riverboat pilot. Before he learned his trade, he enjoyed the movement of water—the ripples and waves seemed pretty. But as he pursued his apprenticeship as a pilot he had to learn which swirls of water meant submerged stumps, which sandbars, which rocks that might tear the bottom out of his boat. He learned to see the flow of water in a new way.

This close attention to detail, and to important underlying processes, is what separates technical process description from daily conversations. Rather than trying to entertain, writers describing technical processes are informing readers about often complex procedures involving complicated equipment. Readers should leave these descriptions more aware of how processes occur, more knowledgeable about how equipment functions, or better able to carry out the process themselves.

When to Use Technical Process Description ————

Writing a technical process description is a very common task. Technical process descriptions are typically found in repair manuals (both to explain steps a person should follow or to describe normal operation of the machine), research reports (describing actions the researchers took), feasibility studies (explaining tests conducted), progress reports (reviewing actions taken by a project team), and proposals (describing work to be done). Most writing contains at least some references to processes, procedures, or changes.

Types of Technical Process Description ————

Although technical process descriptions are used under a wide variety of circumstances, most fall into one of two categories: either they *inform* the reader about a process or they *instruct* the reader in how to perform a process.

Process Descriptions That Inform

A principal function of process descriptions is to explain the way machines operate or natural processes occur. Only by understanding how the machine moves from step to step do we understand its full functioning. Similarly, only by looking at a natural process in motion (like a weather system moving in) do we understand it. Explaining operations is a common purpose for process descriptions.

The machine description below entails such a process description.

As its name implies, the function of this section is to receive the die cut sheets from the collector, break the tabs that were holding the sheets together and convey the now individual cartons to the belt delivery section. The hardware provided to do this consists of individual upper and lower tape sections which are driven at a slightly higher speed than the exit nip of the collector. This causes the leading row of cartons to accelerate as it enters the tab breaker nip, thereby breaking it free from the rows following. One tape section is required for each row of cartons in the across direction. The angular placement of each section with respect to the centerline causes the cartons passing through to be deflected sideways, thus breaking the tabs and creating separation between the rows of cartons as they arrive at the belt delivery section. The top and side view schematically presented in Fig. 22-12 will serve to clarify these functions (Schaffer, 1984, p. 478).

Notice the relative lack of detail in this description. One line states that "[t]he angular placement of each section with respect to the centerline causes the cartons passing through to be deflected sideways, thus breaking the tabs and creating separation between the rows of cartons as they arrive at the belt delivery section." The description does not give the angle of each section or state how far the cartons are deflected to the side. That information is not necessary here. The purpose is to explain the general operation of the machine.

Process Descriptions That Instruct

Another purpose of process descriptions is to provide enough information that a reader could replicate the process described. When this description is taken to its logical conclusion, the result is an instruction manual. Usually process descriptions do not include quite as much detail but still could be used by someone wanting to perform a similar process. The example below is such a process description. It has sufficient detail that another researcher could perform a similar investigation, yet it does not reproduce the complete details of the study.

The study area was divided into 10 travel routes or survey segments that radiated about a centrally located basecamp where the research teams camped nightly throughout the study. Each route was covered once each week by daily surveys that originated from the basecamp and covered an average of 27 km. Each route was traveled the same number of times during the summer so that equal time for observation could be given to each part of the study area (Titus, 1981, p. 12).

Sufficient detail is included that any professional in the field could understand the technique used and perform something similar, but few details are supplied. For instance, no information is provided about how to allocate or map routes or how to make observations. That knowledge is assumed. The writer provides enough information for a practicing professional, not enough for a novice.

Writing a Technical Process Description

While the writing of process descriptions may be common in technical documents, it is not easy. Research must often be painstakingly and exhaustively conducted, and it is usually quite difficult to shape the resulting information into a coherent, useful description. Most good technical process descriptions rely on accurate information, a strong organization, and an appropriate level of detail. Often writers find it useful to describe the purpose of the process, the observation methods, and the equipment involved, with graphic aids used to clarify points.

Accurate Information

The process a writer describes may be one the writer has already performed. This is certainly the case in reports of experiments or of studies. In this case, there is little need to research the process being described; the writer already understands the process well.

But often writers are called on to describe the processes performed by equipment or report on natural processes that may be less familiar to them. In gathering information to write this process description, the writer will rely on two principal tools: first-hand observations and secondary sources.

OBSERVATIONS

Careful observations of a process are not easy. Often several things happen simultaneously, and it may not be easy to determine the relationships of events. One approach is to begin with a set of questions to direct the observations. This often involves asking questions that focus on the chief characteristics of processes: motion and change. Here are several questions that demonstrate this approach.

What is the purpose of the process?
Who or what does the process?
When and where does it occur?
What is the original state of the equipment or product?

What event happens first?

What is the order of subsequent events?

What is the final condition of the equipment and/or product?

Such questions cause writers to focus on the changes occurring as they observe. By looking for motion and change, writers should grasp the central elements of the process. In the case of a writer investigating a complicated paper machine, such questions might result in observations like these:

Process purpose: break cartons into individual units.

Who: operator supervision required for jam prevention

What: machine named "Stacker"

When: operation intermittent

Original condition: large sheets perforated into two rows.

Action 1: front edge of sheet is accelerated to break back tabs.

Action 2: each row diverted from centerline to break side tabs.

Final condition: Input trays rise up to belt level.

A second set of observation questions can be based on the needs of readers. Remembering that they have specific backgrounds and purposes in reading should lead to such questions as:

What is unique about this process?

What specialized equipment is used?

What similar processes are my readers likely to be familiar with?

Notes here might include such information as:

Unique: This process looks much like the paper cutter, but can take larger stacks of input paper, and can operate at somewhat higher speeds.

Specialized equipment: Two streams of paper are created. This requires a splitter or router to move the streams to their appropriate stacks.

Similar processes: Paper cutting follows the same general algorithm.

Directed observations are likely to be more productive than undirected viewing. By focusing on specific questions, the writer becomes more observant.

SECONDARY SOURCES

While observations can be helpful in learning about a process, writers should not ignore written or graphic information that already exists. Machines come with installation, repair, and operation manuals. Each of these can be a good source of information. They may contain a level of detail unnecessary for the process description, but some of their contents will be informative, and they may contain graphics and short descriptions that can be very useful to you.

Strong Organization

The information you gather through observation and secondary sources will probably be quite substantial; you will need a strong and consistent organization to make your raw data into a useful process description. Many people who haven't written them think that the organization of process descriptions is quite straightforward — they are arranged chronologically, with the first step first, the second step second, and so on. Unfortunately, this idea is complicated by two factors: the need to describe equipment and the complexity of many processes themselves.

First, technical processes often involve equipment. Both equipment and process may need to be described, so the writer must determine how to integrate the two descriptions. Is it best to start with a description of the process? Or is a reader better served by a detailed description of all the equipment that is employed in the process? This is a difficult decision to make. The paragraph below focuses on the process of how cartons are stacked as they come out of a paper machine. The writer leaves descriptions of the machinery until later. This may work for readers who are familiar with paper machines; but it may totally confuse readers who are unfamiliar with such machines.

The Stacker Section

The cartons as they proceed through the stacker are dropped into individual pockets where they are side and end jogged into precise stacks. Upon command of a counting device which has been preset to the desired number of cartons per stack, the streams are interrupted and the pockets emptied in preparation for the next sequence. The mechanisms involved to accommodate this action are outlined in Fig. 22-12 (Schaffer, 1984, p. 481).

Some authors create separate sections for equipment descriptions so that the main body of their report can focus on actions. Other authors integrate the equipment descriptions into their process reports, assuming that readers need to understand the equipment and its actions if they are to understand the process.

Second, technical processes are often complicated. Several activities may be occurring simultaneously. At the same time that a machine is taking materials in, it may be working on other materials or pushing a completed product out. The writer, however, can only describe one thing at a time. The writer may start at one end and work to the other, follow a chronological order, or pick some other approach, but all solutions are imperfect. The process is simultaneous, the description is not. Any writer faced with this situation will have significant decisions to make.

Appropriate Level of Detail

Readers of technical process descriptions are almost always professionals in the field. They will have a good general background in the process being described. But they will not already be experts in the particular process you are describing, or they would not be reading your description. This means that general terms or procedures will not have to be described in great detail, but any specialized equipment or unusual procedures will need to be explained. The following description of a psychology experiment strikes just such a balance.

> Hecht et al. had the subject viewing with one eye and looking at a dim red light as a fixation point. The experimenters were able to present the test flash anywhere in the visual field. To clarify the discussion, some method is needed to describe the location where the test flash is presented. The easiest way of describing the location is according to the angle of viewing with respect to a fixation point (Massaro, 1975, p. 149).

The writer feels no need to provide details about the "dim red light," or to define the "visual field." Both would be well understood to anyone doing vision research in psychology. Yet there is a need to describe the location of the test flash, and this is carefully described. The writer assumed general knowledge on the part of readers and supplied specific information only where necessary.

As you prepare to write a process description, you should determine what your readers already know about the process in general. What will be new to them? Which equipment is common in the field, and which may be specific to the process? How similar is this process to others they have encountered? Each answer will tell you a great deal about how much you need to include in your description.

Description of the Purpose of the Process

Not all descriptions include a discussion of purpose, since such purposes might be obvious. For example, we don't need to be told why we want a car to start or why a flat tire should be changed. But purposes are sometimes included in process descriptions if the process is new to the reader, or if the process might be an alternative to a standard approach to a problem. In the description below, notice the author's apparent need to justify a particular approach to paper coating.

> A more recent development of this oxidation method, now finding commercial acceptance, involves the use of an ozone generator from which a high concentra-

tion of ozone is delivered through a nozzle to impinge upon only the side of the melt to be joined to the substrate. This system permits lower melt temperatures to be used, increases the adhesion of the polymer to the substrate and reduces the oxidation on the opposite surface of the polymer so that heat sealing characteristics are also improved (Mainstone, 1984, p. 124).

Because the process being described is new, the author is concerned that readers might not be familiar with it and might not understand why they should concern themselves with it rather than with more familiar methods. Therefore, before describing the process, the author devotes a paragraph to justifying this new process.

Description of Observation Methods

The observation activity itself may need to be described along with the process. This is because the act of observing something may affect the process being observed. For instance, someone describing the working procedures in an office may have observed employees using one process to handle contracts. However, the employees may use a totally different process when they are not under observation. For that reason, such process descriptions normally include some information about how the observations were made. Notice how this is accomplished in the waterfowl description below:

> Nests sites were located by extensive and careful searches along the main shorelines and around all islands. Canoes were paddled as close to the shore as water depth would allow, usually within 3–5 m (Titus, 1981, p. 12).

By explaining what process was used to approach the nests, readers can determine for themselves if the writers had an effective observation strategy, or if the information they compiled was corrupted by their observation procedure. For the writer, this means not only describing the process, but remembering to describe how the process was observed.

Description of Equipment

Equipment descriptions can cause a real dilemma for writers. Most technical descriptions involve descriptions of tools and equipment. But where should these descriptions be placed, and how much should be included? The description below integrates the description of equipment into the description of the process itself.

> The cartons as they proceed through the stacker are dropped into individual pockets where they are side and end jogged into precise stacks. Upon command of a counting device which has been preset to the desired number of cartons per stack, the streams are interrupted and the pockets emptied in preparation for the next sequence. The mechanisms involved to accommodate this action are out-

lined in Fig. 22-12. The indexing sequence is initiated by the clamp which stops the oncoming streams. At the instant the clamp comes on, the transfer belts accelerate, causing all unclamped cartons to move quickly into the pockets. The elevator table, which forms the bottom of the pockets, then moves down to clear the completed stacks from the side joggers and end plate. The belts, which are a part of the elevator system, move the stack on to the final delivery table, after which the elevator table raises to receive the oncoming streams of cartons which have already been freed of the clamp. This entire sequence takes place within five seconds (Schaffer, 1984, p. 461).

Notice all the equipment that is mentioned in passing by this description: "stacker, pockets, counting device, clamp, transfer belts, elevator table, side joggers, end plate, elevator system, delivery table." The description includes a reference to a graphic aid so that readers have some help in visualizing the equipment mentioned, but there are no lengthy descriptions in the text itself. The "counting device" is not precisely named, much less described. We are not told the type of clamp used. The parts of the "elevator system" are never mentioned.

There are good reasons for such descriptions to exclude more detail about equipment. Imagine, if you will, the paragraph above rewritten to include such detail. You will see the problem immediately.

The cartons as they proceed through the stacker are dropped into individual pockets (see illustration 3-12 for a detailed schematic of the pockets). These pockets must be ½ to ¾ inch larger than the cartons they are to receive, and should be at least deep enough to hold fifty cartons. Once cartons fall into these pockets they are side and end jogged into precise stacks. Side jogging is usually performed by a Merstein Jog-All (see illustration 3-13 and 3-14). The jogger operates at a speed of 700 Jogs per minute and consists primarily of a metal plate hammering at the stack of cartons with an impact of 23 psi. The end jogger may also be a Merstein Jog-All. It is placed perpendicular to the side jogger with a ⅜ inch offset so that its plate does not impact the side jogger. It should operate at the same speed as the side jogger.

Few readers would understand the process of stacking if they had to work their way through this level of detail. By including so much detail the author has changed the focus of the passage. It is now a description of equipment rather than a description of a process. The original version of this passage does a much better job of balancing the description of the equipment with the description of the process.

The one difficulty with the first version of the passage is that it assumes a fair amount of understanding on the part of the reader. A reader who did not know what a jogger was would be hard-pressed to understand the stacking process and would benefit from more description. In such cases it may be advisable to include a separate section (possibly a glossary) of equipment descriptions, or to provide a more detailed description of equipment in the introduction. Such separate sections are quite routine in reports of experiments or laboratory reports, where they not only describe equipment but also list materials needed to repeat an experiment.

Graphics

As in most technical writing, graphics play a major role. In general, the standard techniques for use of graphics apply in process descriptions, but there are one or two subtle differences worth noting.

Since the purpose of a technical process description is to explain equipment in operation, there are often marks added to any graphics used here to help readers see such things as direction of flow and motion of parts. The use of arrows and such is a response to the problem of using a static display medium like print to describe a dynamic operation like a machine. The example in Figure 12-1 shows how this marking is often done.

The problem of static graphics is so serious that some technical descriptions have been moved from paper to CD-ROM and other electronic medium that support full-motion display. Because of the cost of CD-ROM, most technical writers will be forced to continue using printed graphics, graphics that need to be carefully marked to help overcome their innate limitations.

Figure 12-1. Diagram Used in a Process Description

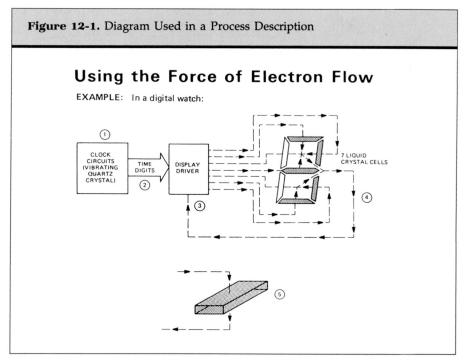

Source: *Contemporary electronic circuits and components.* (1983). Washington, DC: McGraw-Hill Continuing Education Center. Courtesy of McGraw-Hill Continuing Education Center.

A second characteristic of graphics used in process descriptions is their use in a series. Three or four graphics are placed sequentially to show a process in various stages of completion. By seeing the drawings or photos in a series, the reader is able to see the changes that take place during each stage of an operation or as each part of a machine becomes engaged. The example in Figure 12-2 shows how a series of graphics can detail the steps in a

Figure 12-2. Series of Graphics Used in a Process Description

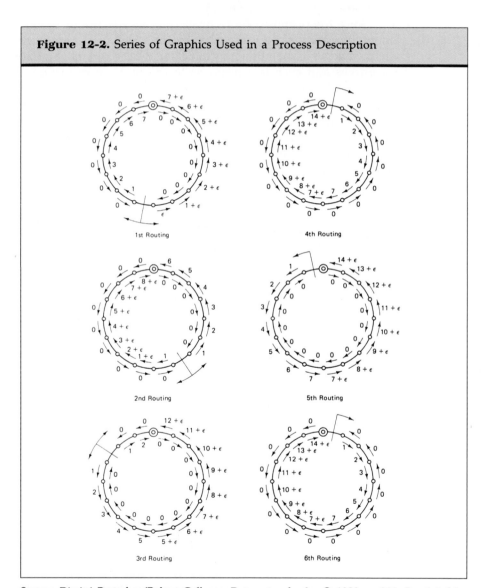

Source: Dimitri Bertsekas/Robert Gallager. *Data networks,* 2e, © 1992, p. 335. Reprinted by permission of Prentice Hall, Englewood Cliffs, New Jersey.

process. In this example, the writers needed to illustrate an approach to computer network message routing. The graphic series shows the changes that occur in the network as messages move.

Guidelines for Technical Process Description ─────────

1. Use an Appropriate Level of Detail

Remember the needs of the reader. Since most readers use process descriptions for general information, descriptions with a high level of detail may not be necessary. This is demonstrated well in the example below. The writer is describing a common strategy for computer networks.

> **IEEE 802.4 Token Bus Standard**
> The IEEE 802.4 standard corresponds essentially to the system we have just described. . . . To allow new nodes to enter the round robin token structure, each node already in the structure periodically sends a special control packet inviting waiting nodes to join. All waiting nodes respond, and if more than one, a splitting algorithm is used to select one (Bertsekas, et al., 1987, p. 266).

The writer explains the issues faced by growing networks, and describes in general terms how the network can respond to new members. For most readers this would be enough. Notice that it would not be enough information for a reader trying to replicate this process. A person newly charged with managing a computer network, for instance, would want more detail about the communication between network nodes, possibly including example computer code or more detailed algorithms. The needs of readers will determine how much detail should be included in descriptions.

2. Describe Time Spans as Needed

Specific information about time is not always included in process descriptions, but since processes involve action and change, there are many occasions where a description of time intervals is necessary. For instance, time information may be required for readers to judge the quality of a process. A machine that produces one product every minute is very different from one that can produce ten products in the same time. Time information can also be useful in describing processes that have subprocesses or major components.

This is demonstrated well in the example below. The writers use time descriptions to help explain stages in the migration process of deer.

Fall Migration

Five deer, all in the Ely region, were radiotracked only occasionally in late fall 1974, but 14 deer in 1975 and 7 deer in 1976 were followed during their migrations to winter ranges. Five of the 7 deer in 1976 had been followed in 1975.

In 1975, below freezing temperatures first occurred in late November and initiated fall migration for 7 radiocollared deer. In contrast, freezing weather in fall 1976 first occurred in late October and triggered migration for 4 of 7 marked deer, roughly 4 weeks earlier than in the previous year.

In fall 1975 and 1976, 11 or 21 radiomarked deer (52%) migrated when the daily minimum temperatures dropped below −7 C. Of the remaining deer, 7 of 21 (33%) migrated later in fall and early winter as temperatures fluctuated and snow depth gradually increased. The remaining 3 deer migrated before freeze-up and their cases are discussed below. In some cases, deer showed a delayed response by migrating just after a major cold period even though current temperatures were increasing. In other cases, deer showed no response during several periods of temperature changes (Nelson, 1981, pp. 19–20).

In this example the author needed to describe the movements of the deer over several years and to correlate their movements with changes in temperature. Dates were important, as were time spans. The statement, "freezing weather in fall 1976 first occurred in late October and triggered migration for 4 of 7 marked deer, roughly 4 weeks earlier than in the previous year," is typical. Because this description essentially relates two sequences of events —weather and migration—time is especially important and occupies much of the process description.

Example of a Technical Process Description ⸺

Figure 12-3 shows an example of a technical process description. In this passage, two authors describe the methods they used to observe loons in northern Minnesota. If done well, descriptions of this sort should accomplish two things: readers should be able to judge for themselves whether the techniques used were valid, and future researchers should be able to repeat the techniques used here, or to replicate the experiment. Because this description must be *instructive* as well as informational, the writers use a great deal of detail in their description.

Figure 12-3. Example of a Technical Process Description

METHODS

From mid-May to mid-August 1975 and 1976, two 2-man teams traveled approximately 6,400 km and spent nearly 6,000 man-hours surveying the loon population of the study area. Approximately half the travel was on routes where outboard motors were allowed, with two 17-foot aluminum canoes equipped with 2-horsepower outboard motors; in the other areas, the canoes were paddled.

The study area was divided into 10 travel routes or survey segments that radiated about a centrally located basecamp where the research teams camped nightly throughout the study. Each route was covered once each week by daily surveys that originated from the basecamp and covered an average of 27 km. Each route was traveled the same number of times during the summer so that equal time for observation could be given to each part of the study area.

Locating Nests

Nests sites were located by extensive and careful searches along the main shorelines and around all islands. Canoes were paddled as close to the shore as water depth would allow, usually within 3–5 m. Concurrently, the shoreline was watched carefully by both team members. It was often necessary to stop and disembark to reexamine a segment of shoreline to be sure no nest was hidden in the vegetation or lakeside debris. Troughlike depressions that led up onto the banks often revealed nests under low-hanging vegetation. Special attention was given to the shoreline whenever a loon was sighted nearby. A single loon or pair of loons low in the water that exhibited a Sneak or Hunched (McIntyre unpublished dissertation) posture was used during the searches as an indication of a nearby nest. If no nest was found in an area where loons were consistently seen, the procedure was repeated during each successive observation throughout the project. There was a high degree of confidence that no sites were overlooked, and in no case were young observed where we were unaware of a nest.

Check of Nests

Once a nest site was determined, an unattended nest was censused by paddling the canoe past the site at the maximum distance necessary for determining the number of eggs present. On routes where motors were allowed, the motor was stopped as the loon territory was approached. If an incubating loon was observed vacating a nest as it was approached for a check, the flushing distance was recorded. When an observation was completed, the research crew vacated the territory as quickly as possible. The time a canoe was within 100 m of a nest site usually was less than 5 min.

If an egg count had been obtained during a previous observation and the presence of an incubating loon was confirmed with binoculars or spotting scope, a nest check was considered complete. Where no previous egg count had been obtained, the nest was approached closely, even if it was occupied. In such a case, a flush and egg count were desired. If a nest was always occupied and the bird would not flush without undue harassment, the number of eggs incubated was estimated by the number of young hatched and/or whether fragments of 1 or 2 eggshells remained after nest destruction of hatching.

Figure 12-3. Example of a Technical Process Description (Continued)

When nest failure occurred, it was often impossible to ascertain the reason. An unknown disappearance was recorded if eggs were missing from an intact nest bowl otherwise undisturbed and free of influences such as wave action or high water. There was no positive way to distinguish between depredated and scavenged nests since each nest was visited only once a week. The time between observations made it impossible to know whether or not nest desertion had preceded a nest destruction, thereby opening the nest for a scavenge, or if predators had destroyed viable eggs.

A nest loss was registered as a predation-scavenge whenever 1 or more of the following were exhibited: (1) crushed egg shells or albumen and egg fragments with shell membranes still intact in or near the nest, (2) egg punctures that implied avian predation (Olson and Marshall 1952, Rearden 1951, Sowls 1955), (3) loss of nest accompanied by an inordinate amount of muskrat or beaver activity at the nest site, (4) nest torn apart, or (5) tracks and/or scat in or nearby the nest. Only when the evidence was convincing was predation-scavenge recorded. Otherwise, the failure was considered to be of unknown origin. A renest was registered whenever a new nest was found in a territory where a previous nest loss had occurred.

Checks of Broods

Verification of the presence and number of young from each successful nest continued with each trip for the duration of the study from the maximum possible distance using a spotting scope or binoculars. Although sightings of broods varied in distance due to weather conditions and locations on the lakes, young were never known to be farther than 1.6 km from the nest site from which they were hatched.

Reproductive success for a breeding pair was determined by the presence or absence of young 2 weeks or older. Such a procedure was considered the best indication of reproductive success since Olson and Marshall (1952) experienced no losses after 2 weeks, and during the 2 seasons of this study only 1 of 71 juveniles was lost after the 2-week period. In that case an 8-week-old juvenile succumbed to a perforated small intestine of unknown cause.

Although difficult to determine, the research teams may well have had an adverse effect on the birds, for we regularly observed the nests and kept some birds off the nest. However, our activities were always undertaken with sensitivity to the nesting and brood-rearing requirements of the birds. We tried carefully to minimize our impact and to affect all pairs equally on all study routes. In only 2 cases did we feel we had caused enough disturbance that subsequent nest losses may have been our fault.

Postponement of nesting or failure to nest were also possible responses to our presence. However, the latter seemed unlikely since all loon pairs produced nests in 1975 when we disturbed the population most heavily with our searching efforts.

Assessment of Recreational Impact

During 1975, all observed recreational use was recorded on 1 leg of the daily nest surveys from the basecamp to the farthest point reached. Observations were randomly altered between trips out and back to place them at different times of the day. The location of each observed recreational party was noted on a map and the amount and type of use within 1.6 km of each nest site was determined.

Figure 12-3. Example of a Technical Process Description (Continued)

In 1976, 0.5 hour per visit was spent on each territory to observe the type and degree of recreational use on that part of the study area. The activities of each party as well as the number of people, canoes, motors used, and campsites were recorded. All of those items were used as objective indicators of the degree of recreational use and potential nest disturbance. Nests were ranked, using the number of canoes recorded within 1.6 km, into high, medium, or low categories or recreational use impinging upon that site. A Human Impact Index of 4 categories, that ranged from very isolated nests to those near heavy human use, was also assigned to each nest. A Disturbance Potential Index, using 4 groupings, was given to each nest site on the basis of whether or not the nest was in sight of the main route of travel through the lake on which it was found and the distance to that route of travel from the nest. Each nest was also assigned a Visibility Index as an indicator of potential human disturbance from shoreline use. Six categories were employed to describe nest visibility (with a loon present) to an observer passing 10 m from shore. Categories ranged from 1 (conspicuous) to 6 (not visible).

Source: Titus, U. R., & Van Druff, L. W. (1981). Response of the common loon to recreational pressure in the Boundary Water Canoe Area, Northeastern Minnesota. *Wildlife Monographs, 79,* 12–14. Copyright © The Wildlife Society. Used by permission.

Activities

1. In small groups, discuss the example process description presented at the end of chapter 12. Consider the following topics:
    a. What level of detail is used? Does this amount of detail tell you something about the expected readers? The description was published in a journal for experts in wildlife biology. What changes would you expect to see if the article were submitted to a general audience publication like *Field and Stream*? Mark places in the report where these changes would be made.
    b. Locate references to time in the example. Discuss why these time references are important in the report.
    c. There seem to be few tools described in the example report. Yet there are some. What are they, and why are they mentioned? How would you characterize the amount of description these tools received? Why do you think the authors chose to describe their tools in that way? As a group, rewrite one of the descriptions so that it could be understood by a novice. Compare your group's revised description with those written by other groups. Do some descriptions seem more effective than others? Discuss why or why not.
2. In laboratory manuals, textbooks, or professional journals, find at least

two illustrations showing machines in operation. What techniques are used to show the direction of motion and sequence of activity? Can you think of ways to improve the drawings?

Examine the illustrations found by your classmates. Which illustration is the most effective? Why is it better than the others? Would the same illustration technique be appropriate for all the examples? Why or why not?

3. As a class, agree on a process you wish to observe. This might be a process such as dispensing food in the cafeteria, accepting shipments at the bookstore, or repairing a computer or other piece of campus equipment. Then observe the process, either singly or as a class. Compare your notes. How much difference is there from one student to another? Whose log would be the most valuable to you as a writer if you were to write a report on the experiment? What makes this log the best?

4. A local factory asks you to describe the operation of one of their machines. List the questions you might prepare before you visited the factory.

5. In the following process description, circle the names of tools or equipment. How much description of these tools is included? What conclusions does that lead you to about the expected audience? Begin a glossary of equipment descriptions. Should the glossary precede the passage or follow it? Why?

Adhesion

One of the most critical factors in an extrusion coating or lamination is the adhesion of the polymer to the substrate. Different end uses will have different demands for the degree of adhesion. For example, some relatively low adhesion requirements will allow the coating process to run at speeds in excess of 600 m/min, while some relatively high adhesion requirements will limit processing speeds to much less than 300 m/min unless some adhesion promoting system is employed.

There are a variety of adhesion promoting systems and their selection usually depends upon the coated or laminated construction required and the substrates and polymers to be employed. Oxidation of the polymer melt in the drawdown distance from the die to the substrate is one such system and has some range of adjustability in that melt temperature can be increased to increase the level of oxidation and the drawdown distance can be increased for longer exposure to air. This method, however, soon runs into limitations since, although adhesion to the substrate is improved, the excessive oxidation of the melt leads to the problem of odor in the polyethylene, which can be unacceptable in food packaging, and the problem of loss of the heat sealing characteristics of the polymer inhibiting the forming and sealing of the package.

A more recent development of this oxidation method, now finding commercial acceptance, involves the use of an ozone generator from which a high concentration of ozone is delivered through a nozzle to impinge upon only the side of the melt to be joined to the substrate. This system permits lower melt temperatures to be used, increases the adhesion of the polymer to the substrate and reduces the oxidation on the opposite surface of the polymer so that heat sealing characteristics are also improved (Mainstone, 1984).

6. Write a technical description of any of the following processes. Assume that readers are familiar with the general area and are reading primarily to better inform themselves about the specific process.
   a. The operation of a furnace
   b. The registration process at your school
   c. The migration process of a bird species
   d. Cell mitosis
   e. The functioning of a computer network
   f. The movement of an automobile engine

Technical Summary

The Role of Technical Summary

When to Use Technical Summary
To Help the Reader Find Documents of Interest
To Provide Main Points of a Document
To Guide the Reading of a Document

Types of Technical Summary
Descriptive Abstracts
Informative Abstracts
Executive Summaries
Glosses

Writing a Technical Summary
Writing a Descriptive Abstract
Writing an Informative Abstract
Writing an Executive Summary
Writing Glosses

Guidelines for Technical Summary
1. Keep All Summaries Concise
2. Do Not Use a Summary to Evaluate the Document
3. Use a Logical Order
4. Do Not Add Information
5. Do Not Omit Important Points
6. Make Glosses Visually Distinctive

Examples of Technical Summary

Activities

The Role of Technical Summary ─────────────

> We are drowning in information but starving for knowledge.
>
> J. Naisbitt

By one estimate, between 6,000 and 7,000 scientific articles are written each day (Naisbitt, 1982). This is an amazing torrent of new technical information. Add to that flood of information new technical reports, business studies, and daily memos, and it is no surprise that even very bright people are drowning. The job for the technical writer is to turn the torrent of information into useful knowledge. This is done many ways, but one of the most effective is the summary.

When to Use Technical Summary ─────────────

Technical summaries are generally used to accomplish three important tasks: to help readers identify articles of interest, to outline the main points of an article, and to guide the reading of longer documents. If your document is long or is only one of many that might be read, consider preparing a summary. Certain technical documents, such as research reports, proposals, and progress reports, almost always include some sort of summary.

To Help the Reader Find Documents of Interest

Selecting published articles of value is the first chore in any research endeavor. Because so many articles exist on any subject, choosing the most valuable can take many hours. If we relied on titles alone to identify pertinent articles, we might miss some, and if we were to read completely every article connected to the subject of interest our search would take far too long. Summaries are a nice middle ground. They contain enough information that readers can make an educated judgment about whether the article will be useful, yet the summary is short enough to be read quickly. The abstract below is an excellent example of how an abstract can help in article selection:

> Grace, J. Cuticular Water loss unlikely to explain treeline in Scotland. OECOLOGIA (1990) 84 (1) 64–68.

> The hypothesis that poor development of the cuticle (and thus an inability to conserve water in winter when the soil is frozen) results in tree failure at their altitudinal limit was tested at a natural treeline of Scots pines (Pinus sylvestries)

in the Caingorm Mountains. Needle samples from 5 altitudes between 300–670 m (spanning open, 20-m-tall stands; open, 5-12-m-tall stands, elfin woodlands 1–4 m tall and isolated 1-m-tall pines) were taken December 1986 and April 1987, and transpiration rates measured. Although no evidence of poor cuticular development was found, needles from old elfin woodlands and young isolated trees at higher altitudes lost water more rapidly than those of trees from lower altitudes. Results are attributed to stomatal dysfunction due to leaf mechanical damage, rather than a thinner or less developed cuticle. Calculations show that a small increase in cuticular transpiration is unlikely to cause frost-draught. (*Forestry Abstracts*, 1991, p. 483).

This abstract briefly explains the study that has been done, summarizes the results, and gives complete bibliographical information if a reader wished to read more. Abstracts like these are so valuable for finding relevant information that they have become a major feature of scientific fields. Such abstract indexes as *Psychological Abstracts*, *Dissertation Abstracts International*, *Physics Abstracts*, *Chemical Abstracts*, and *Ecological Abstracts* are among the most popular.

To Provide Main Points of a Document

A summary may also provide a digest of all the important points made in an article. For many readers this digest is all they need to know. They have neither the time nor the interest to read the details. A specialized form of summary, the executive summary, often fills this role. It sums up the points of interest to decision-makers in a form that they can review quickly. Such a summary might look something like this:

QA Test Recommendations
This report recommends that quality assurance tests be performed by Six Sigma Designs. The three principal reasons for this choice are their consistently high performance ratings, their proximity to our facility, and their demonstrated low cost.

Such a summary might extend to several pages for a very long document, but in each case its principal purpose is to list the main points for readers who may not have the time or need to read more of the report.

To Guide the Reading of a Document

For those who will continue reading the rest of the report, the summary performs a valuable service as well. It serves as an "advance organizer" for their reading. Research into this aspect of summaries was first done by David

Ausubel (1960) in his effort to improve reading. His "advance organizers" can be viewed as a way of preparing readers for what they will read. A short passage tells readers what major pieces of information they will find and the order in which they will occur. With this information in mind, readers read more easily; there are no surprises. Research confirms what we would expect: given the advantage of an advance organizer, readers read more quickly and with greater comprehension.

A well-written summary is a perfect advance organizer. It tells readers what they will find, and often presents the information in the same order in which it occurs in the full report. It should give readers a major advantage then, when they go on to read the document.

Types of Technical Summary

All summaries have the same general value for readers, but some summaries have been formatted to fill specific needs. There are now four principal forms of summary: descriptive abstracts, informative abstracts, executive summaries, and glosses.

Descriptive Abstracts

The descriptive abstract is similar to a table of contents: it describes what a document is about. If the document is a report, the descriptive abstract will note the topics covered. If the document is a journal article, it will list the sections of the article. Because this abstract generally identifies only the major topics of documents, it tends to be the shortest summary. It also tends to be written from an external perspective—as if the writer were describing someone else's report, even though the abstract author may be the report writer. This brevity and tone are visible in the example below.

> **Feasibility of a Portage County Waste-to-Energy Project**
> Studies the feasibility of implementing a waste-to-energy facility in Portage County. Three different volumes of solid waste generation were considered, and three different technologies for waste processing were looked into.

This abstract is so short that it did not even name the three technologies reviewed in the article. The sole function of such an abstract is to help readers make a selection quickly: Is this an article I want to read, or shall I pass it over? Words are used sparingly so that readers can make this decision quickly.

Informative Abstracts

An informative abstract includes more information about the contents of a document. For instance, it might describe any conclusions or findings included in the document. It might also describe the purpose of the report, the scope of an inquiry, and the principal methods used in research. Its purpose is not only to describe the contents of a document, but also to supply much of the central information in the article, so that the reader may substitute reading the abstract for reading the whole document. As a result, an informative abstract is generally longer than a descriptive abstract.

The example below shows this difference. It is an informative abstract for the feasibility study we looked at earlier.

Department of Natural Resources personnel studied the feasibility of implementing a waste-to-energy facility in Portage County. The purpose of the facility would be to extend the life of the Portage County landfill by reducing the volume of waste and the cost of waste disposal. Three different volumes of solid waste generation were considered: only Portage County waste after a recycling program has been implemented; Portage County waste without any recycling effort; and a 150 ton per day (TPD) facility that would process 125 TPD of waste consisting of all the waste from Portage County, and the balance from Waupaca County. Three different technologies of waste processing were looked into: mass incineration, waste distillation, and refuse-derived fuel. With considerations regarding the cost of the landfill, the potential life expectancy of the landfill, the location of the facility, the transportation of waste to the facility, the transportation and storage of the gas, and the funding for the project, the recommendation is to implement the waste distillation option (*Feasibility*, 1988).

This abstract is three or four times longer than the descriptive abstract given earlier, but it should be. It is serving the purpose of providing the most important points to readers who want that information but do not have time to read the whole report.

Executive Summaries

There is a tendency in some companies to call any summary an executive summary—it sounds more impressive, doesn't it? There is a sense in which all summaries are like executives: they concentrate on the main points and leave details for others. But not all summaries are executive summaries. The distinction is one of audience and purpose. An executive summary is by definition a summary intended for upper management. That implies it will not only need to be summarized for quick processing, but should be organized to help with decision-making.

Because the people reading this summary are charged with making decisions, the summary has to contain the information they will need for this purpose. For that reason, an executive summary might not be balanced in the same way that an abstract would be. It might contain all the main ideas of a report, but it might change the order in which they are presented, or include some details but leave out others. The writer must use some judgment, and constantly ask, what information will they need in order to act?

Take the example of a feasibility report. Although the full report may contain recommendations, an explanation of the original problem, and a detailed comparison of alternatives, executives have other needs. They are most in need of recommendations, need little information on the problem (other than why it is important enough to warrant their time and money), and probably care least about alternative solutions not chosen. An executive summary, therefore, would emphasize some areas of the original report and say little about other areas. Rather than exactly mirroring the original, the executive summary is tailored to the needs of the audience.

For an example of this flexibility, let's look again at the waste-to-energy feasibility study. If the informative abstract were rewritten to be an executive summary, it might need the following changes:

> The Department of Natural Resources recommended that Portage County proceed with its waste-to-energy program using the waste distillation process. This decision was based on a cost analysis of the landfill, the potential life expectancy of the landfill, the location of the facility, the transportation of waste to the facility, the transportation and storage of the gas, and the funding for the project. Other processes considered were mass incineration and refuse-derived fuel. The expected result of this process would be to extend the life of the Portage County landfill by reducing the volume of waste and the cost of waste disposal. Three different volumes of solid waste generation were considered: only Portage County waste after a recycling program has been implemented; Portage County waste without any recycling effort; and a 150 ton per day (TPD) facility that would process 125 TPD of waste consisting of all the waste from Portage County, and the balance from Waupaca County.

This executive summary is significantly different from the original informative abstract. Notice that the recommendation is given first, along with its justification. Only later does the executive summary list alternatives and details about waste volumes. The same information is presented in both summaries, but the organization has been changed to respond to the needs of executives.

Glosses

Glosses began as translations of technical terms, often placed in the margin. When many were needed they were placed in a "glossary." More recently the idea of having explanations in the margin has changed to putting a running summary of a document there. Such a gloss repeats main points in an area where they are easy to find and easy to read. If glosses are done well, a reader could skip the document itself and learn the main points just by reading the glosses. Figure 13-1 shows how glosses appear on the page.

In many ways glosses work much as informative abstracts do: they distill main points and present them to a reader. But they have one advantage. Because glosses are normally placed next to the information they summarize, if readers become interested in a point made by a gloss and want to learn more, the detailed information they want is right there, next to the gloss. In this fashion, glosses serve as possible entry points into a text. The disadvantage of glosses is their use of space. Notice in the example above that fully

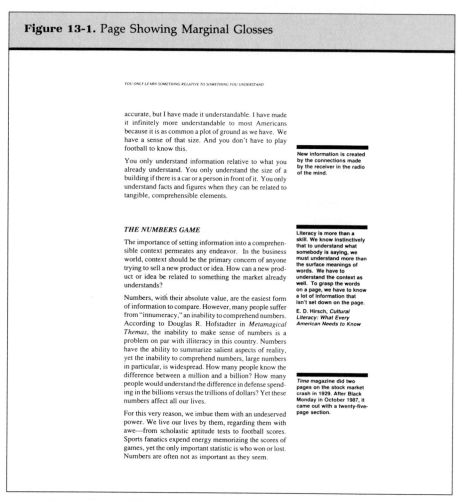

Figure 13-1. Page Showing Marginal Glosses

Source: Wurman, R. S. (1989). *Information anxiety: What to do when information doesn't tell you what you need to know* (p. 173). New York: Bantam. Copyright © by Richard Saul Wurman. Used by permission of Doubleday, a division of Bantam Doubleday Dell Publishing Group, Inc.

one fourth of the page is given over to glosses. The result will be much longer documents and increased duplication expenses.

Writing a Technical Summary ━━━━━━━━━━━

Writing a Descriptive Abstract

Because the principal purpose of a descriptive abstract is to help readers decide if they want to read a document, the first job of a writer is to outline the contents of the document. What are the main topics included? If you were to write a table of contents for your report, what would be in it? A brief list of these topics should be sufficient to start. A topic list for a process description might include items like these:

> Description of the machine
> List of materials used
> Step-by-step fold
> Final outcome

The next step is to write the abstract from an impersonal perspective. This means that even if you are writing an abstract for your own report, you would not write, "I begin by describing. . . ." You would write, "The article describes. . . . " The result might look like this:

> **Folding Machine**
> This report describes the T46 Folding Machine. Standard materials, the folding process, and machine output are explained.

Writing an Informative Abstract

The main difficulty in writing an informative abstract is determining the most important points. As a writer you probably felt that everything in the report was important; if it wasn't, you wouldn't have included it in the first place. Now you must identify the five or ten percent of the report that matters most.

One strategy to use is to reread your document and highlight the main points. In preparing an abstract on the shuttle design, for example, you might underline the following points.

> **The Space Shuttle Design**
> The embryo Shuttle program faced a number of evolutionary design changes before it would become a system in being. The first design was based on

a "fly back" concept in which two stages, each manned, would fly back to a horizontal, airplane-like landing. The first stage was a huge, winged, rocket-powered vehicle that would carry the smaller second stage piggyback; the carrier would provide the thrust for lift-off and flight through the atmosphere, then release its passenger—the orbiting vehicle—and return to Earth. The Orbiter, containing the crew and payload, would continue into space under its own rocket power, complete its mission and then fly back to Earth.

The second-stage craft, conceived prior to 1970 as a space station ferry, was a vehicle considerably larger than the later Space Shuttle Orbiter. It carried its rocket propellants internally, had a flight deck sufficiently large to seat 12 space station-bound passengers and a cargo bay big enough to accommodate space station modules. The Orbiter's size put enormous weightlifting and thrust-generating demands on the first-stage design.

This two-stage, fully reusable design represented the optimum Space Shuttle in terms of "routine, economical access to space," the catch-phrase that was becoming the primary guideline for development of Earth-to-orbit systems. It was, however, less than optimum in terms of the development investment required: an estimated $10–13 billion, a figure that met with disfavor in both Congress and the Office of Management and Budget.

In 1971, NASA went back to the drawing board, aware that development cost rather than system capability would probably be the determining factor in getting a green light for Shuttle development. Government and industry studies sought developmental economies in the configuration. One proposal found acceptance: eliminate the Orbiter's internal tanks and carry the propellant in a single, disposable External Tank. It provided a smaller, cheaper Orbiter without substantial performance loss.

For the launch system, NASA examined a number of possibilities. One was a winged but unmanned recoverable liquid-fuel systems, expendable solid rockets and the reusable Solid Rocket Booster. NASA had been using solid-fuel vehicles for launching some small unmanned spacecraft, but solids as boosters for manned flight was a technology new to the agency. Mercury, Gemini and Apollo astronauts had all been rocketed into space by liquid-fuel systems. Nonetheless, the recoverable Solid Rocket Booster won the nod, even though the liquid rocket offered potentially lower operating costs. The overriding reason was that pricing estimates indicated a lower cost of development for the solid booster.

Emerging from this round of design decision making was the Space Shuttle: a three-element system composed of the Orbiter, an expendable external fuel tank carrying liquid propellants for the Orbiter's engines, and two recoverable Solid Rocket Boosters. It would cost, NASA estimated early in 1972, $6.2 billion to develop and test a five-Orbiter Space Shuttle system, about half what the two-stage "fly back" design would have cost. To achieve that reduction, NASA had to accept somewhat higher system operating costs and sacrifice full reusability. The compromise design retained recoverability and reuse of two of the three elements and still promised to trim substantially the cost of delivering payloads to orbit.

The final configuration was selected in March, 1972 (Presidential Commission, 1986).

Be sure not to get carried away with your underlining. Underlining every phrase is not abstracting. Ask yourself, if you only had two minutes to explain your report to someone else, what would you say?

The other difficulty is to maintain a balance. Don't take all your key points from a single section. An abstract should reflect the entire report. Find key ideas in every section to underline.

Once your key ideas are identified, you are ready to create the abstract. If you are doing your work on a word processor, you now have the choice of erasing all the text that is not underlined, or moving (or copying) the underlined text to another place on the screen. In either case the result would look something like this:

> The embryo Shuttle program faced a number of evolutionary design changes before it would become a system in being. The first design was based on a "fly back" concept in which two stages, each manned, would fly back to a horizontal, airplane-like landing. The first stage was a huge, winged, rocket-powered vehicle that would carry the smaller second stage piggyback; the second-stage craft was a vehicle considerably larger than the later Space Shuttle Orbiter.

> Development investment required: an estimated $10–13 billion. Government and industry studies sought developmental economies in the configuration.

> For the launch system, NASA examined a number of possibilities. One was a winged but unmanned recoverable liquid-fuel systems, expendable solid rockets and the reusable Solid Rocket Booster. Emerging from this round of design decision making was the Space Shuttle: a three-element system composed of the Orbiter, an expendable external fuel tank carrying liquid propellants for the Orbiter's engines, and two recoverable Solid Rocket Boosters.

All the main points are there in the same order as in the original, but they can be further distilled, and they hardly make good reading. The next job is to turn them into a coherent text—an abstract. Here is the text after those editing changes have been made:

> The early Shuttle program faced a number of evolutionary design changes. The first design was based on a "fly back" concept in which two stages, each manned, would fly back to a horizontal, airplane-like landing. The first stage was a huge, winged, rocket-powered vehicle that would carry the smaller second stage piggyback. The second-stage craft was a vehicle considerably larger than the later Space Shuttle Orbiter. Cost was a problem for this design: an estimated $10–13 billion. Government and industry studies sought developmental economies in the configuration. For the launch system, NASA examined a number of possibilities. One was a winged but unmanned recoverable liquid-fuel systems, expendable solid rockets and the reusable Solid Rocket Booster. Emerging from this round of design decision making was the Space Shuttle: a three-element

system composed of the Orbiter, an expendable external fuel tank carrying liquid propellants for the Orbiter's engines, and two recoverable Solid Rocket Boosters.

If an abstract is written well, a reader should be able to read it and get all the main points to be found in the original document. Readers should feel the only reason they need to read the complete document is to get additional background or more detailed information about the topics mentioned in the abstract.

Writing an Executive Summary

There are two main steps to writing executive summaries. The first is to find the main points of the document. These are the same main points that would be included in an informational abstract. Be especially careful to identify only the main points; an executive summary should be no longer than an informational abstract.

The second step is to place the points in order of importance *as viewed by an executive*. This usually means placing conclusions or recommendations first, justification for the decision second, and alternatives third. The assumption is that an executive first wants to know the impact of a report then wants to know how the report justifies that impact, and cares least about alternatives, since those presumably would have been examined in detail by the people writing the report.

Writing Glosses

Glosses are usually approached in one of two ways. One approach is to find the main point in each section of the report and restate it in the margin. In this way, the gloss serves as a kind of heading for the section. Readers see all the main points contained in the document, and, if they are interested in one of the points, they know where to find additional information—beside the gloss.

A second approach is to use direct quotations from the main report. These are often quotations that emphasize key points. By putting them in the margin as glosses, the writer gets additional attention for these quotations. This approach is most common in popular magazines, but it is also used in books on technical or scientific subjects. The example in Figure 13-2 comes from a recent technical article.

The writer must be careful with quotations. It is difficult to select quotations that sum up or reinforce key points in the main passage. If not selected with care, the quotations placed in the margin might not represent the main point of the article at all, and so might give a very different impression about the subject of the article than the one the writer wished to create.

Figure 13-2. Glosses Drawn from the Main Text

Selecting Color Scanners
Drew Robison

Overview

June 1991

Scanning an image is a powerful, easy and fast way to incorporate images into any presentation or application. Selecting the proper color scanner is, however, a difficult task. With a wide variety of cost and capability between individual scanners, and without a means of testing or comparing these scanners, most organizations and academic institutions must rely on a combination of limited product knowledge and blind luck when making a purchase. IAT's technical support group recently researched five color scanners that were rated highly by some noted periodicals.

Implications

It is important to determine the quality of image that is desired. If your organization needs an image for a high-resolution output device rather than a low-resolution output device, then this will certainly affect your decision making when selecting a scanner. These questions and many others must be asked. Finding the scanner that best suits your organization's present needs, yet flexible enough to support future expansion, requires thorough planning.

With a wide variety of cost and capability between individual scanners, and without a means of testing or comparing these scanners, most organizations and academic institutions must rely on a combination of limited product knowledge and blind luck when making a purchase. IAT's technical support group recently researched five color scanners that were rated highly by some noted periodicals.

Specifying your needs

Some of the questions that should be addressed before making a decision follow:

　　1) **Resolution** -- Resolution is usually defined in d.p.i. (Dots Per Inch). This tells you how many dots are in one inch of the scanned image. The minimum resolution should be

1

Source: Robinson, D. Selecting color scanners (p. 1). In *IAT scanner.* Chapel Hill, NC: The Institute for Academic Technology. Courtesy of the Institute for Academic Technology.

Guidelines for Technical Summary ─────────────

1. Keep All Summaries Concise

A descriptive abstract is normally one paragraph in length. Informative abstracts may be longer than descriptive abstracts, but are still less than a page long.

An executive summary must also be brief. If a recommendation seems controversial, it is tempting to present an exhaustive justification for it, even in the summary, but only the main reasons for the decision should be in the summary, and they should be simply stated. The main document is the place for detailed rationales.

Glosses should be no longer than a sentence or two. They must be short enough to fit in a narrow type area, with plenty of space above and below.

2. Do Not Use a Summary to Evaluate the Document

Abstracts and other summaries are nonjudgmental. An abstract does not include phrases like "poorly explains" or "neglects to mention." The principal function of an abstract is to disclose succinctly the information in a document. Readers can decide later if the information in the document is valid or the presentation is well done.

An informative abstract may include conclusions, but only those conclusions expressed in the document. The example abstract above states that cost was a problem. This opinion should come directly from the article and should not reflect the personal opinion of the abstract writer.

An executive summary may include comments and conclusions about the report if they guide an executive. Such phrases as "the main consequence of the report" or "of particular interest will be" are acceptable if they direct the attention of the executive to significant places in the report.

3. Use a Logical Order

Most abstracts are organized similarly to the original report. By following the general outline of the original document, you give the reader an idea of the organization and development of the original. The same is true for glosses.

The organization of the executive summary reflects the priorities of executives, not the order of the main document. Even if the first third of the document consists of lengthy considerations of alternatives, if such are not of primary importance to an executive, they should be placed later in the summary.

4. Do Not Add Information

Summaries should not include information not contained in the original. Because the purpose of an abstract, summary, or gloss is to accurately reflect the purpose and content of a document, you need to be sure that you do not inadvertently include details or conclusions that are not in the original document.

5. Do Not Omit Important Points

This is a particularly crucial guideline for glosses. Select sentences or quotations for glosses carefully. Would a reader who read only the glosses reach the same conclusions about the report as someone who read the whole report?

6. Make Glosses Visually Distinctive

Set the gloss in a slightly larger type size than the rest of the report, or use boldface or italics. There should be enough space around the glosses that they appear separate from the document and from each other.

Examples of Technical Summary ─────────────

Figures 13-3, 13-4, and 13-5 show three abstracts, one of the most common and versatile forms of technical summary. These abstracts of research reports were published in indexes designed to help researchers and other interested people find exactly the sources they needed.

Figure 13-3. Example of a Technical Summary

112: 26523d Upper bounds on 'cold fusion' in electrolytic cells. Williams, D.E.; Findlay, D.J.S.; Craston, D.H.; Sene, M.R.; Bailey, M.; Croft, S.; Hooton, B.W.; Jones, C.P.; Kucernak, A.R.J.; et al. (Harwell Lab., UK At. Energy Auth., Didcot/Oxfordshire, UK OX11 0RA). *Nature (London)* **1989,** 342(6248), 375–84 (Eng). Expts. using 3 different calorimeter designs and high-efficiency n and γ-ray detection on a wide range of materials fail to sustain the recent claims of cold fusion. Spurious effects which, undetected, could have led to claims of cold fusion, include noise from n counters, cosmic-ray background variations calibration errors in simple calorimeters and variable electrolytic enrichment of Tr.

Source: *Chemical abstracts.* (1990). *112, 377.*

Figure 13-4. Example of a Technical Summary

90-07391P Radon risk communication: The effectiveness of an integrated media campaign versus communicating when a house is being sold. [en] Presented at 83. Annual Meeting of the Air & Waste Management Association, Pittsburgh, PA (USA), 24–29 Jun 1990. by A. Fisher, J.K. Doyle, G.H. McClelland, W.D. Schulze (U.S. EPA, PM-221, Washington, DC 20460, USA) in 83. ANNUAL MEETING OF THE AIR & WASTE MANAGEMENT ASSOCIATION, p. 69.

A television station presented newscasts and public service announcements about radon, and advertising the availability of reduced-price radon test kits in a local supermarket chain. The large number of test kits sold was a success from a marketing perspective, but not from a public health perspective—very few high readings were mitigated. In contrast, a study of housing sales showed much more testing and mitigation when risk communication accompanied the housing transaction, rather than being directed toward the general public. The paper examines the relative effectiveness of these alternative approaches to risk communication, emphasizing the implications for developing and implementing radon programs.

Source: *Pollution abstracts.* 21(5), 209.

Activities

1. The purpose of the summaries presented at the end of this chapter is to identify articles of interest, provide main points of an article, and guide the reading of longer articles. Compare the three example abstracts. Rank them according to how well they achieve these purposes. Are there things that all three do well? That all three do badly?

2. Test the value of abstracts as an advance organizer. As a class, pick an article in a professional journal that includes an abstract. After reading the abstract, each student should try to outline the report. What information do you predict it will contain? In what order will it appear? Now check the article itself. How accurate were your predictions? How would this information have affected your reading of the article? Discuss differences between students in your class.

3. Assume you are trying to hire someone. As a class, make a list of information a manager would need to make a decision about which candidate to select. Order the list from most to least important. How would this list be incorporated into an executive summary?

4. Take a descriptive abstract and rewrite it as an executive summary. What information did you add? What information did you delete? How much reordering was needed?

Figure 13-5. Example of a Technical Summary

8501 BOUWER, H. **Linkages with ground water.** In *Nitrogen management and ground water protection [Edited by R.F. Follett]*. Amsterdam, Netherlands; Elsevier Science Publishers (1989) 363–372 ISBN 0-444-87393-7 [En, 27 ref., 2 fig.] USDA-ARS, U.S. Water Conservation Lab., 4331 E. Broadway Road, Phoenix, Arizona, USA.

In nonirrigated areas, downward flow or deep percolation rates (Darcy flux) in the vadose zone typically range from approximately 1 mm/yr or less for semiarid or arid areas to approximately 500 mm/yr or more for humid areas. Irrigation areas in dry climates can have deep percolation rates as a few cm/yr to a few dm/yr for relative efficient irrigation systems and as much as 1 or 3 m/yr for inefficient irrigation systems. Assuming an NO_3-N leaching of 50 kg/ha/yr (approximately one-fourth of the fertilizer N applied), these fluxes give calculated NO_3-N concentrations in vadose zone water of 10 to 100 mg/l, which agree with values found in practice. To prevent high NO_3-N concentrations in vadose zone water below irrigation areas, best management practices for irrigation must be combined with best management practices for nitrogen fertilizer. Rates of downward movement of nitrate in the vadose zone are equal to the pore velocities of the water. Theoretically, these are calculated as Darcy flux divided by volumetric water content. Because of preferential flow, however, actual velocities can be considerably higher, i.e., 2 to 5 times higher for soils with cracks, rootholes, wormholes, etc. Downward velocities of nitrate in the vadose zone below agricultural land may range from a few dm/yr to a few m/yr. Longitudinal dispersion may be influenced more by spatial variability of infiltration rates than by hydrodynamic effects. Nitrate concentrations in water pumped from the aquifer eventually can approach those in the water from the vadose zone. Where nitrate concentrations exceed the maximum limit for drinking, treatment of the water may be necessary. Of the various techniques, the method that combines anion exchange and denitrification holds promise.

Source: *Soil and fertilizers.* (1990). *53*(7), 1073.

5. Glosses are still relatively uncommon. Find a technical journal in your field that uses glosses. Is there anything about the journal that might make it more willing to try such a technique? Find a journal article that does not use glosses and write the glosses for the first two pages of the article. How did you select information to put in the glosses? Compare your glosses with those of others in the class. Identify the glosses that best capture the content of the article.

6. In his report to the Presidential Commission, Richard Feynman, Commission member, presented his observations about the *Challenger* accident. Figure 13-6 is a section of his report on avionics. Write a descriptive abstract, an informative abstract, an executive summary, and a set of glosses to accompany this report. Compare your summaries with those of others in the class. How would you determine which summaries were the best?

Figure 13-6. Excerpt from Report on the *Challenger* Accident

Appendix F
Personal Observations on Reliability of Shuttle

by R. P. Feynman

. . . .

Avionics

By "avionics" is meant the computer system on the Orbiter as well as its input sensors and output actuators. At first we will restrict ourselves to the computers proper and not be concerned with the reliability of the input information from the sensors of temperature, pressure, etc., nor with whether the computer output is faithfully followed by the actuators of rocket firings, mechanical controls, displays to astronauts, etc.

The computing system is very elaborate, having over 250,000 lines of code. It is responsible, among many other things, for the automatic control of the entire ascent to orbit, and for the descent until well into the atmosphere (below Mach 1) once one button is pushed deciding the landing site desired. It would be possible to make the entire landing automatically (except that the landing gear lowering signal is expressly left out of computer control, and must be provided by the pilot, ostensibly for safety reasons) but such an entirely automatic landing is probably not as safe as a pilot controlled landing. During orbital flight it is used in the control of payloads, in displaying information to the astronauts, and the exchange of information to the ground. It is evident that the safety of flight requires guaranteed accuracy of this elaborate system of computer hardware and software.

In brief, the hardware reliability is ensured by having four essentially independent identical computer systems. Where possible each sensor also has multiple copies, usually four, and each copy feeds all four of the computer lines. If the inputs from the sensors disagree depending on circumstances certain averages, or a majority selection is used as the effective input. The logarithm used by each of the four computers is exactly the same, so their inputs (since each sees all copies of the sensors) are the same. Therefore at each step the results in each computer should be identical. From time to time they are compared, but because they might operate at slightly different speeds a system of stopping and waiting at specified times is instituted before each comparison is made. If one of the computers disagrees, or is too late in having its answer ready, the three which do agree are assumed to be correct and the errant computer is taken completely out of the system. If, now, another computer fails, as judged by the agreement of the other two, it is taken out of the system, and the rest of the flight canceled, and descent to the landing site is instituted, controlled by the two remaining computers. It is seen that this is a redundant system since the failure of only one computer does not affect the mission. Finally, as an extra feature of safety, there is a fifth independent computer, whose memory is loaded with only the programs for ascent and descent, and which is capable of controlling the descent if there is a failure of more than two of the computers of the main line of four.

There is not enough room in the memory of the main line computers for all the program of ascent, descent, and payload programs in flight, so the memory is loaded about four times from tapes, by the astronauts.

Because of the enormous effort required to replace the software for such an elaborate system, and for checking a new system out, no change has been made in the hardware since

Figure 13-6. Excerpt from Report on the *Challenger* Accident (Continued)

the system began about fifteen years ago. The actual hardware is obsolete; for example, the memories are of the old ferrite core type. It is becoming more difficult to find manufacturers to supply such old-fashioned computers reliably and of high quality. Modern computers are very much more reliable, can run much faster, simplifying circuits, and allowing more to be done, and would not require so much loading of memory, for their memories are much larger.

The software is checked very carefully in a bottom-up fashion. First, each new line of code is checked, then sections of codes or modules with special function are verified. The scope is increased step by step until the new changes are incorporated into a complete system and checked. This complete output is considered the final product, newly released. But completely independently there is an independent verification group, that takes an adversary attitude to the software development group, and tests and verifies the software as if it were a customer of a delivered product. There is additional verification in using the new programs in simulators, etc. A discovery of an error during the verification testing is considered very serious, and its origin studied very carefully to avoid such mistakes in the future. Such unexpected errors have been found only about six times in all the programming and program changing (for new or altered payloads) that has been done. The principle that is followed is that all the verification is not an aspect of program safety, it is merely a test of that safety, in a non-catastrophic verification. Flight safety is to be judged solely on how well the programs do in the verification tests. A failure here generates considerable concern.

To summarize, the computer software checking system and attitude is of highest quality. There appears to be no process of gradually fooling oneself while degrading standards so characteristic of the Solid Rocket Booster or Space Shuttle Main Engine safety systems. To be sure, there have been recent suggestions by management to curtail such elaborate and expensive tests as being unnecessary at this late date in Shuttle history. This must be resisted for it does not appreciate the mutual subtle influences, and sources of error generated by even small changes on one part of a program on another. There are perpetual requests for changes as new payloads and new demands and modifications are suggested by the users. Changes are expensive because they require extensive testing. The proper way to save money is to curtail the number of requested changes, not the quality of testing for each.

One might add that the elaborate system could be very much improved by more modern hardware and programming techniques. Any outside competition would have all the advantages of starting over, and whether that is a good idea for NASA now should be carefully considered.

Finally, returning to the sensors and actuators of the avionics system, we find that the attitude to system failure and reliability is not nearly as good as for the computer system. For example, a difficulty was found with certain temperature sensors sometimes failing. Yet 18 months later the same sensors were still being used, still sometimes failing, until a launch had to be scrubbed because two of them failed at the same time. Even on a succeeding flight this unreliable sensor was used again. Again reaction control systems, the rocket jets used for orienting and control in flight, still are somewhat unreliable. There is considerable redundancy, but a long history of failures, none of which has yet been extensive enough to seriously affect a flight. The action of the jets is checked by sensors, and, if they fail to fire the computers choose another jet to fire. But they are not designed to fail, and the problem should be solved.

Source: Presidential Commission. (1986). *Report to the President on the space shuttle Challenger accident.* (Vol. 2). Washington, DC: U.S. Government Printing Office.

Technical Analysis

14

The Role of Technical Analysis

When to Use Technical Analysis

Types of Technical Analysis
Classifications
Partitions

Writing a Technical Analysis
Writing a Classification
Writing a Partition

Guidelines for Technical Analysis
1. Use a Justifiable Principle of Analysis
2. Don't Use the Same Approach in Every Document
3. Name the Approach You Are Using
4. Use Labels for Each Group or Part
5. Keep the Number of Groups or Parts Reasonable

Examples of Technical Analysis

Activities

The Role of Technical Analysis ─────────────

Consider these situations:

- You have been observing a manufacturing process. You have ten pages of notes. How do you order them for presentation?
- You have been reading about new products that might be useful in a research lab. You want to write an overview of the products. How do you organize your comments about these products?
- Each morning employees are encouraged to drop suggestions for quality improvement into a suggestion box. There are fifty-eight suggestions in the box. Can they be grouped?
- Ground water pollution is a local problem. Everyone wants to do something, but no one even agrees on what pollution is. Where can you start?

Technical analysis is a way to make sense of an otherwise overwhelming amount of information. By using analysis, writers establish a whole, identify the parts that make up the whole, and clarify the relationship between the whole and the parts.

When to Use Technical Analysis ─────────────

Technical analysis is a writing technique you will use every time you incorporate large quantities of information into a feasibility study, a proposal, a research report, or almost any other kind of technical document. Each of these writing tasks requires you to turn raw information into something that can be easily comprehended; this is analysis.

There is no such thing as a "classification report" or a "partition report"; analysis is always included as part of a larger document. Analysis is so important in technical writing that standard sections of many documents are set aside for this purpose. The "Conclusions" section of a research report or feasibility study, for example, consists mainly of an extended analysis. Of course, the technique of analysis is used in many ways throughout every document, not just in these special sections.

Types of Technical Analysis ─────────────

Technical analysis is based on two standard techniques: classification and partition. Classification consists of organizing collections of facts into groups. Partition entails breaking a single large item or concept into its major parts.

Each approach is a way of making sense of large quantities of information. The difference is one of direction. Classification is largely "bottom up" in that it takes individual items and forms groups with them. Partition is "top down" in that it starts with a single item and breaks it into smaller parts.

Classifications

In the suggestion box example above, classification would be a method of grouping the fifty-eight suggestions. If they could be grouped, they would be easier to summarize and easier to understand. For instance, it might turn out that fourteen of the suggestions referred to a single machine, three were suggestions about a particular supervisor, twenty-two mentioned inventory problems, and the remainder were irrelevant. So, one way to group them might have been according to subject. Of course, they might have been grouped according to the name of the person making the suggestion, the day or time of the suggestion, their apparent importance, or even by length of the suggestion. The point is to find some meaningful approach to grouping, and use it to convert a jumble of information into useful categories or classifications.

The value of grouping information in some meaningful way is so apparent that the process occupies a major portion of a recent book on the subject, *Information Anxiety: What to Do When Information Doesn't Tell You What You Need to Know* (Wurman, 1989). The author of that book suggests that information can be grouped in any of five principal ways: category, time, location, alphabet, and continuum. In the suggestion box example, this would translate into grouping by subject, time of suggestion, location in the factory, an alphabetized listing of subjects or authors, and a listing from best to worst or longest to shortest.

To see just how many ways information can be grouped, consider this list of twelve cities. How could they be classified?

San Francisco	Oshkosh
Tokyo	Miami
Berlin	Seoul
New York	Mexico City
Atlanta	Portland
Seattle	San Jose

They could be grouped by a category such as size (small, medium, and large), or arranged by age (youngest to oldest), or grouped by location, or listed alphabetically, or put on a scale (warmest to coldest or richest to poorest). Any of these five approaches would give us some way to take a long list and give it some meaningful form.

Partitions

Partition consists of breaking a large item down into its constituent parts. We take a problem and break it into its major causes. We describe a machine as a set of subassemblies. Our system of government is described as having three branches. In each case, in trying to create a description, we sense that the best approach is to break something into parts.

The purpose of this kind of analysis is usually clarity and completeness. Some subjects are so complex that the only way to fully describe them is to examine sections of them in detail. In explaining a computer to novices, for example, writers normally divide the computer up into major pieces: the CPU (central processing unit), memory bank, input devices, output devices, and storage devices. Then the function of each component is explained in more detail. By describing the parts of the computer, writers hope they are bringing clarity to the whole while simultaneously helping readers see how the machine functions.

Writing a Technical Analysis ━━━━━━━━━━━━━━━

Writing a Classification

The first step in classification is finding some principle for grouping. On what basis will you group items? Let's take a simple example. Suppose you are writing an assembly manual for a small machine. You will want to include a list of parts so that readers can check to be sure they have everything they should have. You could just list all the parts in random order, but that would be confusing. So you decide to group "parts" into subunits. On what basis do you divide the parts?

One option might be size. You group all the parts under an inch together, then all the parts one to two inches, and so on. Another approach might be to divide the parts according to the time when they will be needed in the assembly process: parts needed for the first step, then parts needed for the second step, and so on. Parts could also be grouped by location — where they will be used — according to subassemblies: wheel parts together, drive mechanism parts together, and so on. Figure 14-1 shows a more common approach, a division based on type of part. All nuts are grouped together, all bolts are grouped together, and so on.

How do you select the best approach to division? Some approaches may be standard in your field. Descriptions of wildlife, for instance, almost uniformly group animal behaviors into food gathering, reproduction, and self-defense. Your description would normally follow that approach as well, unless you have a good reason to divide behavior into some new categories (most fun to watch, or most like humans).

Figure 14-1. Division Based on Type of Part

1	4153224	Hardware Package *7037-1*
6	4080345	*1* Screw #10×⅜" Comb P Hd
10	4153500	*2* Screw, Self-tapping #8×½" Comb P Hd
4	4080301	*3* Screw #10×¾" Comb Tr Hd
4	4080151	*4* Bolt, Carriage #10×¾"
2	4080371	*5* Screw #8×1" Comb Tr Hd *7037-1*
20	4080352	*6* Screw, Self-tapping #10×1-¼" Comb Tr Hd
8	4080303	*7* Screw #10×1-½" Comb Tr Hd
1	4080360	*8* Screw #10×1-½" Rd Hd *Brass*
4	4080324	*9* Screw ¼"×1-½" Comb Tr Hd
2	4080355	*10* Screw ⁵⁄₁₆"×3-¼" Comb Tr Hd
2	4080373	*11* Screw #10×3-¾" Comb Tr Hd
1	4080433	*12* Screw, Wing ¼"
2	4080226	*13* Nut, Hex ⁵⁄₁₆"
19	4153095	*14* Nut, Hex #10
1	4080205	*15* Nut, Hex #10 *Brass*
3	4153094	*16* Nut, Acorn #10 *7037-1*
4	4153349	*17* Nut, Barrel ¼"×⁹⁄₁₆"
2	4080431	*18* Washer, Star #10
3	4080422	*19* Washer, Flat ⁵⁄₁₆"
2	4080418	*20* Washer, Insulation
1	4080537	*21* Flanged Sleeve
2	4080412	*22* Spacer ⁵⁄₁₆"×1-⁹⁄₁₆"
2	4156513	*23* Pin, Hinge ¼" × 1-⅛"
3	4156509	*24* Pin, Hitch
2	4080228	*25* Nut, "J" #8 *7037-1*
1	4080536	*26* Wire Cap
4	4080470	— Cap, End 1" w/Hook
1	4080524	— Spring

Source: Instructions for "Char-Broil" gas grill. (n.d.). Columbus, GA: Char-Broil. Courtesy of Char-Broil.

Where there are no models to guide you, you may want to brainstorm a list of possible approaches. First list as much as you can about the subject. What information have you already collected, or do you already know? With this list in front of you, some major groups will seem obvious. But also consider your audience and purpose. A feasibility study might have information about a potential action listed as "advantages and disadvantages." Or, if there are many advantages, they might be grouped by importance: most significant advantages, moderate advantages, and incidental advantages. A progress report might list activities by time: ahead of schedule, on time, and behind schedule. A report to management might group points in a proposal under "economic consequences and personnel consequences." In all these cases, considering the readers and purposes will help identify useful classifications.

Once you have an approach to analysis, you want to use it not just to organize the information you have already collected, but as a tool to find more. For instance, in the example of the suggestion box, if a writer had grouped the suggestions by time, it might have become apparent that most of the suggestions were coming from a particular shift. This could have led to interesting questions and further research. If the suggestions had been grouped according to location, the results might have led you to focus on a particular machine and additional questions about why that machine was mentioned so frequently. Every approach to a problem brings a special perspective with it. Use the perspective to investigate further, to be sure there is nothing you are missing.

After you have found a useful approach to your analysis and have investigated your subject fully, preparing the report should come fairly easily. In doing your analysis you have been taking information and grouping it according to some principle. This organization can be used with little alteration in your report. Each group becomes its own section.

One tool that many writers find helpful in writing classifications is the use of graphics. The most useful graphics for classifications are generally those that give a schematic rather than an analogous presentation of your material; that is, tables and diagrams rather than drawings. If your information is primarily numerical, a table is often the most efficient way to present it; the columns and rows you choose will allow you to establish the pattern of classification you think best. In many cases a diagram can also be used to illustrate a classification, since it offers the opportunity to display the exact conceptual relationships among things within a group. The diagram in Figure 14-2 shows the reader exactly how various Indo-European Languages are grouped and how these groups relate to one another.

Writing a Partition

The first step in writing a partition is finding an effective way of breaking a large item into parts. There are often many choices in doing this. Take the example of pollution. The topic is so large that it seems impossible to speak about it usefully as a whole. Writers routinely break it into parts. How they break it into parts may vary. One writer might describe types of pollution: water, air, and so on. Another writer might break the subject according to causes: industrial, agricultural, residential, and so on. Several approaches are possible, and each leads to a very different description.

In choosing a method of partition, you might begin by listing the major approaches you have already noticed being used by other writers. If communication protocols are always broken into synchronous and asynchronous,

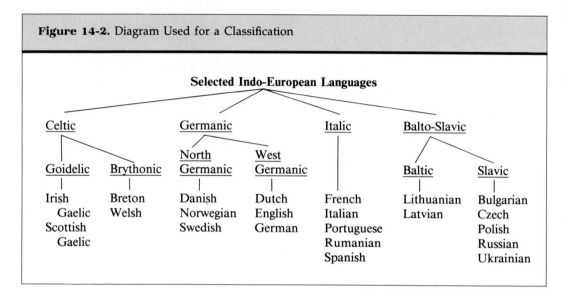

Figure 14-2. Diagram Used for a Classification

you probably will want to use the same divisions in your material. The larger purpose of the document will also help. If the document is intended to educate maintenance workers, for instance, equipment maintenance might be broken into daily, weekly, and monthly activities. A proposal attempting to sell a service might break "advantages" into short-term and long-term advantages. Readers should also be considered for approaches. For instance, if they are accustomed to descriptions that break costs into capital, supplies, and labor, that would be an important partition for you to use.

Once you have decided how to divide a subject, make a list of each major division and the information that belongs with each. It might look something like this:

Project costs

Immediate	Short-term	Long-term
site prep.	personnel hiring	salaries
equip. purchase	personnel training	maintenance
moving costs	conversion	utilities
		rent
		depreciation

Such a listing is a good test. If parts of the project do not seem to fit on the list, the partition being used isn't right for this subject. In this example, some cost that was not immediate, short-term, or long-term would mean that this partition was invalid.

Once the means of partition has been identified and tested, you will want to choose an order of presentation. There may be several divisions, but only one can be presented at a time. Pick an order—usually most important to least important—and write the description.

As with classifications, partitions are often improved by graphics. Diagrams are particularly useful for showing exactly how the individual items produced by the partition relate to one another. Partition also makes good use of labeled drawings, when the thing being partitioned is a concrete object such as a machine. Through the use of one or more exploded drawings, for example, a writer can show how the whole can be broken down into its constituent parts. Figure 14-3 shows an example of this.

Figure 14-3. Exploded Drawing Used for a Partition

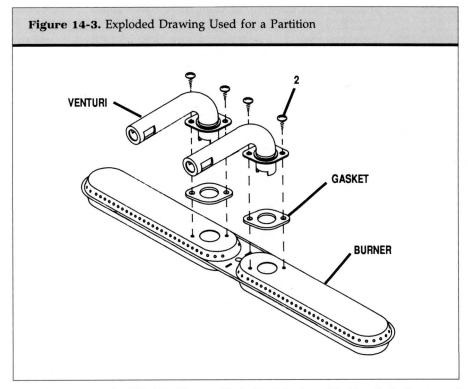

Source: Instructions for "Char-Broil" gas grill. (n.d.). Columbus, GA: Char-Broil. Courtesy of Char-Broil.

Guidelines for Technical Analysis _____

1. Use a Justifiable Principle of Analysis

Your principle of classification or partition—the basis on which you join individual items into groups or divide an item into its parts—should be both valid and appropriate for the situation. You should be confident that your approach is the most meaningful that could be taken, given the context of your document. For instance, in a feasibility report, a likely approach to analysis is cost. You could group solutions by cost, or group problems by cost. In this instance, analyzing solutions on the basis of color scheme or warranty period would be harder to justify.

2. Don't Use the Same Approach in Every Document

Cost might be one approach to classification, but if you use it over and over, you may be missing very important alternative approaches. Remember the discussion some of the engineers had after the *Challenger* disaster? It occurred to them that the perspective they had taken in debates just prior to launch had caused them to ignore some crucial questions. The way they had framed the discussion forced one set of answers and caused them to miss others. Choose your approach to classification or partition carefully. Each document, each decision, is new and should be approached with a fresh perspective.

3. Name the Approach You Are Using

Your readers should be told early in your document what approach you will be using. You may also want to explain why you selected the approach you did, especially if the approach is unusual.

4. Use Labels for Each Group or Part

Readers should have no difficulty knowing what the groups or parts are called, or understanding when your document moves from one group to another. Such an organization simplifies the work of the reader.

5. Keep the Number of Groups or Parts Reasonable

George Miller's famous studies of memory revealed that people generally remember seven plus or minus two things. This research was based on recall of meaningless data, like random numbers or shapes. For data that have meaning to us, we can remember far more. (For instance we can remember the names of more than seven friends.) But for situations in which we are learning new material, seven is not a bad upper limit. Writers should observe that boundary when categorizing information. If they describe a machine as having twelve or thirty distinct functions or parts, they are stretching the memory capacity of their readers. There may be some occasions when the material requires more than seven groups, but in general, if you find yourself with a large number of groups, you should reconsider how you are grouping your information. It may be time to consider a new principle of classification.

Examples of Technical Analysis _____

Figures 14-4, 14-5, and 14-6 contain examples of classification and partition. In examining the possible causes of the *Challenger* accident, the Presidential Commission reviewed all the changes that had been required in the shuttle. The list was quite substantial. The author of this section of the Commission's report chose to group those changes to make them easier for readers to interpret. He noticed that four causes could account for all the changes. With that in mind, he could organize his information and present it in an effective way. His classified listing is presented in Figure 14-4.

This analysis is immediately useful to us. While it is clear that many, many changes were made in shuttle flights, we can quickly see that there were four main causes of these changes. The author has made it easy for the reader to understand the source of the changes and the general nature of the changes.

The author had many options in how he performed this analysis. He chose to group the flight changes by cause. He could also have grouped them by other factors, such as length of resulting delay (less than one month, one to three months, over three months), resulting cost impact (small, medium, or large), or impact on mission safety (little, moderate, great). Any of these approaches would have allowed him to take the list of shuttle mission changes and explain it by creating meaningful groups.

Each approach would have resulted in a different view of the material, and each would have had a significant impact on conclusions readers might have drawn. Would they focus on the causes of the changes or the effects of those changes? Would they come away thinking about costs or risks? The basis for analysis has a crucial impact on how the material itself is viewed.

Figures 14-5 and 14-6 contain two brief sections of a manual for safety engineers. The first section uses partition to list the purposes of the manual. In the second section the author classifies all the causes of dam failures into eight categories.

Figure 14-4. Example of Classification

The following manifest changes in the general categories mentioned in section b occurred in 1985:

1. Hardware Problems
   - TDRS (cancelled 51-E, added 61-M)
   - Syncom (added to 61-C)
   - Syncom (removed from 61-C)
   - OV-102 late delivery from Palmdale (changed to 51-G, 51-I, and 61-A)
2. Customer Requests
   - HS-376 (removed from 51-I)
   - G-Star (removed from 61-C)
   - STC-DBS (removed from 61-E)
   - Westar (removed from 61-C)
   - STC-DBS (removed from 61-H)
   - EOS (removed from 61-B)
   - EOS (removed from 61-H)
   - Space Telescope (swap with EOM)
3. Operational Constraints
   - No launch window for Skynet/Insat combination (61-H)
   - Unacceptable structural loads for TDRS/Insat (61-H)
   - Landing weight above allowable limits (61-A)
   - Landing weight above allowable limits (61-E)
   - Landing weight above allowable limits (71-A)
   - Landing weight above allowable limits (61-K)
4. External Factors
   - Late addition of Senator Garn (51-D)
   - Late addition of Congressman Nelson (61-C)
   - Late addition of PVTOS experiment (51-I)

A detailed explanation of an example from each of these four categories of manifest changes to illustrate both the immediate impact of the change-cause as well as the downstream effects is given below.

Source: Presidential Commission. (1986). *Report to the President on the space shuttle Challenger accident* (Vol. 2, p. 167). Washington, DC: U.S. Government Printing Office.

Activities

1. The example in Figure 14-6 shows that there are many reasons why dams fail. How did the author group those reasons? Devise at least two other ways of grouping the reasons. Why would one method be selected over the others? What effect would the different groupings have on readers?

Figure 14-5. Example of Partition

PURPOSE OF MANUAL

This manual provides a guide for use by professional personnel in performing safety evaluations of existing dams and appurtenant structures. The principles and concepts of examination and evaluation, causes of failures, and examples of adverse conditions are discussed. The manual sets forth guidelines for:

- Scope and frequency of examinations
- Team selection and responsibilities
- Preparing and updating Data Books
- Reviewing design, construction, and operations
- Making onsite examinations
- Developing conclusions and recommendations
- Preparing reports

Source: U.S. Department of the Interior. Bureau of Reclamation. (1983). *Safety evaluation of existing dams* (p. 1), Denver, CO: U.S. Government Printing Office.

2. Computer descriptions almost always begin with a division of computers into CPU, memory, I/O, and secondary storage. What other approaches to analysis could be used (such as cost or complexity)? What would they tell readers that the hardware approach doesn't? What would they miss? Outline a report using such a new organization.

3. Assume you were asked to analyze one of the subjects below. For each, list all you know about the subject. Then list several approaches to analysis you could choose. For each approach, list additional information about the subject that comes to mind. Compare your lists with those of others in the class. Why do some approaches generate longer lists?

  a. Costs for a new dormitory
  b. Impact of a tuition increase
  c. Uses of personal computers
  d. On-the-job safety
  e. Freeway design
  f. International trade

4. Rewrite the example description of manifest changes (Fig. 14-4) using a different method of classification. For instance, the changes could be classified by amount of delay (short, medium, long), cost to implement (high, medium, or low), or some other means. Compare your way with those of others in the class. Which way seems to work best? Which way brought the most surprising conclusions?

5. The report in Figure 14-7 summarizes reviews of O-rings in the space

Figure 14-6. Example of Classification

CATEGORIES AND CAUSES OF FAILURE

Failure	Cause
Foundation deterioration	Removal of solid and soluble materials
	Rock plucking
	Undercutting
Foundation instability	Liquefaction
	Slides
	Subsidence
	Fault movement
Defective spillways	Obstructions
	Broken linings
	Evidence of overtaxing of available capacity
	Faulty gates and hoists
Defective outlets	Obstructions
	Silt accumulations
	Faulty gates and hoists
	Gate position and location
Concrete deterioration	Alkali-aggregate reaction
	Freezing-thawing
	Leaching
Concrete dam defects	High uplift
	Unanticipated uplift distribution
	Differential displacements and deflections
	Overstressing
Embankment dam defects	Liquefaction potential
	Slope instability
	Excessive leakage
	Removal of solid and soluble materials
	Slope erosion
Reservoir margin defects	Perviousness
	Instability
	Inherent weaknesses of natural barriers

Source: U.S. Department of the Interior. Bureau of Reclamation. (1983). *Safety evaluation of existing dams* (p. 18). Denver, CO: U.S. Government Printing Office.

Figure 14-7. Report on Space Shuttle O-Ring Reviews

Appendix H

Flight Readiness Review Treatment of O-ring Problems

This Appendix describes case-to-case field joint and nozzle-to-case joint O-ring anomalies experienced in flight and documents to what extent these anomalies were discussed in subsequent Flight Readiness Reviews. The sequence of Flight Readiness Review briefings on the Solid Rocket Boosters included up to five successive briefings at Marshall: SRM Pre-board, SRB Board, Shuttle Projects Office Board, and Center Board. The Flight Readiness Review briefings were culminated with a Level I Headquarters briefing, nominally 10 days prior to launch and typically referred to as "the" Flight Readiness Review. The second Shuttle flight, STS-2, launched November 12, 1981, experienced erosion of the primary O-ring in the right SRM aft field joint. The eroded area was at the 900 location and was a maximum of 0.053 inches deep. However, this erosion was not discussed in the STS-3 Flight Readiness Reviews, even though other anomalies mentioned in these reviews were identified as "Not a Crit. I/II failure." The field joint seal was, of course, classified Criticality 1R. This erosion was the deepest experienced in flight in a case field joint, until STS 51-L.

* * *

STS-6 (flight 6) was the first use of the lightweight SRM case; it was also the first flight following the case joint seal criticality change from 1R to 1. At the STS-6 SRB Preboard on January 4, 1983, Marshall Safety, Reliability, and Quality Assurance mentioned the criticality change as "in approval cycle." At the Marshall Center Board briefing on February 25, 1983, responding to an STS-6 Vehicle Configuration Review action item, Thiokol discussed the gap size and O-ring squeeze for the lightweight case segments, also referring to the criticality change: (Chart 1). Subsequently, Mr. Lawrence Mulloy, the SRB Project Manager, also mentioned the pending criticality change in his Flight Readiness Review briefing to Level I on March 17, 1983. It was identified as not being an open item, even though the waiver for flight as a Criticality 1 was not approved by Level I until March 28, 1983: (Chart 2).

* * *

STS-6 was launched on April 4, 1983. When the SRMs were dismantled, blowholes through the putty in both nozzle joints were found. The O-rings were affected by heat, but were not eroded. The blowholes were at 3180 on the left nozzle and 2510 on the right nozzle. These observations were not discussed in any of the STS-7 Flight Readiness Reviews.

* * *

STS 41-B (flight 10), experienced O-ring erosion in both the right hand nozzle joint and the left SRB forward field joint. The field joint O-ring erosion was centered at the 3510

Figure 14-7. Report on Space Shuttle O-Ring Reviews (Continued)

location and the area of maximum erosion extended over 3 inches with a maximum depth of 0.040 inches. The nozzle O-ring erosion was at the 14.40 location. The erosion was localized, extending only over a 0.75 inch span to a depth of 0.039 inches. These anomalies were described by Thiokol in the STS 41-C SRB Board briefing on March 2, 1984: (Charts 3,4). In the same briefing, O-ring erosion was presented as a technical issue. For the first time in any Flight Readiness Review, the O-ring problems on STS-2 and STS-6 were mentioned. The idea that putty blow holes and O-ring erosion were related was also discussed for the first time: (Charts 5, 6, 7). On March 8, 1984, at the Shuttle Projects Office Board for STS 41-C (flight 11), the notion of "acceptable" erosion is first mentioned: (Chart 8). In assessing the erosion, the concern was expressed in terms of the following: erosion of the primary O-ring could lead to erosion of the secondary O-ring and subsequent joint failure. However, the joint had been reclassified as Criticality 1, not 1R, indicating the primary O-ring should have been considered a single point failure. Also noteworthy in the briefing was the first mention of an analysis showing that the "maximum possible" erosion for joint O-rings was .090 inches and the rationale that this was acceptable since a test showed that O-rings with a simulated erosion of .095 inches could seal at three times motor pressure: (Charts 9, 10, 11, 12, 13). The concept of "acceptable erosion" was advanced again in the Flight Readiness Review briefing to the Marshall Center Board on March 20, 1984: (Chart 14). At the same briefing, Thiokol discussed the O-ring erosion in some detail. The briefing was substantially the same as that given at the Shuttle Projects Board. Similar charts were used by Mr. Mulloy at the Level I Flight Readiness Review on March 27, 1984: (Charts 15, 16, 17, 18, 19). During the Level I briefing, Mr. Mulloy received an action item to review the case and nozzle joint seals. The action item was identified as "Outside the Flight Readiness Review," which indicated that the action did not have to be completed before the STS 41-C flight: (Chart 20). Mr. Lawrence Wear, the SRM Element Manager, subsequently directed Thiokol to establish a plan and a test program to investigate the issue. In his tasking, he asked Thiokol to determine if the O-ring erosion was acceptable, and if so, why: (Charts 21 and 21A).

* * *

STS 41-C (flight 11), launched April 6, 1984, experienced erosion of the primary O-ring in the right-hand nozzle joint. The erosion, located at 3190, was 0.034 inches deep and extended over a 1.8 inch span. The primary O-ring in the left-hand aft field joint was affected by heat in areas adjacent to blow holes in the putty, but there was no measurable erosion. A Thiokol presentation to the STS 41-D SRM Preboard Flight Readiness Review on May 30, 1984, described these anomalies: (Charts 22, 23, 24, 25, 26). The nozzle joint O-ring erosion was also discussed as a technical issue in the same briefing: (Charts 27, 28). Similar charts were prepared by Thiokol for the Marshall Center Board on June 8, 1984, but they were not presented. The STS 41-C O-ring erosion was not mentioned in the STS 41-D Level I Flight Readiness Review on June 18, 1984.

Source: Presidential Commission. (1986). *Report to the President on the space shuttle Challenger accident.* (Vol. 2). Washington, DC: U.S. Government Printing Office.

shuttle. The report is organized by flight. What other ways could the information have been grouped? List the new groups that would result from this new form of classification. How would the report have been changed if one of the alternative classification approaches had been used? Given the intended audience for the report, which approach do you feel would have been most effective — the original, or yours? Why?

6. Write a report for high school students describing your college major.

a. First write a report that begins by classifying the courses you have taken.

b. Write the same report, but this time write it based on a partition of your typical day.

15

Technical Argumentation

The Role of Technical Argumentation

When to Use Technical Argumentation

Types of Argumentation
  Logical Appeal
  Emotional Appeal
  Ethical Appeal

Writing a Technical Argument
  Thesis Statement
  Background Information
  Logical Argument
  Evidence
  Anticipation of Objections
  Clear Conclusion

Guidelines for Technical Argumentation
  1. Project a Credible Image
  2. Anticipate Objections Early
  3. Do Not Argue Matters of Taste
  4. Do Not Preach to the Converted

Examples of Technical Argumentation

Activities

The Role of Technical Argumentation ────────

As discussed in chapter 3, there are three basic purposes in technical writing: to inform, to instruct, and to help others make a decision. Writing to help others make a decision means to help readers decide whether to carry out a course of action. In helping others make a decision, technical writers rely heavily on argumentation to convince readers to adopt the writer's point of view.

Though the terms "persuasion" and "argumentation" are often used interchangeably, they are not the same. Persuasive writing encompasses any writing that tries to persuade readers. In this regard, almost all writing is persuasive. Authors of definitions, descriptions, and other informative materials want to persuade readers that they are reliable and know what they are talking about. Argumentative writing, on the other hand, refers only to those documents that propose an identifiable argument.

When to Use Technical Argumentation ────────

Technical argumentation is used whenever the purpose of a document is to help the reader make a decision. A memo advising the purchase of a new laser printer, a proposal urging adoption of a new method for waste-water treatment, a report of a new procedure for aligning large cylinders for welding, a feasibility report recommending a site for a recycling center all have an argumentative edge. Proposals and feasibility studies can, in fact, be understood as extended arguments. And because many research reports are written specifically to settle a question that must be answered before a decision can be made, these documents often contain technical arguments within their "Conclusions" section.

Types of Argumentation ────────

Traditionally, there are three forms of appeal in argumentation: the logical appeal, the emotional appeal, and the ethical appeal. These do not necessarily represent separate arguments; in fact, the three forms of appeal are often used together in combination. Knowing when to make which appeal can mean the difference between an effective argument and an ineffective one. Too often, inexperienced technical writers expect a logical appeal to carry their arguments, when a logical appeal combined with an ethical appeal or even, occasionally, an emotional appeal would have been more effective.

Logical Appeal

An argument based on logical appeal appeals to reason. It relies on a clear, appropriate progression of thought or evidence. Because technical argumentation generally relies heavily on logical argument, it depends heavily on facts to support its conclusions.

However, an argument that is logical and well supported by evidence will accomplish nothing if it is not persuasive. That is, how the argument is presented is almost as important as the validity of the argument. If the audience is convinced that the writer is credible, reliable, intelligent, and reasonable, half the battle of argumentation is won. The next two forms of appeal will help you become more sensitive to your ultimate purpose in argumentative writing, which is to convince your readers and get them to agree with you, or at least to look at something in a different way.

Emotional Appeal

An argument based on emotional appeal engages the reader's heart. It attempts to evoke an emotional response in the reader. Although an overt emotional appeal is inappropriate in technical writing, correctly assessing the readers' emotional attitude toward a topic (acid rain, nuclear power) is often an important step in developing a persuasive argument. For example, in attempting to persuade readers to improve access for the handicapped, a writer might describe the frustrations experienced by blind employees and visitors caused by not having brailled elevator keys.

Ethical Appeal

An argument based on ethical appeal asks the reader to consider the writer's credibility. It attempts to convince a reader by demonstrating that the writer is knowledgeable (has a reputation in a particular field), by directly communicating certain values, and by adopting particular tones in communicating those values. For example, in arguing against purchasing a piece of equipment, a technical writer might argue that one of the subcontractors is a member of the Ku Klux Klan and, assuming that his or her readers find racism abhorrent, recommend that the subcontractor be boycotted.

Writing a Technical Argument

The challenge of writing good technical arguments is primarily one of presentation. The advantages of a new method for beading aircraft tires might be obvious to you, but you must make those advantages equally obvious to the reader. You might have a brilliant argument, and all the facts might prove your point, but you must marshall those facts in a reasonable and convincing manner.

It is not enough just to present the facts. You must also consider the reader's reaction to your presentation of those facts. In every piece of writing you want to say what you have to say as well as you can. In technical argumentation, however, how you present your argument is almost as important as the validity of your argument. As Robert Sternberg, author of *The Psychologist's Companion: A Guide to Scientific Writing for Students and Researchers*, has pointed out, "whereas it is usually easy to distinguish well-presented good ideas from well-presented bad ideas, it is often impossible to distinguish poorly presented good ideas from poorly presented bad ideas" (1988). Because it is difficult to tell whether a sloppy, disorganized piece of writing reflects the quality of the idea or the quality of the presentation, readers tend to consider poorly presented ideas as poor ideas.

Although your readers might not be hostile to your ideas, you still must convince them. Anything that interferes with the logic of your argument, with the train of thought in your presentation, with your attempt to convince your readers, is what William Zinsser in *Writing to Learn* refers to as "noise" (1989). According to Zinsser, noise refers to "all the random interferences with what man or nature is trying to say." In technical argumentation, noise is anything that reduces the effectiveness of your argument: a misspelled word; the lack of a sound, logical, and well-supported argument; an opinion masquerading as a fact; an inappropriate tone; an inaccurate assessment of the audience.

Most well-developed instances of good technical argumentation consist of several standard elements: a thesis statement, background information, a reasonable argument, sound evidence, anticipation of objections, and a clear conclusion. These parts are rarely labeled and often overlap; for example, an argument is rarely presented independent of the evidence on which it is based. But if each of them is sound and well executed, the argument as a whole will be more effective. Finally, many arguments are enhanced by the strategic use of graphics. A photograph may show the dilapidated condition of a building much more clearly than mere words, and diagrams, drawings, and tables are often invaluable when it is important to present dense information clearly, succinctly, and persuasively.

Thesis Statement

The thesis is the point you are making about the subject. In argumentative writing, the thesis is the statement of the main idea of the argument. Too often novice writers fail to state the thesis explicitly, assuming that the argument as a whole will make it clear what the point is. But this is rarely an advantage; it just leaves the readers wondering whether they have understood the discussion.

Occasionally a writer may decide that a thesis statement is best placed in the middle or at the end of an argument. This may be the best approach if the thesis is so controversial or threatening that it might be rejected out-of-hand if it were baldly stated at the beginning. In these cases writers often lead up to the thesis statement gradually, using the logical development of ideas to get the reader in the right frame of mind and cushion the blow.

Background Information

Because not all readers will be aware of the situation that led to the report, the study, or the argument, it is usually necessary to provide a context for the document. After assessing your audience, you should know whether this is necessary, and if so, how much background information you should provide.

Logical Argument

The argument itself must be sound and reasonable; it should not contain any logical fallacies. Each proposition should follow logically from the one before it, without any leaps of faith or unidentified assumptions. If any link in the chain of your argument is flawed, the whole thing may fall apart.

Below is a discussion of some of the most common logical fallacies. Be careful to avoid them in your writing.

HASTY GENERALIZATION

The most common fallacy is the hasty generalization. A hasty generalization is a generalization or conclusion based on insufficient or unrepresentative evidence. It is often recognizable from use of such words as "all," "always," and "everybody." It is usually necessary to qualify statements when you write, and to be wary of unqualified statements when you read. Although good readers usually recognize the more obvious examples of hasty generalization ("young people are conformists"; "teachers like to give bad grades"), it is the less obvious ones that give readers the most trouble. Below are two examples of hasty generalizations. There is no support offered for either of the two assertions.

> Conformity is caused by many things, but the greatest cause is a lack of strong beliefs.

How could the writer possibly know this? Has she interviewed thousands of conformists?

> It still amazes me to realize that most people in the U.S. have no interest in sports.

How does one come to know anything about most of the people in the United States?

NON SEQUITUR

Another common fallacy is the non sequitur (Latin for "it does not follow"). There are usually two parts to a non sequitur, an opening statement and a conclusion that is presented as a logical conclusion of that opening statement. A non sequitur occurs when one of the parts does not logically follow from the other, as in the statement, "I'm old enough to fight for my country; therefore, I'm old enough to drink."

BEGGING THE QUESTION

Begging the question is a form of non sequitur: the writer acts as if an arguable proposition were already agreed upon.

> Since the HAL 2001 is the best computer on the market, the marketing division should trade in their old models.

In this case, the writer first needs to prove that the HAL 2001 is the best computer on the market.

FALSE ANALOGY

Though analogies can be dramatic and can often help clarify and explain complex ideas, their usefulness is limited by the number of points of similarity. An argument is analogous to a house in that both must be carefully constructed, but it would be foolish to extend the analogy and imply that an argument has windows and chimneys. Some analogies fall apart before they begin. These analogies are called false analogies. Several years ago, a Wisconsin state senator, urging his colleagues to vote against legalizing bingo, based his entire argument on this analogy:

> You can't eat without having dirty dishes, and you can't have bingo without corruption.

While the sentence might have balance and a nice ring to it, it lacks the logic required of a reasonable analogy.

Similarly, it is fashionable among some child development specialists to explain the stages of human development in terms of analogies. One explanation likens children to roses:

> Let's suppose you'd never seen a rose and you knew nothing about roses and how they grew. Then, one day, you saw a rose growing in a field. With your

eyes only on the delicate and beautiful flower, you reach down and grasp the stem. What a shock! You jerk back your hand. You didn't know about the *thorns*!

Children are like roses. Much of their unpleasant ("thorny") behavior is normal. When parents realize this, they are less apt to feel frustrated and inadequate. Furthermore, they can be consoled by the knowledge that pleasant stages of development come between the "thorny" periods.

This explanation goes on to name each thorn, beginning with colic at age three to six weeks and running through each sharp and painful period until finally the bud is reached, by which time the thorns are gone and the child has reached "sweet 16." Anyone who has reached the age of 17 will recognize the falseness of the analogy.

EITHER/OR FALLACY

To suggest that only two alternatives exist, when in fact there are more, is to be guilty of the either/or fallacy. The following statement appears in a public utility brochure:

The choice is between clean, inexpensive nuclear energy and dirty, expensive fossil fuels.

The writer has conveniently omitted mention of other types of energy, such as solar, geothermal, and wind.

Evidence

Evidence is the facts and opinions that writers use to support their claims. A *fact* is information that can be verified through experience or observation. Evidence is often presented in the form of tables, graphs, photographs, drawings, diagrams, statistics, examples, and definitions.

To convince readers, technical writers also rely on what is called "appeal to authority": they cite the opinions of authorities in order to enhance their own credibility. If you rely on the opinions of technical or professional experts, be sure to give their credentials and qualifications.

A good rule of thumb is never to underestimate the intelligence of your readers and never to overestimate their knowledge. Always err on the side of giving too much evidence rather than too little. Evidence is *sufficient* when it provides enough reliable facts, details, examples, and opinions to support whatever point or claims you are making. Evidence is *insufficient* if it is inaccurate, if it is based on misunderstanding of material, if there is no logical connection between the support and the claim, or if it is unclear.

Anticipation of Objections

Regardless of the audience, most readers will not know as much about the subject as the writer. However, readers will almost assuredly be at least somewhat knowledgeable about the subject; thus, they will probably know when the writer has failed to acknowledge arguments for the other side.

Inexperienced writers sometimes believe that acknowledging an opposing viewpoint will weaken their argument. To the contrary, according to psychotherapist Carl Rogers, acknowledging the opposition is an extremely effective strategy in argumentation (1951). According to Rogers, a person will refuse to consider an alternative argument if that argument seems threatening; therefore, for that person to become receptive to your argument, the threat must be eliminated.

The Rogerian strategy is to let readers know that their positions are understood, recognized, and respected. Tell readers where their arguments are valid, and convince readers that writer and reader have the same goals. The objective of the technical writer in argumentation is to get readers to understand the writer's position in the same way that the writer has understood the readers' position. This means that the writer must know what the opposition is, must have some sense of what readers already know, and must have a good idea of how much they need or want to know.

Clear Conclusion

Do not end abruptly. A curt conclusion, because there is no sense of completion, will leave the reader feeling that the writer has rushed through the argument. A conclusion that trails off will weaken a strong argument by leaving the impression of an insubstantial argument. Although there are no clear rules for good conclusions, all conclusions should clearly relate to what has been discussed in the document. In the report of plating, for example, the conclusion tells the reader what experiments need to be conducted before the process becomes available.

Guidelines for Technical Argumentation _____

1. Project a Credible Image

Although readers of technical writing will often want to believe you, they will not unless they perceive you to be trustworthy and competent. As a rule, if readers feel you are honest and know your subject, they will be inclined to be persuaded. To this end, you must be honest in your dealings with your readers.

2. Anticipate Objections Early

It is difficult to hold a reader's attention if the reader is thinking about objections she or he has about the point you are making. Therefore, effective writers anticipate a reader's objections. Anticipating objections makes the writer appear thorough and objective, and helps blunt any objections. In the report on plating, for example, the writer anticipated possible objections to the proposed method of plating parts by describing the drawbacks of existing methods.

3. Do Not Argue Matters of Taste

Inexperienced writers often slip into arguing about taste when they ought to focus on more substantial or relevant topics. For example, trying to convince readers that the bald eagle is the most beautiful bird in the world is probably futile and certainly irrelevant, but a writer *can* argue that for certain reasons, the eagle should be protected from extinction, and these are the steps to take. Matters of taste depend, for the most part, on personal preference.

4. Do Not Preach to the Converted

Don't waste time and energy trying to convince people who already agree with your position. You may have just identified a new line of reasoning that supports a commonly held conviction or new evidence that strengthens the case. When addressing readers who already agree with you, however, you need only present these as *information*, not as *argument*. Otherwise, your readers may be offended by your argumentative tone and may end up disagreeing with you.

Examples of Technical Argumentation

Figures 15-1, 15-2, and 15-3 contain examples of technical argumentation. The report in Figure 15-1, written for an audience assumed to have an interest in and knowledge of plating methods, describes an improved method for plating various surfaces. Although the order of the parts might differ from argument to argument and from document to document, this report contains the typical ingredients of an argumentative piece of technical writing. The report begins with a thesis statement, offers background information, provides evidence, anticipates objections, and ends with a conclusion.

Figure 15-1. Example of Technical Argumentation

A simple, inexpensive method of plating parts with glasslike amorphous metal alloys has been developed. It is hoped that use of this technique will ultimately result in significant savings of strategic metals, such as chromium.

Whereas ordinary metals are composed of crystals, amorphous metals are noncrystalline, like glass. For this reason, they are often stronger and more resistant to corrosion and abrasion—characteristics that give them great potential value as protective coatings. However, there has been no easy way to produce them either in bulk or as films. Thus extra strength and corrosion resistance are ordinarily achieved by supplementing ordinary metals with the addition of so-called strategic metals, such as chromium, cobalt, and molybdenum.

These metals are expensive and frequently are imported. "If we can mass produce amorphous metal coatings, the use of strategic metals could be much reduced," says Kay Hays, developer of the new coating process and supervisor of Sandia's metallurgical cleaning and coating technology division. "For instance, we could substitute a low carbon steel that has a corrosion resistant coating for a material like stainless steel that contains up to 18 percent chromium."

The coating technique developed by Dr. Hays involves converting gases containing the desired metals to a plasma so that the metal alloy grows on the workpiece as a glasslike, amorphous film. The deposition occurs at room temperature.

With plasma deposition, it is possible to reach the desired metal composition by adjusting the ratio of alloy constituents. "To form the films, you just adjust the feed gas flow to get the required proportions," says Dr. Hays. The type of gas used to carry the metal and metalloid to the plating surface is also important. Although the current work of Dr. Hays is aimed primarily at developing coatings containing nickel, chromium, and phosphorous, she believes the same general technique should permit deposition of many different amorphous metal coatings on a variety of materials, from steel strip to plastic mouldings.

The process appears superior in cost and simplicity to such conventional coating techniques as sputtering, evaporation, electroless chemical deposition, and ion implantation.

The porosity and adhesion of the amorphous metal coatings are now being investigated, but these characteristics are not expected to present obstacles to utilization of the process.

Source: U.S. Department of Commerce. National Technical Information Service. (1986, May). Coating with plasma provides protective metal film. *Tech Notes*. Washington, DC: U.S. Government Printing Office. (NTN No. 86-0481)

Figure 15-2. Example of Technical Argumentation

Recycling doesn't eliminate the solid-waste problem, but it greatly reduces it. We need to exceed even our most ambitious recycling goals. The American Paper Institute's recently announced target of 40 percent paper recycling by 1995— laudable by today's standards—will barely keep pace with the expected increase in paper use by then! Because of the concurrent projected growth in paper products, the current recycling goal is unlikely to cut even a single ton from the 60 million tons of paper presently dumped or incinerated. Research indicates that paper fibers can be used up to a dozen times. If we bury the paper, we waste landfill space as well as fibers that could have been used repeatedly (and we're also causing more trees to be cut for new fiber). If we burn the paper, we use the energy for combustion, but we waste the opportunity to reuse the fibers.

Source: Reprinted, with permission of the editors, from Davis, A., & Kinsella, S. (1990, May-June). Recycled paper: Exploding the myths. *Garbage, 2*(3), 48–54.

The text in Figure 15-2 is a short excerpt from a corporation newsletter arguing for recycled paper. In this section of the article, the writers are responding to the concern that recycled paper will end up as garbage anyway. The purpose of the article is to convince readers that recycling is, in fact, environmentally sound.

Notice that the writers use several important argumentative strategies. First, they are realistic about their claims: "Recycling doesn't eliminate the solid-waste problem, but it greatly reduces it." Second, they anticipate and respond to objections that the reader might raise: What's wrong with just burying the paper? Why don't we always burn the paper for energy? Finally, they offer concrete facts to support their argument: "60 million tons of paper [are] presently dumped or incinerated"; "paper fibers can be used up to a dozen times."

The memo shown in Figure 15-3 was written five months after a near-serious accident at the Davis-Besse nuclear power plant in Toledo, Ohio. It was written by Bert Dunn, Manager of Emergency Core Cooling Analysis (ECCS) at Babcock and Wilcox Company, and directed to James Taylor, Manager of Licensing in the Nuclear Power Generation Division of Babcock and Wilcox. The purpose of the memo was to convince Mr. Taylor to change the operating procedures to allow for termination of high-pressure injection under certain conditions. The memo has been widely analyzed because it describes and attempts to prevent a situation that not only occurred but caused a very serious problem. Babcock and Wilcox was also the private vendor for the Three Mile Island nuclear power plant, and the very problem

Figure 15-3. Example of Technical Argumentation

THE BABCOCK AND WILCOX COMPANY
POWER GENERATION GROUP

TO: Jim Taylor, Manager, Licensing

FROM: Bert M. Dunn, Manager
      ECCS Analysis (2138)

SUBJ: Operator Interruption of High Pressure Injection

DATE: February 9, 1978

This memo addresses a serious concern within ECCS Analysis about the
potential for operator action to terminate high pressure injection
following the initial stage of a LOCA [loss-of-coolant accident]. Successful ECCS operation during small breaks depends on the accumulated
reactor coolant system inventory as well as the ECCS injection rate. As
such, it is mandatory that full injection flow be maintained from the
point of emergency safety features actuation system (ESFAS) actuation
until the high pressure injection rate can fully compensate for the
reactor heat load. As the injection rate depends on the reactor coolant
system pressure, the time at which a compensating match-up occurs is
variable and cannot be specified as a fixed number. It is quite possible, for example, that the high pressure injection may successfully
match up with all heat sources at time t and that due to system pressurization be inadequate at some later time t2.

The direct concern here rose out of the recent incident at Toledo.
During the accident the operator terminated high pressure injection
due to an apparent system recovery indicated by high level within the
pressurizer. This action would have been acceptable only after the
primary system had been in a subcooled state. Analysis of the data from
the transient currently indicates that the system was in a two-phase
state and as such did not contain sufficient capacity to allow high
pressure injection termination. This became evident at some 20 to 30
minutes following termination of injection when the pressurizer level
again collapsed and injection had to be reinitiated. During the 20 to 30
minutes of noninjection flow they were continuously losing important
fluid inventory even though the pressurizer indicated high level. I
believe it fortunate that Toledo was at an extremely low power and
extremely low burnup. Had this event occurred in a reactor at full power
with other than insignificant burnup it is quite possible, perhaps
probable, that core uncovery and possible fuel damage would have
resulted.

Figure 15-3. Example of Technical Argumentation (Continued)

The incident points out that we have not supplied sufficient information to reactor operators in the area of recovery from LOCA. The following rule is based on an attempt to allow termination of high pressure injection only at a time when the reactor coolant system is in a sub-cooled state and the pressurizer is indicating at least a normal level for small breaks. Such conditions guarantee full system capacity and thus assure that during any follow on transient would be no worse than the initial accident. I, therefore, recommend that operating procedures be written to allow for termination of high pressure injection under the following two conditions only:

1. Low pressure injection has been actuated and is flowing at a rate in excess of the high pressure injection capability and that situation has been stable for a period of time (10 minutes).
2. System pressure has recovered to normal operating pressure (2200 or 2250 psig) and system temperature within the hot leg is less than or equal to the normal operating conditions (605 or 630 F).

I believe this is a very serious matter and deserves our prompt attention and correction.

BMD/lc

cc: E.W. Swanson
     D.H. Roy
     B.A. Karrasch
     H.A. Bailey
     J. Kelly
     E.R. Kane
     J.D. Agar
     R.L. Pittman
     J.D. Phinny
     T. Scott

Source: United States. (1979, October). *Staff report to the President's Commission on the accident at Three Mile Island*. The role of the managing utility and its suppliers (pp. 224–225). Washington, DC: U.S. Government Printing Office.

Mr. Dunn had foreseen resulted in the serious accident at that plant little more than a year after the memo was written.

You have probably noticed that Mr. Dunn relies heavily on facts, logic, and certain assumptions about Mr. Taylor. The logic is helped by the orderly presentation of the details surrounding the incident at the Davis-Besse plant. The argument begins with two major premises:

1. Full injection flow must last from ESFAS actuation until the high-pressure injection rate can compensate for reactor heat load (because successful ECCS operation depends on the coolant system inventory as well as the injection rate).
2. The time of such compensation is unpredictable (because injection rate depends on reactor coolant system pressure).

The conclusion that is reached from these premises is that high-pressure injection should be terminated under two conditions only.

The assumption about Mr. Taylor is that he and Mr. Dunn share many of the same concerns—namely the concern for safety and the need to provide "sufficient information" to reactor operators. Note that Mr. Dunn begins his memo by emphasizing the seriousness of the matter and reemphasizes the seriousness in the conclusion. Only after he feels he has been convincing does Mr. Dunn then present his recommendations.

Even though the memo presents the facts logically and appeals to what Mr. Dunn believes to be common concerns, the memo was not a fully effective piece of argumentative writing. The purpose of the memo was to convince Mr. Taylor to change the operating procedures to allow for termination of high-pressure injection under certain conditions. Mr. Taylor was not convinced and did not change the operating procedures.

Activities

1. Do you find the report arguing for a new method for plating parts with metal alloys (Figure 15-1) convincing? If so, why? If not, how could the argument be made more effective?

2. Find two articles, one arguing for and the other against the same issue. Working with two or three other students, write a report on why one argument is stronger than the other.

3. Look through professional journals in your field. Find an article that is argumentative. In a one-page memo to your instructor, explain what the argument is about and the author's position. Are there any indications that the writer has considered the feelings and attitudes of the reader?

4. Find an article in a professional journal that argues against your position on an issue. Do you still disagree after having read the article? Why? What might the writer have done to move you closer to his or her point of

view? Write a letter to the author of the article explaining your position and why you were not convinced.

5. Find a short argument (two pages or less) in a technical journal. Change the audience of the article, and rewrite the argument to reflect an awareness of the intended audience.

6. Write a memo to your immediate supervisor arguing your position on one of the following questions:

   a. Should a cellular telephone be installed in all company-owned vehicles?

   b. Should a new security system be installed?

   c. Should the company establish a day-care center?

After your first draft, ask yourself the following questions: What type of appeal have I used? Why is it more effective than another type of appeal? Have I given sufficient background information? Have I projected a credible image? Have I anticipated objections to my position?

IV

TECHNICAL
RESEARCH

Collecting Information

Determining What Information You Need
  Using Tagmemics
  Using Burke's Pentad
  Framing a Research Question

Sources of Information

Yourself
  Brainstorming
  Free Writing
  Invisible Writing

The Library
  The Card Catalog
  Periodical Indexes
  Reference Books
  The On-Line Catalog
  Computerized Indexes
  On-Line Databases

The Field
  Interviews
  Letters of Inquiry
  Questionnaires
  On-Site Visits
  Observations
  Controlled Experiments

Activities

Technical writing entails communicating information. Regardless of the writing task, whether you are writing a simple memo suggesting the purchase of a new fax machine or preparing a complex manual describing a nuclear generator, you need to have information on hand and you need to present it effectively. Why, exactly, should the company buy a new fax machine? How, exactly, does a nuclear generator work? This emphasis on information in technical writing means that the information for every technical document is held to the highest standards: it must be absolutely complete and absolutely accurate.

At some point in the process of writing almost every document, then, you must conduct research: collecting whatever information is needed, determining what that information means, and effectively presenting the information within your document. This section of *Communicating Technical Information* explores how information for technical documents is collected, analyzed, organized, and integrated into technical documents.

Determining What Information You Need ———————

Before you begin to collect information, you must determine what you need to find out. For many short writing tasks, such as memos and brief reports, you might have most of the information in your head or on paper before you begin writing. For longer writing tasks, such as literature reviews, research reports, proposals, feasibility studies, progress reports, and instructions, you will likely not have all the information you need. The initial problem for many writers is knowing what information they need. This chapter provides strategies for helping you determine what information you need. Knowing that will make the rest of the process of collecting information much more efficient.

Long before computers and other advanced forms of technology were even thought of, speakers and writers faced the same problem we have today: "How can I discover what is really central in this subject? How can I determine what matters most?"

Previously writers used lists of specific questions to describe an object thoroughly and effectively. Such devices for self-questioning are known as "heuristic" techniques. Aristotle was one of the first to create such a list for orators. His *Topics* helped speakers prepare for Athenian debates. Some of the questions were tied to specific topics, others were for general analysis of topics. These general questions asked for such things as definition, partition, comparison, analogy, antecedent, consequence, cause and effect, genus, and species (D'Angelo, 1975).

Since then a number of question sets have been created to help a writer be insightful, complete, and original. Some of these lists have been computerized, and a few additional computer prompting techniques have been invented. We will look at the traditional questioners first.

Using Tagmemics

Beginning during World War II, three professors at the University of Michigan developed a theory of observation and description called tagmemics. Central to the theory is the belief that to understand something we have to know three things: what makes the thing unique, what possible forms it can take, and where it fits in a larger world (Becker, Pike & Young, 1970).

This became known as the particle-wave-field approach to observation. "Particle" questions are contrastive: they look at a thing in order to show differences when compared. "Wave," or variation, questions check for the range of traits the subject can have and still be the subject. "Field" questions view the topic in a larger context.

These kinds of questions can help people think more clearly about their own experiences and can also help researchers become better observers. For example, suppose you are writing a proposal for establishing a new wildlife preserve designed to protect an endangered species. You are concerned about detailing everything you can about the species in order to determine what special dangers exist for it or what special traits it may have that will help its survival.

You start with the species and begin asking "particle," or contrastive, questions about it. What features make it different from other species? What is unique about the size or health or habitat of the species? What is its principal means of defense? How could you identify this species in a group of similar species?

"Wave," or variation, questions ask, what range of behaviors are noted in the species? What is the range of forage? What are its sizes or colors? "Wave" questions also consider a subject over time and might include questions such as, how has the habitat of this species changed in the last hundred years? In the last ten years? What changes are possible in the future?

"Field" questions that try to view the species in a larger context might include issues such as what distinguishes the male or female of this species from the male or female of other species. What makes the habitat unique? How has its adaptations varied from the adaptations of other species under duress?

Asking and answering all the particle, wave, and field questions can be time-consuming but should leave you with a much clearer understanding of your subject. You can be reasonably sure you haven't missed any important aspects of the subject, and you may discover a question or two leading you in new directions for research.

How can you use tagmemics when writing a document? You would probably want to use it several times. The first might be before you have done much research at all. You might begin by asking yourself questions that help you look closely at your subject. Can you determine what is unique about the endangered species? Can you define it in some detail? Can you see that something is changing? Something tied to a larger world? This first use of tagmemics should help you think more clearly about your subject and think of new approaches to it.

A second time to use tagmemics is after you have begun your research. You have been reading for several days and have collected a fair amount of information. How does that information fit together? If your reading has led you to believe that the endangered species is totally unique, you may not be able to say much about how it has changed over time or about its relationships with other species. Now that you look back on it, is the species that unique? Are your sources being fair in not mentioning how the species ties to others? Have you been looking in the wrong places for information? Do you feel comfortable taking this approach to your subject?

What if you are less able to determine how the endangered species is unique but readily able to say how the species interconnects with other species? Perhaps the more you study the species, the more it seems to be tied to others. You find yourself unable to explain the species without describing these others, too. This should give you an important insight into what direction you will take with your research and a major point you will have to make to your readers.

As you use tagmemics, you should find that it helps you determine how much you really know about your subject and which aspect of your subject is primary: its uniqueness, its range of forms, or its connection to other subjects.

Using Burke's Pentad

Another set of questions that can help you analyze your subject is Kenneth Burke's pentad (Burke, 1945). To a certain extent you have used much of Burke's pentad if you have ever tried to answer the "five W's and an H" question journalists use: who, what, when, where, why, and how. In the case of an automobile accident, a journalist might ask, Who was involved? What happened? When did the accident occur? Where was the accident? Why did it occur? How did it occur?

Burke reframes these questions slightly, using the terms "agent" (who), "action" (what), "agency" (how), "scene" (when and where), and "purpose" (why). This gives five primary questions, or a pentad of questions. Burke then adds two more aspects: circumference and ratio. Circumference refers to the general background of the event: in the case of the auto accident, it might be the American love of cars, current problems with overconsumption of alcohol, and questionable safety standards for American products. Ratio simply asks, of all the areas of the pentad, which is most important? Which is central? If you were to draw an event as five petals on a flower, which of the five would be the largest and which the smallest? Would agent (who) be crucial in the accident—a person with a history of drunken-driving arrests? Or might it be the scene—a hilly road covered with fog? As the writer tries to draw the relative sizes of the pentad petals, he or she is forced to make judgments about what is crucial and what incidental.

Let us apply Burke's pentad to the writing of a report on the problem of a large number of injuries reported by people using a quarter-inch electric drill.

1. **Agent:** Who is being injured? What are the injury rates for various age groups? Are the rates the same for all income or education levels? All races or nationalities? What do we know about the people being hurt?
2. **Action:** What happened to the people? Are the injuries similar?
3. **Agency:** How are they getting hurt? What were they working on at the time of the accident?
4. **Scene:** When and where are the injuries occurring? Are they occurring on the job or at home? Do they seem most common at a particular time of day? Near the end of the day, when people are tired? Near the time some other activity is taking place?
5. **Purpose:** For what task was the drill being used? Is that the same function the manufacturer expected?
6. **Circumference:** What is the situation in which the injuries are occurring? How does it connect to other home or workplace problems? Are these injuries occurring in other countries, or is it uniquely American? Has this always been a problem with electric drills in the United States, or are we seeing something new?
7. **Ratio:** What is key to understanding these injuries?—The people involved, or the general workplace atmosphere? Should we focus most on the reasons why people are being injured, or should we look first at the drill itself? Which area tells us the most about the problem? Where should we begin to look for solutions?

The idea of a "ratio," the attempt to rank reasons for a problem in some order, is especially helpful with complex problems. Without a ratio, we gather increasing amount of information about a subject and may create a long list of research questions, but we may still be confused about what is central to a problem. In researching a problem, it is appropriate to try to identify the main cause. Identifying the ratio is one way of forcing ourselves to look for the main point. We may not find it right away—the key point is seldom obvious—but trying to find the key point can guide our research in a specific direction and help us avoid wandering aimlessly through endless facts.

Framing a Research Question

To conduct your research efficiently, you must have a clear idea of what question you are researching. Determining what information you need will ultimately lead to a specific research question. For example, if you are writing a proposal for establishing a new wildlife preserve designed to protect an endangered species, exploring your topic might lead you to ask what the major obstacles are in establishing the preserve. Without a clear question, your research will be unfocused; unfocused research is time-consuming and ultimately might prove fruitless.

Make sure your research goals are practical. Often the research project as originally conceptualized is impossible to carry out and perhaps not even necessary. Writers need to examine their research goals and strategies and, if necessary, to narrow or limit their focus. This step may need to be performed several times in the research project.

For example, ideally you might want to observe all sorts of people performing a process under all sorts of conditions before writing instructions, but time and resources may make it more expedient to determine what the most important variables are and make sure they're covered in a small sampling.

Sources of Information _____

For any research situation, there are basically three places you can find information: yourself, the library, and the field. Although the research process and the sources used vary from field to field and from task to task, typically researchers start by determining what they already know about a topic, go to the library to discover what others have found, and then, if necessary, conduct field research to find out about any missing pieces.

The information collected in this way can be classified as either primary research or secondary research. **Primary research** is first-hand research and consists of "undigested" information. Typical examples are interviews with accident victims to determine what happened in a car crash, a controlled experiment to establish the interaction between two drugs, or observations of an assembly process to discover the sequence of steps involved. As these examples demonstrate, primary research is usually thought of as field research, but it can be conducted in the library, too. Say researcher x conducts an experiment and publishes his findings. When researcher y reads this research report and decides to use it in her own report, she can do one of two things: she can use researcher x's interpretation of the data as he presented it in his "Discussion" section, or she can go to the raw data itself (in the "Results" section). The former is a secondary source, the latter a primary source.

Secondary sources give "digested" information, information that has already been processed and interpreted. Any information that has been gathered by one person and reported by another is considered secondary information. Much of what appears in journals, for example, is secondary information.

Both primary and secondary sources have strengths and weaknesses. Information gathered through primary research may be biased or unrepresentative; it is almost always time-consuming and costly. Using someone else's raw data as a source of primary information is risky because it is difficult to know, for example, how accurate an observation was, how biased an interview was, whether or not the respondents to a questionnaire were broadly representative. Because secondary sources are at least one step removed from the original source, their disadvantages include the possibility of an error in transmission, a misinterpretation of material, or bias on the part of the reporter. As you gather information, evaluate your sources carefully and use a combination of primary and secondary sources.

Yourself

If you are writing a memo suggesting your solution to a problem, you are the source of the information. On the other hand, if you are preparing a manual or providing instructions, you might not see yourself as a resource. However, before you look for information anywhere, it is important to find out just what you do know. This will not only provide a possible source of information, but it will help you find out what you do *not* know. Here are three strategies for finding out what you might or might not know.

Brainstorming

One approach to finding out what you already know is brainstorming. Brainstorming means simply to open your mind and allow your thoughts and ideas to come out. What do you know about the subject? Can you remember anything you have read about the subject, remembered from class, or heard someone mention in the employee's coffee lounge at work? All that is grist for your intellectual mill.

The central feature of brainstorming is speed. You write or speak as quickly as you can, jotting down notes as you go. The hope is that if you go fast enough, you can overcome the "censor" that lives in most of us. This censor can be helpful later in determining which conclusions are more valuable, but if we begin censoring too early, we end with almost no ideas: all our effort goes into thinking why our ideas are wrong, instead of finding new ideas.

You may have worked in groups that brainstorm. The group uses a large sheet of paper, and everyone puts as many ideas as possible on the sheet. No one is allowed to say anything negative about any of the ideas until later. By the time the group is done, there may be more ideas on paper than anyone would have imagined possible.

Free Writing

You may have done brainstorming in the form of "free writing." As quickly as you can, for about five or ten minutes jot down as much as you can about your subject. During that five or ten minutes do not stop writing or correct anything you have written. Normally by the end of five minutes you have written well over a page, and although some of the writing may be silly, there are often surprising and useful ideas on the page. The hope is that you will uncover not only the information at the surface of your knowledge but also information you forgot you knew. You can do this with a pen or pencil, or you can sit at a computer terminal.

Invisible Writing

If you have access to a word processor, consider a variation of free writing: invisible writing. While you spend the same amount of time (five or ten minutes) and do the same things, you turn off the computer screen (but remember to leave the rest of the computer on). The word processor will still work and will record what you write, but you will not be able to see what is on the screen. With nothing to see on the screen you will not be distracted by what you have written, so that you can concentrate on getting your thoughts down.

In between the typos and unfinished sentences, you will almost always find ideas you did not expect. Since your censor wasn't able to keep a lid on your ideas, a few new ones got out.

To improve the value of invisible writing, you might start with some initial idea. You are concerned about why migration patterns for a particular species are changing. You might want to start by writing a beginning sentence first, with the screen on. Just a few words will do; all you want is a starting place. Now shut off the screen and start writing. Write as fast as you can and do not stop for anything. With luck, by the time you finish you will have pages of ideas about migration trends. If one of the ideas seems especially useful, rewrite that one at the top of the screen and try invisible writing again, this time trying to elaborate on that idea. You don't want to put too much focus on your writing, as that, too, will limit you, but having a starting point may help you get below the surface of your topic.

The Library ————————————————————————

An important source of information is the library. For example, if you are preparing a literature review, the library will be the predominant if not sole source of information. However, while the library is an important resource, it is rarely the inevitable and ultimate goal of research. For most technical writing tasks, the library is the place to gather background information, to search the literature as a preliminary step before gathering first-hand information.

Although technical writers are usually quite familiar with methods of library research, they must also be familiar with the wide variety of libraries.

- **Public libraries** are primarily city, state, and regional libraries. While they typically serve a general audience, do not overlook these libraries as potential sources of information. For example, they may have city, state, and county data, or they may be a U.S. Government depository.
- **Academic libraries** are usually affiliated with a specific college or university. Many technical writers put the library out of their mind as soon as they graduate from school. This is a mistake. Academic libraries are often excellent sources of information, especially those libraries that are depositories for government documents.
- **Professional libraries** are usually highly specialized. While they exist throughout the world of work, they are especially prevalent in businesses, in industries, and in medical centers. These are excellent sources for information. If you are new to a company, regardless of the size of the company, ask if there is a library. Many companies, even small ones, maintain a library.

Most libraries today have two generations of research sources available. The traditional methods of finding information — the card catalog, indexes, and reference works — are still available and still useful. But for most of your research you will need to have access to an enormous number of references in order to find the relatively few that are truly relevant. You will need so many, in fact, that these conventional manual methods may not be adequate. You will almost certainly need a more powerful information-gathering tool. That tool is the new electronic information technology. These electronic tools can search through immense lists in a fraction of the time that a team of human researchers could do the same. They are so efficient and powerful that researchers can — in fact, must — be selective: filtering one's sources has become an important step in the research process.

The Card Catalog

The card catalog lists all the books currently in a library. Because you have likely been locating books with card catalogs ever since you first went to the children's library back home, probably you will first think of moving at once to the card catalog to begin the information-collecting process. Certainly the card catalog will be part of your research strategy in some instances, but because books are often not the most current of sources, you may find greater value in materials not indexed in the card catalog.

Periodical Indexes

Periodical sources are usually shorter and more focused than those found in books, enabling you to concentrate your research on directly relevant information. To locate these items of information, you will use general and specialized indexes found either in the library's reference room or in an electronic database.

The general readership magazine that typically appears in general indexes, such as the *Readers' Guide to Periodical Literature*, commonly fails to report in any detail the work and interpretations of acknowledged experts in technical fields. If you are interested in the opinions of the person on the street concerning the impact of computers, locate sources in the *Readers' Guide*; if you are researching the impact of the computer on productivity, look to specialized indexes.

Specialized indexes provide the advanced sources of information required for most technical writing tasks. Most of the journals indexed here provide abstracts to accompany their articles; these will often provide valuable assistance to understanding complex reading material.

Reference Books

Reference books include encyclopedias, technical dictionaries, handbooks, almanacs, and many other general and specialized reference tools. The reference section of a library is a good place to find basic information and can often provide leads to other sources.

The On-Line Catalog

While not all libraries currently have an on-line, or computerized, catalog, many do, and the on-line catalogs at many libraries are similar. Often people can call these up via a modem at home or at work, so the writer can look at what's available at lots of libraries, even those in different cities. If the writer finds a source that looks promising, he or she should be able to obtain these sources through interlibrary loans.

On-line catalogs provide access to books, non-print materials, many government documents, and journal titles held in a library; however, generally they do not include titles of individual articles. Though you might not find many books on your topic, in order to conduct as complete a search as possible, you should check the on-line catalog.

DETERMINING A SEARCH STRATEGY

The main menu in your library's on-line catalog will probably look something like this:

> Choose the type of search you wish to perform
> 1 — By Subject
> 2 — By Author
> 3 — By Title
> 4 — For other searches enter a key word to see possible matches
> Choice:

For much library research, the most logical choice at this point is to search either by subject or by key word. Though it might be tempting just to choose selection 1 and type in your subject, it's often best to begin with the "key word" selection, the most powerful search tool in the on-line catalog.

After you choose the "key word" selection, which for most computers will be similar to the four choices listed above, the terminal will respond with a list of key words to choose from. If your search produces a large number of citations (generally twenty or more), the terminal will respond to your choice with a screen similar to the following:

> Your search has identified many citations. What would you like to do next?
>
> 1. Narrow search
> 2. View citations sequentially
> 3. Start a new search

If you do not think there are enough citations to warrant narrowing your search, you can choose to view the citations sequentially. If you do this the titles will appear in reverse chronological order (most recent first). When you call up the citations, the record for each will include bibliographic details in a format similar to a traditional catalog card: call number, location, number of copies, and so on. However, there is one major difference: almost all on-line catalogs include status information. As soon as a book is checked out or in, its availability status is noted on the record screen. So, not only will you find the information you want, you will also find out if the books you want are available, and, if not, when they are due back.

FILTERING SOURCES

If you decide to narrow your search, you will be given several options for limiting your list, such as the following:

1. Limit by publication year
2. Limit by language
3. Limit by personal author
4. Limit by subject
5. Limit by material type
6. Limit by other searches

Basically, the computer takes the citations it has already collected and performs a second, more rigorous search on them to weed out any that don't meet your criteria.

Computerized Indexes

Like its printed cousins, a computerized index (also called an electronic index) covers a broad, but still limited, selection of periodicals. Computerized indexes have significant advantages over more traditional print indexes. One advantage is that the computer search strategy is much faster than the print search strategy. The computer can scan thousands of citations in just a minute or two, saving days of tedious searching through print indexes. Another advantage is that the computerized search strategy allows you to manipulate terms to broaden or narrow your search in ways that are simply unavailable with print indexes.

One disadvantage of electronic indexes is their limited time span. Though some computerized indexes include entries before 1970, electronic indexes generally begin coverage with works published around 1983. Thus, if you need access to sources older than this, you will also have to work with conventional, printed indexes.

Often, however, the one disadvantage of electronic indexes can actually be an advantage. The limited time span of electronic indexes acts as a filter for researchers who are not interested in sources published much before 1983.

Different libraries, of course, have different systems and different capabilities. One standard electronic index system is published by the H. W. Wilson Company. These indexes, called "Wilsondisc," are on CD-ROM (Compact Disk Read-Only Memory). Accessible through a special laser-reading machine, CD-ROMs often contain massive amounts of information on one compact disk.

SELECTING INDEXES

Because there are many electronic indexes, your first task is to choose the appropriate indexes to search. Listed below are some of the more common electronic indexes that you might expect to find in a university library, in a medical center library, and even in many public libraries.

Applied Science and Technology Index covers engineering, chemistry, mathematics, physics, computer technology, data processing, and related disciplines.

Biological and Agricultural Index covers biology, agriculture, biochemistry, environmental science, ecology, microbiology, and nutrition.

Business Periodicals Index covers management, marketing, economics, transportation, and computers, as well as specific industries, businesses and trades.

ERIC (Education Database) includes references and abstracts of documents and journal articles in education from 1969 to the present.

Government Printing Office Index contains materials issued by all U.S. federal government agencies, covering a wide range of topics such as energy research, agriculture, economics, tax reform, and health.

General Science Index covers the physical, life, and health sciences, and includes astronomy, earth science, conservation, nutrition, and oceanography.

Index to Legal Periodicals covers corporate law, labor relations, taxation, real estate law, estate planning, and criminal law, and includes statutes and case notes.

Readers' Guide to Periodical Literature covers popular literature, and includes current events, business, fashion, politics, crafts, food, education, sports, history, and science.

Social Science Index covers the interdisciplinary social sciences and includes economics, politics, international affairs, anthropology, psychology, and sociology.

Because each library is structured differently, it is a good idea to confer with a librarian who specializes in computer searches. The librarian may actually do the search for you.

Though it is not always easy to decide which indexes will be the most helpful, the most efficient method is to read carefully the descriptions of the indexes found in the library and choose the ones that seem most likely to contain the subject you are looking for. If, for example, you are investigating some aspect of computing, you will find that the *Applied Science and Technology Index*, the *Business Periodicals Index*, and the *Library Literature Index* include references to computing. On the other hand, if you are investigating an environmental issue, the *Biological and Agricultural Index,* the *Government Printing Office Index,* and the *General Science Index* might be most helpful.

Because there is often overlap among indexes, be careful not to choose more indexes than you can reasonably use. A rule of thumb often used by technical researchers is to limit the number of indexes to four.

FILTERING SOURCES

Most electronic indexes offer users several search strategies for each database. For example, the Wilsondisc CD-ROM Index offers either the Browse Search or Wilsearch strategy.

Browse Search is typical of the search strategies available with computerized indexes. With Browse Search, which is the simpler method of the two search strategies, the user enters a term or phrase that best describes the topic, and the system displays an alphabetic list of terms that includes the topic and those closest to it. However, scanning citations individually is extremely time-consuming: this search strategy usually results in an enormous list, especially if you have several indexes to look at and many more terms to look for.

Because you will want to pare down your list of citations to those articles that show the greatest promise of relating to your topic, we recommend that you forgo the type of strategy represented by Browse Search strategy in favor of a more efficient method. In addition to scanning thousands of citations in a matter of seconds, one of the computer's strongest features is that it allows manipulation of terms, a feature unavailable in traditional searches. The manipulation of terms is called Boolean strategy.

Boolean strategy allows you to combine terms in order to narrow or broaden a search. Different databases use different terms to accomplish the same thing. In Wilsearch, for example, the user narrows a search by entering the subject, up to seven terms, and the word "and." The computer calls up only those publications that have all of the terms linked by "and" in the title or description. For example, a researcher looking for information about the government regulation of pesticides would probably request "pesticides *and* government *and* regulation." This way, only titles with all three key words would be called up; most articles outside the researcher's field of interest would be ignored.

To broaden a database search on this same system, a researcher can use the word "or" to indicate alternatives to the computer; most database systems have a similar command. For example, a researcher who wants to find out about the relative effects of diet, exercise, and family history on the incidence of osteoporosis might first try to search by using the phrase "osteoporosis *and* diet *and* exercise *and* family history." However, the likelihood of finding much through this search is very small; it will only identify articles that contain *all* of these terms in the title or description, which is probably only a handful of articles. So this researcher might broaden the search by using the phrase "osteoporosis *and* diet *or* exercise *or* family history." Now the computer will highlight any items in the index that are about osteoporosis and either diet, exercise, or family history.

Using an effective and efficient search strategy will not only provide you with appropriate citations, it will also save you considerable time.

On-Line Databases

A third form of electronic information commonly available is on-line databases. On-line databases are similar to the electronic indexes described above; they provide lists of citations. However, unlike electronic indexes, on-line databases provide researchers access to massive collections of information accessible by communication between computers. This form of electronic information allows researchers to gather information from remote computers offering databases on almost any subject. The range of databases is immense. There are very highly specialized databases, such as one at Johns Hopkins University that contains almost everything known about genetically caused diseases. There are also databases that are multidisciplinary, such as *Dissertation Abstracts International* (DAI) and the *Educational Resources Information Center* (ERIC). An increasing number of companies are now providing access to multiple databases through a single searching language. Dialog Information Services, for example, makes available more than 320 databases in almost all disciplines.

DETERMINING A SEARCH STRATEGY

Although several new on-line systems, such as Dialog's *Knowledge Index*, Bibliographic Retrieval Services' *BRS-After-Dark,* and *Compuserve*, were designed for users with little or no training in information retrieval, few libraries currently offer unassisted use of on-line databases. Thus, you will probably need help from one of the highly trained research librarians.

The librarian will first ask you to fill out a form such as a "request for database services." The two most typical questions on a "request for database services" form are the following:

Subject of your search: Please describe in detail the subject of your search. (Define any technical terms that you use.)

Key Words: Please underline any words in the above description that are key words, and identify any terms under which you have found this subject in printed abstracting and indexing services (e.g., Psych. Abstracts).

For example, a researcher interested in the causes of Lyme disease might write the following:

> I am researching the <u>causes</u> of <u>Lyme disease</u>, especially which <u>animals</u> carry the <u>spirochete</u> and how the disease is <u>transmitted</u> to <u>humans</u>.
>   deer, pets, mice, birds, cats, vector, host, tick, Burgdorferi, Dammini, horse

These words will be used by the librarian to construct a search request much like the one used for the electronic index described above. Be careful when filling out the form because computer time is expensive. It is especially important to think of appropriate synonyms for what you want. In order for the librarian to be successful, the researcher must be a partner in the search, providing key words and concepts, especially for highly technical subjects.

FILTERING SOURCES

Searching an on-line database will almost assuredly result in a large number of citations. The problem is how to filter sources to get a reasonable number of reliable citations. There are several common strategies.

Limiting by Chronology.   One way to filter sources is to retrieve those that are most recent. Because citations are always listed in reverse chronological order, you might consider limiting your search to citations that cover the past two or three years, depending on your subject.

Limiting by Title Descriptions.   An effective method of filtering sources is to eliminate those citations that do not look promising from the title descriptions. For example, you will probably want to eliminate anything published in a foreign journal because it would be difficult to obtain and might have to be translated. You might also be able to weed out the few irrelevant items that have managed to slip through the search request.

Limiting by Availability.   Few libraries have all the references that result from the on-line database search. For those citations that are not in your library, you will have to use interlibrary loan. Some databases do contain full texts; however, because retrieving full texts is very expensive, most libraries prefer to use interlibrary loan. Because information in abstracts is often insufficient for a literature review, and because waiting for interlibrary loan will often delay your report, the material you use will often be determined by what is available.

The Field _____

For most of your technical writing tasks, it is likely that after you have figured out what you already know about a topic and have searched the literature for information that others have collected, you will still have to go out into the field and collect some first-hand data. Typical writing tasks that would require you to collect data from others are describing services available to a customer, interpreting lab analysis, providing instructions, initiating a new procedure, submitting a bid, and preparing a proposal.

Interviews

Sometimes a technical writing situation clearly dictates that you conduct an interview. For example, if you were preparing a brochure describing a new diagnostic service that repairs computers, facsimile machines, and voice-processing equipment over the telephone, you would likely interview company personnel responsible for providing the service. Similarly, if you were writing a progress report on construction of a new plant, you would want to interview those most directly involved in the project. Although no two interviews are exactly the same, there are some guidelines that will help you conduct a successful interview.

1. *Choose the interviewee carefully.* Make sure the person is the right person and is knowledgeable and credible. It may be difficult to find the right person to interview. If you are describing services your organization offers, your choices are limited and thus easier to make. However, if you are preparing a feasibility study for a construction project, you might have to interview people not affiliated with your company. For example, you might need information about percolation tests, local and state codes, and zoning. Choosing the wrong person to interview will result in time wasted and possibly even misinformation.

2. *Prepare a list of questions beforehand.* The people you interview are usually expert in their field; at least, they are likely to know more than you, and they are helping you. Simple etiquette and common courtesy demand that you be on time and be prepared with precise questions.

Generally, simple yes or no questions ("Do all branches offer the diagnostic service?") can usually be answered without an interview. However, avoid very broad and general questions ("How does the remote diagnosis work?"). Most interviewees do not like wasting time conveying information that can be found in company publications. Therefore, know the purpose of the interview, and conduct as much background research as possible before asking for an interview. If you have done your homework, you will be prepared with specific, pointed questions ("What kinds of equipment can be diagnosed and repaired over the telephone?" "How does a customer arrange for service?").

3. *Be tactful.* Some interviewees will be friendly, some will be unfriendly, some will be talkative, some will be relatively uncommunicative. Your challenge is to be tactful, friendly, and systematic. The following suggestions will help maintain a cordial yet business-like atmosphere:

- Generally stick to the list of prepared questions.
- Ask for clarification if you're not sure of an answer.
- If some of the questions are controversial, leave them for the end of the interview.
- Be a good listener. In an interview it is more important to be a good listener than a good talker.

4. *Tape-record the interview to ensure accuracy.* However, because many people do not like to be taped, make sure you get permission first. Unless you have tape-recorded the interview, as soon after the interview as you can go over your notes to make sure you understand everything.

5. *Thank the interviewee.* Send a thank-you note to let the person interviewed know you appreciate the courtesy. This is not only a matter of politeness. It will help keep the channels of communication open in case you need to contact that same person again in the future.

Letters of Inquiry

If you cannot conduct a personal interview, an alternative is a letter of inquiry. Letters of inquiry range from a one-sentence request for information about a new product to a lengthy and detailed request for information on a new method for flushing nuclear waste. Even more than in an interview, you must frame your questions carefully because you will not have a chance to ask follow-up questions immediately, and there is always the chance that the reader will not understand your question. Because a follow-up letter is costly and time-consuming (both for you and for the recipient), the initial letter of inquiry should be free from ambiguity. Following are some guidelines that will help clarify your letter.

1. *Identify yourself and your position.* Before they are willing to be candid and helpful, most people like to know who's asking for the help. This will also let the reader know whom to call if anything needs clarification.

2. *Be concise.* Don't take up more of the reader's time than you have to. As a rule, state what you want to know early in the letter. Be as specific as you

can. Unless you really want *all* the information a company has on remote diagnostic services, ask only for the information that you actually want ("I am requesting information on the cost of your new remote diagnostic service").

3. *Give background information as needed.* In order to make your request understandable, you might have to give some background information ("Our company is investigating the feasibility of remote diagnostic services for our voice-processing equipment").

Questionnaires

The most immediate advantage of a questionnaire over an interview and a letter of inquiry is that a questionnaire will allow you to gather information from a large number of people over a wide geographical area. If you are looking for a large number of opinions or for statistical information about a large number of people, a questionnaire might work well for you.

However, do not be misled by the apparent ease of constructing and administering questionnaires. Questionnaires are difficult to design well, and, because many people are reluctant to complete questionnaires, you might not get a representative cross-section of respondents. If you cannot get the information you need from any other source, such as an interview or a library, ask for help from someone who has some experience designing questionnaires. Preparing a flawed questionnaire will do more harm than good by calling into question both you and your work.

If you decide to use a questionnaire, here are some guidelines that might make your task easier.

1. *Introduce the questionnaire carefully.* Though you will not want to begin with a lengthy explanation for the purpose of the questionnaire, some concise, clear explanation is necessary. That information should explain who you are, what your position is, why you are asking for this information, what you might do with it, and — very important — when you need the information back. Be specific. Do not say, "Return as soon as possible." Give a date. Also, whenever possible, send a self-addressed, stamped envelope.

2. *Choose your format carefully.* There are a number of formats to choose from. A combination of formats is often most effective. When it is important to be able to tabulate and calculate with a large number of responses, the most convenient options are usually either-or questions:

	True	False
1. I use the microwave most often to prepare snacks.	____	____
2. I use the microwave most often to prepare meals.	____	____

multiple-choice questions:

1. I use the microwave to . . .
   (Please check all those that apply.)
   _____ prepare snacks
   _____ prepare meals
   _____ reheat leftovers
   _____ warm beverages

or questions asked with a Likert scale:

	Often					**Never**
1. I use the microwave to prepare snacks.	1	2	3	4	5	6
2. I use the microwave to prepare meals.	1	2	3	4	5	6

If you want more complete, and possibly more thoughtful, answers you can ask for written answers, either brief (sentence length) responses or longer and more thorough responses.

As a rule of thumb, respondents are less likely to take the time to reply to written answers; therefore, unless you need the kind of information you get with written answers, use a more objective format.

3. *Make the questions precise and unambiguous.* Remember that you won't be able to correct any misunderstandings until it's too late.

4. *Always have the questionnaire reviewed before using it.* Before sending out the questionnaire, ask a colleague to review its quality. Once a questionnaire with a mistake goes out, you won't have the opportunity to correct misunderstandings until it's too late. Take every precaution possible to be sure that your questionnaire is perfect *before* you use it.

5. *Select respondents carefully.* Choosing people to receive the questionnaire and ensuring that they respond is essential if you want meaningful data. For the information to be helpful, respondents should be representative, which means you need a good number of them and they can't all be the same. To ensure a high response rate, you might contact them ahead of time, whenever possible.

Figure 16-1 shows a brief questionnaire that was sent to a large number of companies. The names and addresses of the respondents were provided by the marketing division of each company. The primary purpose of the questionnaire was to determine the level of need for services and what kind of services the company might anticipate. Like many questionnaires, this one also had a secondary purpose—in this instance, to announce the new remote diagnostic service. Included with the questionnaire was a brief cover letter. (A cover letter is usually necessary, especially when mailing questionnaires.) Both the cover letter and the questionnaire were brief because the longer the questionnaire, the less likely the busy reader is to complete and return it.

Note that the questions are clear, concise, and short. Although some

---

Figure 16-1. Example of a Cover Letter and Questionnaire

February 20, 199X

Dear Customer,

Breakdowns of computers, facsimile machines, and voice processing equipment can now be repaired over the telephone, without a service technician in sight.

Our new remote diagnostics service enables our engineers to repair machines anywhere in the world, usually within 20 minutes, by two-way electronic communication. A customer's malfunctioning machine ''reads'' its operating instructions and last 20 transactions by phone to a similar machine at a diagnostics center. Engineers identify the problem via error codes and make software changes, which are automatically relayed to the customer's machine.

In order to help us serve you better, we would appreciate your response to the enclosed questionnaire. While the questionnaire might take a few minutes of your time, we believe that your answers will help us improve the efficient and effective service you have come to expect from us.

Thank you,

The Xavier Corporation

Figure 16-1. Example of a Cover Letter and Questionnaire (Continued)

Customer Needs Survey

1. Does your company/organization use one or more of the following? Check all relevant items.

   computers _____

   facsimile machines _____

   voice processing equipment _____

2. Approximately how many of the following does your company/organization have?

   computers _____

   facsimile machines _____

   voice processing equipment _____

3. How frequently do the items have to be repaired? Once a week? Once every two weeks? Once a month? Once every six months?

   computers _____

   facsimile machines _____

   voice processing equipment _____

4. What is the name of the individual responsible for maintenance of

   computers?_____

   _____

   facsimile machines?_____

   _____

   voice processing equipment?_____

   _____

questions can be answered by a yes or no, others require more thought on the part of the respondent.

On-Site Visits

If you cannot get the information you need from interviews, letters of inquiry, or the library, you might need to visit a site to get information first-hand. Although not as common a method of collecting data and conducting research as interviews or letters of inquiry, on-site visits can be a highly useful method of gathering information. If you are preparing a report on a conflict that has arisen at a construction site, you will probably not be able to rely on those at the site to gather impartial information. You will likely have to visit the site yourself. If you do visit a site, here are some guidelines to help you.

1. *Take copious notes.* Do not be selective in recording impressions and details. The more information you gather, the more you will have to work with later on.

2. *Consider using a tape recorder or a laptop computer.* For some people, a tape recorder is easier and quicker than a pen or pencil. Although relatively new as a tool for collecting information during on-site visits, a laptop computer can be easier and faster than either a tape recorder or a pen and pencil. A laptop has the added convenience of allowing you to manipulate information more readily once you are back in your office.

Observations

In presenting your observations of the world around you, your writing must convince your readers of two things: (1) that you carefully recorded what you observed, and (2) that you are a qualified observer under the conditions you reported and of the phenomena you observed. When you report on some aspect of the world around you, the quality of your information depends on a variety of conditions more or less beyond your control. The information provided by the field worker in wildlife and by the advertising student conducting a sidewalk survey is affected by weather, time of day, and the availability of subjects to observe. Readers will understand and use your information if they know not only what you observed, but the conditions under which you observed it. Thus, the successful presentation of observations requires thoughtful preparation in the field.

While gathering information, be certain to record all details that could have affected your observations. In addition, you must be prepared to list and, depending on your audience, describe in detail any equipment that aided you in your observations. Finally, report any conditions that might have altered the quality of your observations, and discuss the expected effect of these conditions.

Controlled Experiments

When the subject of your writing is an experiment whose conditions you established and controlled, recording the conditions of the experiment is even more critical because your goal is to allow readers to reproduce the experiment itself. Without an adequate description of the preparations for the experiment, the reader's results are quite likely to be different from yours, and your credibility will be damaged.

Insofar as the laboratory conditions are yours to determine and may be recorded before the actual observation takes place, there is no excuse for inadequate description in your lab notebook, which becomes the ultimate record of detail and the source for any document that may result from the experiment.

Activities

1. As a group, practice using tagmemics and Burke's pentad as investigation tools.
   a. Do a tagmemic analysis (particle-wave-field) for the topics listed below. What part of the question set seems easiest to answer? What conclusion does that lead you to? Outline a descriptive paper that might result from your questions.
      (1) Your best friend
      (2) Technical writing classes
      (3) A form of pollution
      (4) A major city near you
      (5) Part-time jobs
   b. Use Burke's pentad to investigate these topics. First, have each person carefully analyze the ratio of the topic. Then, see if others in the group arrive at the same ratio. Complete the investigation as a group.
      (1) Your part-time job
      (2) Factory automation
      (3) Groundwater pollution
      (4) The last party you went to
      (5) Medical terminology
      (6) The computers in a college lab
2. Try brainstorming any of the topics below. Brainstorm first in a group.

With the help of three or four people, list every idea you can on a sheet of paper. Then try invisible writing to see what you can say about the subject without the distractions of words on the screen.

   a. Behaviors of a species

   b. The best movie ever produced

   c. A law of physics that needs changing

   d. A new legal voting age

3. Write a one-paragraph report for your classmates describing the general references and journals that professionals in your field read.

4. The student government at your school is preparing a set of comprehensive recommendations to be presented to the administration. You have been asked by the student government to prepare a report on the state of computerization of your library. Does the library have an on-line catalog? If it has electronic indexes, what are they? If your library has on-line databases, is the service limited in any way?

5. Choose one of the topics listed below, and prepare a report to be delivered orally to the class. For this assignment you will have to conduct on-site visits. Remember to take copious notes, and consider using a tape recorder or, if possible, a laptop computer.

   a. Waste disposal

   b. Water quality

   c. Food service

   d. Power plant emissions control

17

Analyzing Information

Checking Completeness
  Conventional Standards
  Heuristic Techniques
  Statistical Significance

Developing Conclusions
  Standard Procedures
  Spreadsheets
  Spreadsheet-Based Graphics

Checking Conclusions for Reasonableness
  Evaluating Sources
  Checking Procedures
  Checking Data Sufficiency
  Considering Alternative Conclusions

Activities

It would be nice if the research process could be described as a simple set of steps—first read for a while or do a study, then write it up. This formula is very neat, but not very helpful. It doesn't tell the researcher how long to read or what to study. Nor does it give the researcher any idea of when to stop gathering information. This formula also omits a crucial step—making sense of the information that has been gathered.

Let's examine information analysis in relation to several typical writing tasks. Suppose you are asked to review the materials that have been published about water pollution in a particular soil type. How do you know when you have enough information to give an informed summary? How do you know you have understood all the implications of the research? Or perhaps you are doing an attitude survey. How can you be confident you are interpreting the results correctly? The same situation might arise in the course of doing a feasibility study. You may believe you have arrived at a feasible solution, but how can you be sure you are not missing something?

All of these situations can be summed up as three questions. During the course of doing research there will be moments when you will need to stop and ask, (1) Have I taken a complete look at this problem? (2) What is the best conclusion I can come to about the data in front of me? (3) Is my conclusion reasonable? This chapter addresses all three questions.

Checking Completeness

Humans commonly form conclusions on the basis of partial information. We select restaurants without knowing much about them. Students often select a college based on incomplete information. Companies may make purchasing decisions without knowing all they could know. Even people conducting research may miss some things because they run out of time or money. Sometimes being incomplete is unavoidable.

But there are times when the research must be perfect, and when time and money and skill are available to make it perfect. In these cases, how can a writer be sure he or she has all the information available before forming a conclusion? Several techniques are available.

Conventional Standards

One check for completeness that experienced writers rely on is a thorough understanding of the kind of inquiry undertaken. For example, over many years feasibility has come to be defined in three ways: operational feasibility, technical feasibility, and economic feasibility (see chapter 23). By understanding the conventional standards for feasibility decisions, writers know then their research is complete — when they have examined all three kinds of feasibility.

Literature reviews also have their conventional standards. For each field of study there is an index that covers the primary works in the field. Also, there are expectations about the span of years that a researcher will cover (see chapter 20 for detailed descriptions). A writer who follows these standard practices can be fairly sure of being complete.

Thus, one way a writer can check for completeness is to ask, do I understand the standard approaches taken to the kind of research I am doing, and have I followed them? If the answer is yes, one measure of completeness has been reached.

Heuristic Techniques

Chapter 16 described a number of questioning techniques, or "heuristics." Each technique is useful in exploring a new subject, and each can also be used to check for completeness. For instance, tagmemics explores a subject by asking a series of questions tied to particle, wave, and field, or isolation, change, and comparison. Those questions could be used in the beginning stages of research, but they could also be used near the end to ensure completeness. To conduct this check we might ask questions such as these:

What makes *x* special?

How is *x* important?

What are the major features of *x*?

What are some of the forms *x* can take?

What is the most unique form of *x*?

Is *x* different from what it used to be?

Will *x* change in the future? How?

What makes *x* different from *y*?

What else relies on *x*?

What causes changes in *x*?

If you had spent a week reading about a parasite that afflicts white-tailed deer, you could ask yourself each of these questions at the end of the week to see how well you understood your subject. Any question you couldn't answer would be an indication of a general area of research you may have missed.

A second simple heuristic technique is Burke's pentad. This is the technique that asks the five w's (who, what, when, where, why), plus ratio and circumference. For a writer researching a particular activity like data communication, Burke's pentad suggests approaching the research by asking about the people involved in the data communication (*who*), by defining *what* it is, by asking *where* and *when* it occurs, by discovering *why* it is done as it is done, and by reviewing *how* data are communicated. Each of these is an important avenue to research. And, for a full exploration of the subject, each of these should have been reviewed by the time the research was complete. If any has not been looked at, the writer still has some work to do.

Statistical Significance

The completeness checks above are primarily for qualitative information. They essentially say a writer is done gathering information when he or she has done all that is commonly done in a particular kind of report; or, if there are no set practices, a writer is done after exploring all the approaches suggested by standard heuristic techniques.

For quantitative information, the standards are different. For instance, in a study on employee responses to various computer screen displays, the question of completeness would depend on statistics. How many employees should be tested as part of the research? That is totally a function of the research design being employed. Some statistical approaches might be possible with as few as six subjects, whereas other approaches might require twenty or more subjects. This kind of research is complete when enough subjects have been studied to supply a statistically significant result. That number should be known by the writer before the study is begun, since it is an essential part of the overall research design and will determine scheduling and other matters.

Developing Conclusions

As data are gathered, some conclusions will become obvious. At other times a writer will need to decide how to interpret the data. There are several approaches a writer can take to form valid conclusions.

Standard Procedures

In some cases conclusions are reached following standard procedures. A good example is the feasibility study. Companies often have a set standard for economic feasibility, such as a three-year payback period. If a particular approach has a payback period of less than three years, then it is economically feasible. Such standards are available in many situations.

Statistical inference is also made at standard levels. For much research a significance level is set at $p < 0.05$, in some cases at $p < 0.01$. Writers experienced in a certain kind of research quickly learn these standards and base conclusions on them.

Spreadsheets

Often research will turn up masses of numbers. Research in sociology might produce hundreds of income levels. A psychology test might produce a table of reaction times. A discussion of tensile strength may end with a list of numbers.

Although numbers are precise and can clarify findings, numbers themselves may be confusing. For example, Figure 17-1 shows a table of information on gross hourly incomes for various industries. By looking at the set of numbers, what conclusions can you draw?

These figures are from the U.S. Census Bureau and have been adjusted to remove the effects of inflation. Even with all the work the Census Bureau has done, it is not easy to determine a pattern in these numbers. Some industries seem to have done better in the past twenty years, while others have done worse. Analyzing these numbers will not be easy.

Enter the spreadsheet. While originally designed for basic accounting, the spreadsheet is a simple computer program that puts numbers or labels in rows and columns and can process numbers of all sorts, including the data in Figure 17-1. For instance, one way to help understand the wage figures above might be to sort them. We could list the industries from best-paying to worst-paying. In fact, we could do this twice, once based on 1970 earnings and a second time based on 1990 earnings, to see if there is much change. Manual sorting of the data is time-consuming, but all spreadsheets have

---

Figure 17-1. Unsorted Table

Hourly Wages (in 1977 US Dollars)

Industry	1970	1980	1990
Manufacturing	5.23	5.34	5.42
Mining	6.01	6.74	6.86
Construction	8.17	7.30	7.15
Transportation	6.01	6.52	6.55
Wholesale trade	5.37	5.11	5.29
Retail trade	3.81	3.59	3.47
Finance, insurance	4.79	4.25	4.50
Services	4.38	4.30	4.51
AVERAGE	5.04	4.89	4.91

Source: U.S. Bureau of the Census. (1991). *Statistical abstracts of the United States* (p. 413). Washington, DC: U.S. Government Printing Office.

Figure 17-2. Spreadsheet Table after Being Sorted

Hourly Wages (in 1977 US Dollars)
Sorted by 1970 Hourly Wages

Industry	1970	1980	1990
Construction	8.17	7.30	7.15
Mining	6.01	6.74	6.86
Transportation	6.01	6.52	6.55
Wholesale trade	5.37	5.11	5.29
Manufacturing	5.23	5.34	5.42
Finance, insurance	4.79	4.25	4.50
Services	4.38	4.30	4.51
Retail trade	3.81	3.59	3.47
AVERAGE	5.04	4.89	4.91

Source: U.S. Bureau of the Census. (1991). *Statistical abstracts of the United States* (p. 413). Washington, DC: U.S. Government Printing Office.

built-in sorting capability, so that the numbers could be sorted quickly. Figures 17-2 and 17-3 show the same numbers in Figure 17-1, after they have been sorted by a spreadsheet.

With the help of spreadsheet sorting we can now answer questions about individual industries much more easily. We can see at a glance not only which industries paid best and worst but also how they fared in the past twenty years. For instance, while construction paid the best and retail trade the worst in both 1970 and 1990, most of the other industries changed relative positions. Mining and transportation tied for second place in 1970, but mining improved much more than transportation. Manufacturing wages improved in relative position, as did services, while wholesale trade and finance wages faded. A simple sort helped us quickly make some comparisons between industries.

Besides repositioning information to facilitate our analysis, spreadsheets also do many calculations automatically. We have already seen how they can add up a column of numbers. We could just as easily have used a spreadsheet

---

Figure 17-3. Spreadsheet Table after Being Sorted

```
        Hourly Wages (in 1977 US Dollars)
           Sorted by 1990 Hourly Wages

   Industry              1970    1980    1990

   Construction          8.17    7.30    7.15
   Mining                6.01    6.74    6.86
   Transportation        6.01    6.52    6.55
   Manufacturing         5.23    5.34    5.42
   Wholesale trade       5.37    5.11    5.29
   Services              4.38    4.30    4.51
   Finance, insurance    4.79    4.25    4.50
   Retail trade          3.81    3.59    3.47
   AVERAGE               5.04    4.89    4.91
```

Source: U.S. Bureau of the Census. (1991). *Statistical abstracts of the United States* (p. 413). Washington, DC: U.S. Government Printing Office.

to calculate the exact amount of change in wage rates between 1970 and 1990. This involves entering a simple formula (column C minus column B), which is then applied to the figures. Figure 17-4 shows what those calculations look like on a spreadsheet.

Spreadsheets were designed specifically to perform that kind of calculation; they do it quickly and easily. Now that the mathematics has been done for us, we can see exactly how much change in wages occurred in the past twenty years. To make these changes even more meaningful, we can sort the table one more time, this time sorting by amount of change. Figure 17-5 shows the results.

With the third sort we can see that whereas construction paid the best in 1990, it fell the most in wages paid. Wages in the mining and transportation industries, tied for second place in 1970, increased the most over the period. Retail trade, which paid the worst in both 1970 and 1990, was also the

Figure 17-4. Spreadsheet Table after Calculations

```
        Hourly Wages (in 1977 US Dollars)

    Industry                1970    1990    Change

    Construction            8.17    7.15    -1.02
    Mining                  6.01    6.86     0.85
    Transportation          6.01    6.55     0.54
    Manufacturing           5.23    5.42     0.19
    Wholesale trade         5.37    5.29    -0.08
    Services                4.38    4.51     0.13
    Finance, insurance      4.79    4.50    -0.29
    Retail trade            3.81    3.47    -0.34
    AVERAGE                 5.04    4.91    -0.13
```

Source: U.S. Bureau of the Census. (1991). *Statistical abstracts of the United States* (p. 413). Washington, DC: U.S. Government Printing Office.

second-worst loser in terms of average wages. This chart provides a new way of looking at information buried in the original numbers.

With a spreadsheet we can manipulate numbers quickly and easily until they have been repositioned and recalculated enough to be meaningful. Clearly, the spreadsheet is a powerful tool for analysis and interpretation.

When should you use a spreadsheet? First, spreadsheets are useful only for numbers. Second, they are more useful for analyzing a large amount of data than for analyzing two or three numbers. If you encounter only a few numbers in your research, a spreadsheet is not needed to analyze those numbers; you can do it mentally just as well.

A spreadsheet is valuable when you face an array of numbers and have one of two problems: no clear pattern is visible in the numbers, or you want to manipulate the numbers to reveal trends. Each approach will help you form conclusions about research data.

Figure 17-5. Spreadsheet Table after Calculations and Sorting

---

Hourly Wages (in 1977 US Dollars)

Industry	1970	1990	Change
Mining	6.01	6.86	0.85
Transportation	6.01	6.55	0.54
Manufacturing	5.23	5.42	0.19
Services	4.38	4.51	0.13
Wholesale trade	5.37	5.29	-0.08
Finance, insurance	4.79	4.50	-0.29
Retail trade	3.81	3.47	-0.34
Construction	8.17	7.15	-1.02
AVERAGE	5.04	4.91	-0.13

Source: U.S. Bureau of the Census. (1991). *Statistical abstracts of the United States* (p. 413). Washington, DC: U.S. Government Printing Office.

Spreadsheet-Based Graphics

Spreadsheets, which have the capacity to sort, reposition, recalculate, and manipulate numbers, are also used for graphics. When the spreadsheet work is done, the numbers should be easier to understand. They are still numbers, however, and for people and some applications, this is a problem.

In 1981 Jonathan Sachs and Mitch Kapor took the next step and wrote a spreadsheet, Lotus 1-2-3, that not only manipulates numbers but also converts them to graphics. The user sees not only the numbers themselves, but also these numbers reconstructed as bar graphs, pie charts, line graphs, and more. Besides having the computer calculate the difference between two numbers, the user can have it draw the difference.

The ability to graph numbers is crucial in presenting information (see chapter 8). But graphics can also be useful in analyzing data and forming conclusions. To understand the usefulness of graphics in interpreting data, study the following data on alcohol consumption and the four graphs that follow.

This spreadsheet table combines two sets of information: the effects of

Figure 17-6. Spreadsheet Table

Frequency of Alcohol Consumption

	Age			Income (in thousands)		
	20-34	35-54	>55	<$6-	$6-$15	$15-$30
Less than once per year	22	28	45	46	38	27
Less than once per week	36	35	28	29	32	38
Once or twice per week	18	14	9	13	14	13
Three or more per week	24	23	18	13	16	21

Source: U.S. Bureau of the Census. (1991). *Statistical abstracts of the United States.* Washington, DC: U.S. Government Printing Office.

age and income on drinking. We can discern several trends by looking at the table: younger people seem to drink more than older people, and poor people drink less than middle-income people. However, the differences would be much clearer if the data were shown pictorially. Most graphics programs will do just that with spreadsheet information. For example, let's take one age group, 20- to 34-year-olds, and make a bar graph of their drinking habits. Figure 17-7 shows the result.

Now we see clearly that there is not an orderly transition from those who never drink to those who frequently drink. The bars make it obvious that a large group of people drink only on special occasions. We also see that only one fourth of the people in this group could be called "regular" drinkers. We could also derive this information from the spreadsheet, but the bar graph makes the data easier to interpret.

Another graph that clarifies information is a pie chart, which shows not raw numbers but percentages of a whole as pieces of a pie. We can list the numbers of people for each age group; or we can say that of the total pie of heavy drinkers, one slice represents young people, a second slice represents middle-aged people, and a third slice represents older people. Figure 17-8 shows a pie graph generated by the spreadsheet program.

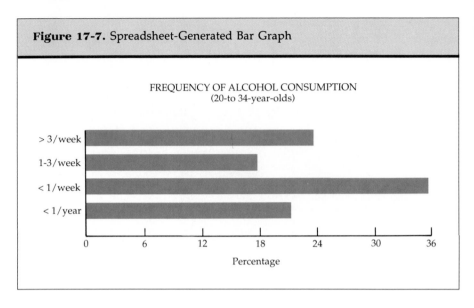

Figure 17-7. Spreadsheet-Generated Bar Graph

FREQUENCY OF ALCOHOL CONSUMPTION
(20-to 34-year-olds)

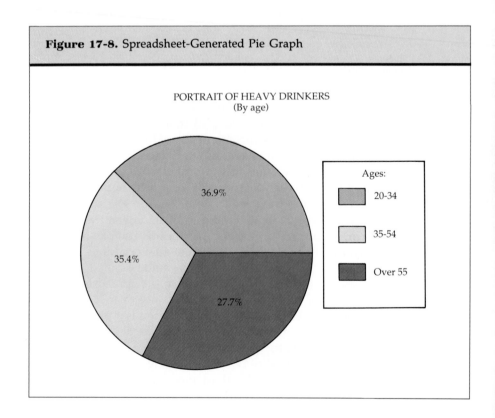

Figure 17-8. Spreadsheet-Generated Pie Graph

PORTRAIT OF HEAVY DRINKERS
(By age)

From this pie chart, it seems that the slices for each of the three age groups are roughly equal. The percentages vary slightly, but there is no preponderance in any one age group. We could have obtained roughly the same information by looking at the spreadsheet, but the graph reinforced and clarified what the spreadsheet showed.

Sometimes line graphs can be used to plot two different trends to see how they offset each other. In our case we might be interested in knowing how the trends look for heavy drinkers, as measured by both age and income. One quick way to discover this would be to create a line graph with one line for income and another for age. Where the lines meet should tell us what ages and income are equally likely for heavy drinkers. This is something our spreadsheet cannot show us, but a graph can quite easily. Figure 17-9 shows a spreadsheet-generated line graph for heavy drinkers.

The line graph shows the relative effect of two trends quite clearly. The lines also let us project out for ages not on the spreadsheet. As people age, what happens to their alcohol consumption? What happens as they get

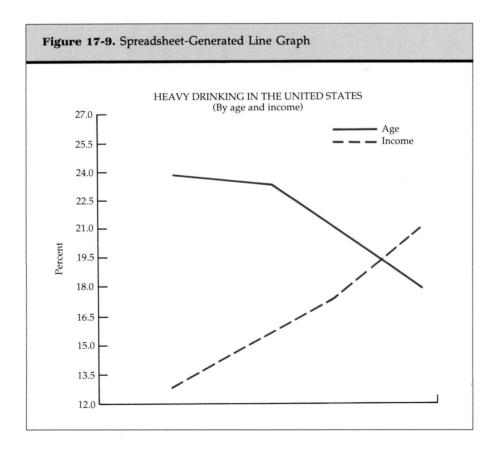

Figure 17-9. Spreadsheet-Generated Line Graph

wealthier? The lines created by the graph make general trends clear. So line graphs can be used not only to make information easier to see than it might be on a spreadsheet but also to allow easy viewing of multiple trends and long-term trends.

Graphs also help clarify information by showing the combined effects of two different trends. In our example, we now know the effect of age and income, but what is their combined effect? Is a young, poor person more likely to drink than an older, wealthy person? The answer to such questions can usually be found through the use of a spreadsheet-generated layer graph. It adds the effect of one trend on top of the other and colors them so we can see the combined influence of two or more trends. The spreadsheet-generated layer graph in Figure 17-10, for example, shows the combined effect of age and income on heavy drinking.

In this case the combined effects produce an almost equal result. There is slightly more heavy drinking when we add high income to old age than there

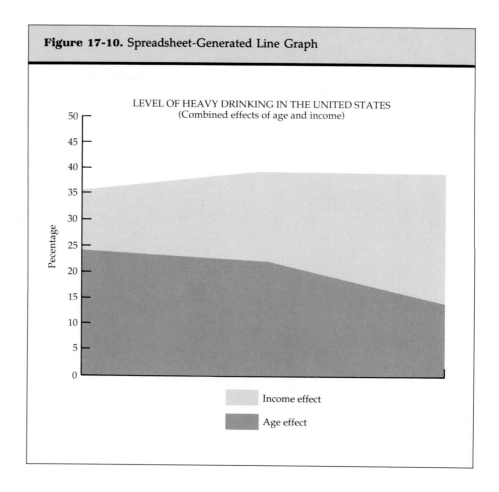

Figure 17-10. Spreadsheet-Generated Line Graph

is when we add low income to youth, but the difference is not very great. It appears that the trends almost totally offset each other. Layer graphs let us see this where spreadsheets would only do it with difficulty.

Are graphs essential to an understanding of numerical information? No. By studying spreadsheets carefully enough, we could obtain most of the information that we would otherwise get from graphs. However, graphing multiple trends does show certain combined effects that we could detect only with difficulty by spreadsheet analysis, and even simple bar graphs reinforce any conclusions we might make about the data. Graphs make complex blocks of numbers much easier to understand.

Checking Conclusions for Reasonableness

Being certain that a conclusion is valid is not always easy. Many errors can occur during the process of collecting information or forming a conclusion. Here are four reviews a writer might use to check the validity of a conclusion.

Evaluating Sources

A conclusion will be no more valid than the information it is based on. Therefore, one valuable check consists of evaluating the quality of sources used. This can be done informally by asking more experienced people how they feel about a particular journal. Some journals are particularly rigorous in their review procedures, while others accept most of the articles they receive. Knowing this, a writer should have some sense of how strong a conclusion can be based on a single journal article. If found in a strongly refereed journal, one article might be sufficient evidence; if found in a lesser journal, one article might not be enough.

A more formal evaluation process can be based on the number of times a particular article or author is cited in other articles. This will not be a total evaluation of quality, but if an author is rarely cited by other authors, it is reasonable to question whether the results given by that author should be used as the sole foundation for your own studies.

Checking Procedures

If the results of a study seem unusual, there are two ways in which procedures could be checked. The first is to redo a study to see if the results come out the same. It may be that the first results were based on flawed procedures or errant materials. The second way is to question the kind of procedures used. Were they the procedures that are normally used? Were they not common practice? Any error in the procedures used will invalidate conclusions based on the results.

Checking Data Sufficiency

Statistical procedures usually explicitly state the number of subjects or samples required for validity. Data sufficiency may be more subject to judgment if statistics are not being used. In this case the question is, would a reasonable person form the same conclusion based on this evidence?

Take the case of observations. It may be very difficult or time-consuming to make many or lengthy observations. Therefore, the question for anyone making conclusions based on observations would be, was enough observed at an appropriate time and in an appropriate manner to support this conclusion? That is a good question to put to a senior colleague before forming a conclusion and releasing recommendations.

Considering Alternative Conclusions

If reliable sources have been used, standard procedures followed, and sufficient data collected, the question becomes, are there other conclusions that could be made about the data? If so, what are they, and which is most valid?

Finding alternative conclusions requires being open to alternatives. We all know people who will never change their minds about anything. They know what they know and will not consider alternatives. Re-seeing information from a new perspective is a trait of the highly intelligent. But even for the highly intelligent it may not be easy. However, several techniques can be used to re-see and consider alternatives.

One technique people routinely use is to take time. A distance of a day or a week can make a great difference in how things look. A night's sleep almost always helps. Another is talk. Discussing a problem with others not only allows them to suggest alternatives but, in the process of presenting the situation, the researcher often sees new approaches. Just putting the problem into words creates new perspectives. A third approach is a kind of negative advocacy. The writer tries not to prove the conclusion, but to prove a different conclusion based on the same data. The effort required to defend an approach that would not otherwise be taken can sometimes help a writer see yet another interpretation of the data.

Activities ━━

1. Pick one of the following report subjects. Then use tagmemics to make a list of topics you would expect to see covered in such a report in order for it to be complete.

a. Deforestration

b. Ozone depletion

c. Safety precautions for medical practitioners

d. Engine maintenance

2. Photocopy the "Conclusions" section of a technical report. Working in small groups, compare the conclusions for reasonableness. For the best reports, why did you accept the conclusions as reasonable? For the weaker reports, suggest additions that could be made to support the original conclusions, or identify other conclusions that could be reached.

3. The table in Figure 17-11 compares cigarette smokers. List at least ten observations you can make from the data. Use a spreadsheet to calculate changes over time. List at least five additional observations you can make now.

4. Line graphs are a good way to compare two trends over time. List at least four trends you could graph from the table in Figure 17-11. Use a graphing tool to draw at least one pair of lines. What do you now notice about the trends? Would you have made the same observations without the graphs?

5. List the journals available in an area of inquiry such as environmental education or quality control. Which are the best journals in the field? How do you know?

Figure 17-11. Cigarette Smoking (% of Total Population)

Number of cigarettes per day	1970		1980		1990	
	Males	Females	Males	Females	Males	Females
1–4	11.5	11.8	8.7	10.2	8.3	10.1
15–24	18.4	12.9	15.2	12.1	15.0	13.2
25–34	5.1	3.0	5.3	3.2	5.2	3.1
>35	6.4	2.5	6.7	3.4	5.9	2.5

18

Organizing Information

Content-Driven Organization
Chronological
Spatial
Order of Importance
General to Specific
Specific to General
Problem – Cause – Solution

Format-Driven Organization

Using Templates

Activities

You've conducted research about your subject, learning as much as you can about it. You've analyzed the information using the most appropriate tools available. The time has come to organize all that information. Organizing information means arranging the information you have gathered in a way that allows the reader to understand — and, if necessary, agree with — the information you are presenting.

The proper organization is the one that is the most effective for your audience and your purpose. A good organization allows readers to find whatever information they want quickly and easily, but it also allows writers to present information as convincingly as possible. As you begin developing the organization for a document, you will find that organization can be thought of as *content-driven* or as *format-driven*. Content-driven organizations are determined mainly by the information you have gathered; format-driven organizations have evolved within various companies and professions.

Content-Driven Organization

Content-driven organizations are determined primarily by the information the writer has gathered. That is, some subjects, on the basis of the material, the audience, and the purpose, lend themselves logically to particular structures. For example, if you are providing a set of instructions for setting up a printer, you will likely present those instructions in the sequence in which the operations will be performed; if, as in the example below, you are preparing a brochure in which you describe the six most important dietary principles, you might choose to list the principles in the order of their relative importance.

The following sections describe six common methods for organizing technical information.

Chronological

In a chronological organization, events are presented in the order in which they occur. Typical writing tasks that follow a chronological organization are instruction manuals that describe a process, and, as in the narrative of the *Challenger* accident in chapter 1, descriptions of events. Figure 18-1 shows the chronological organization used for a technical process description: The authors describe instructions for installing and testing a modem.

Figure 18-1. Chronological Organization of an Instruction Manual

---

1. Assembling Materials and Equipment
   a. A personal computer
   b. Your modem
   c. The 9.5 volt (AC), 2-amp, 18-watt power cord with a transformer at one end
   d. A modular telephone cable
   e. Either a modular telephone jack or a multi-line adapter
   f. A "straight-through" RS-232C interface cable
   g. A screwdriver
2. Detailed Installation
   a. Run CINSTALL before installing your modem
3. Connecting the Modem to the Computer
   a. Serial ports
   b. The serial cable
   c. Making the connections
4. Connecting the Telephone Cables
   a. Two-line telephones
   b. Multi-line telephones
   c. If your telephone doesn't have the right type of connector
   d. Connecting the telephone to the modem
   e. Connecting the modem to the telephone line
5. Testing Your Installation
   a. Testing the modem
   b. Testing the telephone connection
   c. Finding the modem
   d. Testing the telephone connection

Source: *Evercom 24E user's manual.* (1990). Fremont, CA: Everex.

Spatial

In a spatial organization, objects or processes are presented according to the physical arrangement of their parts. Spatial organizations are commonly used in descriptions of a mechanism or a place. The most typical writing techniques that use spatial organization are technical description and technical process description; the most typical documents that use spatial organization are research reports, feasibility studies, and progress reports. The outlines may be organized from left to right, from east to west, from top to bottom, from inside to outside or the opposite of any of these. Regardless of the organization you choose, the important thing is to pick a direction and stick with it.

Figure 18-2. Spatial Organization of a Technical Description

---

   1. Lawn Area (Farthest West)
     a. multipurpose grass areas
     b. rows of old plane trees
     c. warehouse courtyards
     d. 10-foot wide allees
     e. turfed stairs
   2. Parterres (Center)
     a. didactic gardens
     b. labyrinth
     c. neoclassical pavilion
     d. rows of grapes
   3. Romantic Garden (Farthest East)
     a. grotto
     b. canal
     c. circular lake
     d. restaurant
     e. terraces

Source: Clemons, M. (1989, January). Gardens of memory. *Landscape Architecture, 79,* 30.

Figure 18-2 shows the spatial organization used for a technical description of a park. The writer helps the reader visualize the park by carefully moving in only one direction, from west to east.

Order of Importance

When using an order-of-importance organization, the writer ranks the importance of each item and then presents them in order: commonly from most important to least important, less commonly from least important to most important. Organizing a report or a proposal by order of importance is useful when the writer is making a series of points or recommendations and those points or recommendations are not all equally important. This type of organization is especially helpful when the purpose of the document is to inform readers or help them make a decision. Typical writing tasks that would use an organization ranking items by importance are documents in which the writer reports solutions to a problem, as in research reports and progress reports, or is trying to convince the reader to follow a particular course of action, as in proposals and feasibility studies.

Figure 18-3 shows the order-of-importance organization used in a medical pamphlet on the role of diet in maintaining health. Because the writer of the pamphlet wanted to emphasize the fact that some dietary principles are

---

Figure 18-3. Order-of-Importance Organization of a Pamphlet

---

    1. Reduce fat and cholesterol intake.
      a. avoid egg yolks
      b. avoid whole milk dairy products
      c. avoid organ meats
    2. Up your intake of complex carbohydrates.
      a. spaghetti
      b. beans
      c. peas
      d. potatoes
      e. whole grain bread
    3. Be aware of how much salt (sodium) you are ingesting.
    4. If you must, drink alcohol in moderation.
    5. Emphasize vitamin A- and C-rich vegetables and fruits.
    6. Emphasize cruciferous vegetables.
      a. cabbage
      b. broccoli
      c. brussels sprouts

Source: The 5 most important dietary principles. (1989, fall/winter). *Positive Health Decision*. Berlex Laboratories.

more important than others, she prefaced the description of the five principles with the following statement: "Beginning with the primary principle, the six most important dietary principles are. . . . "

General to Specific

A general-to-specific organization is good for situations in which the audience might be interested in having the general statement placed first and does not want to read the entire document to find out the recommendation. Proposals and progress reports often use a general-to-specific organization.

Figure 18-4 shows part of the organization of a technical report on a new method for holding thin semiconductor samples for sectioning. The purpose of the report is to convince the reader that the new method is superior to current methods. The writer begins the report with a general statement about the new method and then proceeds to offer a description of the method.

Specific to General

When using specific-to-general organization, also called simple-to-complex organization, the writer begins with specific items of information and from them builds a general picture. Figure 18-5 outlines a proposal for a new office

---

Figure 18-4. General-to-Specific Organization of a Persuasive Report

---

1. Brief overview of the new method
2. Need for a new method
3. Disadvantages of current technologies
   a. Hinders full potential
   b. Results in surface damage
4. Advantages of the new method
   a. Sample is easily removed after polishing
   b. Results in damage-free polished product
   c. Eliminates chipping of the finished edge
   d. Eliminates torsional stress
5. Description of the device

Source: U.S. Department of Commerce. (1986, May). *Apparatus for sectioning demountable semiconductor samples*. Tech Notes. Washington, DC: U.S. Government Printing Office.

Figure 18-5. Specific-To-General Organization of a Proposal

---

1. Physiological Factors in Fatigue
   a. energy spending rates
   b. reserves
   c. planning the rest periods
2. Variations in Performance
   a. starts out cold
   b. warms up to the task
   c. increases in productivity
   d. another warm-up period after lunch
   e. brief period of high productivity
   f. productivity decreases to end of work period
3. It's Not All Muscular Fatigue
   a. routine jobs
   b. high-concentration jobs
   c. that "tired" feeling
4. The Payoff
   a. workers will take breaks anyway (informally and inefficiently)
   b. schedule routine breaks

Source: Bosticco, I. L. M, & Andrews, R. B. (1960, January-February). Is worker fatigue costing you dollars? *Technical aids for small manufacturers*. Washington, DC: Small Business Association.

procedure. To build up to their recommendations, the writers first present detailed descriptions of worker fatigue and then lead up to the general conclusion: because worker fatigue is costly to business and industry, managers should schedule short breaks routinely throughout the day.

Problem – Cause – Solution

In a problem – cause – solution organization, the writer first states the problem, then states the cause, and then offers a solution. This organization is often used for proposals and feasibility studies. Figure 18-6 shows the problem – cause – solution organization used in a proposal for minimizing household hazardous waste.

Figure 18-6. Problem – Cause – Solution Organization of a Proposal

1. What's Toxic?
    a. paints and solvents
    b. vehicle fluids
    c. pesticides
    d. household cleaners and polishes
    e. miscellaneous items, such as batteries, art supplies, pharmaceuticals, and some cosmetics
2. Disposal Problems
    a. injuries to sanitation workers
    b. damage to sanitation equipment
    c. poisonous leachate at landfills
    d. toxic air pollution
    e. contaminated sewage sludge
    f. contaminated drinking water
3. What Should be Done
    a. use alternatives to oil-based paint
    b. use alternatives to toxic cleaning agents
    c. use alternatives to dangerous pesticides
    d. separate toxic material from ordinary garbage
    e. collect toxic material separately
    f. dispose of toxic material safely

Format-Driven Organization ━━━━━━━━━━━━━━━

Formats are standard plans for the organization and arrangement of information. They are essentially conventionalized, codified, idealized versions of the organizational patterns discussed above. Although they are built from organizational patterns, formats don't always follow a single pattern; more typically, they combine two or three. For example, instructions conveniently place any urgent warnings at the beginning of the document, where they are most visible. In other words, this format combines a chronological organization with an order-of-importance organization.

Formats have evolved in many companies and professions. An example of a company's format is shown in Figure 18-7. This format is required to be used by those taking meeting notes at one of the Limerick Generating Stations of the Philadelphia Electric Company.

The following format for a research report, recommended by the journal *Comparative Biochemistry and Physiology*, is an example of a format-driven organization that exists within a profession.

Abstract

Introduction/Presentation of the Problem

Methods and Materials

Results

Discussion/Conclusions

Literature Cited

Formats such as these are very common. One advantage is that they present information in a form that has been found useful over time. Because readers expect the material to be presented in a certain form, they can more easily scan the work to find the information of interest to them. Another advantage of formats is that they are logical; they follow one or more of the organizational principles discussed above. For example, an executive might first read the "Action Items" section of the *Meeting Notes* before either reading the "Discussion" section or checking to see who attended the meeting. Because the document follows a standard format, searching for particular parts is simple.

The format for the research report combines two organizational principles: order of importance and chronological order. The format begins with the "Abstract," which is the most important part of the report and will be read by most people, and ends with the "Literature Cited" section, which is the least important. The rest of the format follows the chronological order of the research process: define the problem, develop a strategy for investigation, obtain results, and draw conclusions.

Figure 18-7. Format for Notes Taken During a Meeting

CONTROL NO. _____

<u>MEETING NOTES</u>

<u>SUBJECT</u>:   Job No. _____ Sub_____ ,
              Limerick
              Generating Station   <u>DISTRIBUTION</u>
              Peach Bottom   *Attendees plus:

              _____

<u>DATE</u>:   _____

<u>LOCATION</u>:   _____

*<u>ATTENDEES</u>:   _____
                                           _____
                                           _____

<u>DISCUSSION</u>:

PLEASE LIST ACTION ITEMS AT THE BACK

The range of formats in technical writing situations is quite wide. Some formats contain only broad outlines of the information to be included in the document, such as the format above recommended for biochemistry and physiology. Other formats, such as the one required for the city of Appleton, Wisconsin, community development block grant proposals, are comprehensive; they include not only the broad outlines of the structure but the nar-

---

Figure 18-7. Format for Notes Taken During a Meeting (Continued)

```
                                Responsible
      Action Items:               Parties     Due Dates


        1. _____
           _____
           _____   _____   _____


        2. _____
           _____
           _____   _____   _____


        3. _____
           _____
           _____   _____   _____


        4. _____
           _____
           _____   _____   _____











                                Prepared by:


                                (Name of Originator)

      Written Response Req'd: Yes or No


      / / (  ):
      CFD/31
```

Source: Limerick Generating Station, Philadelphia Electric Company. [Meeting notes].

rower areas in between (Fig. 18-8). This format represents the funding agency's order of priorities for the information.

Although the organizations of these documents are limited to a great degree by the conventions of the company or profession, often the writer will still have some flexibility. For example, in the description of the project in the

---

Figure 18-8. Required Format Specified in a Request for Proposal

---

    A. Activity
    B. Amount of funds requested
    C. Name of organization
       1. Local address
       2. Local contact person
       3. Headquarters address
       4. Headquarters contact person
    D. Description of project
    E. National objectives
    F. Describe how the objectives of the project will be measured
    G. Detailed budget
       1. Personnel costs
       2. Training materials & supplies
    H. Neighborhood
    I. Public service
    J. Description of organization
       1. Organization
       2. Board of directors
       3. Resumes
       4. Other projects completed

---

community development block grant proposal, the writer can decide what information to present and in what order. While formats often determine the *overall sequence* of a document (e.g., what a proposal looks like), the author is usually left to decide the organization of the parts *within* the document. For example, there are often formats for technical writing tasks, such as proposals, but rarely for technical writing techniques, such as a description within the proposal.

The "Technical Writing Tasks" section of this book presents typical formats for technical documents. Ultimately, however, unless you are required to use a specific format, the organization of a piece of writing should be determined by the material, the audience, and the purpose. If you are to write a series of similar documents, it will be efficient to develop your own formats.

Using Templates

An outline is a kind of blueprint, showing a condensed description of the contents of a document. Having an outline helps the writer clarify his or her ideas and maintain focus. It also helps the writer see the relationship between the parts of a document, allowing the writer to move parts around to be more effective, to emphasize important sections, to delete others, and yet to add still others. The outline is a guide that helps the writer develop ideas efficiently and effectively.

Templates are a kind of outline. They are outlines based on standard formats for such typical documents as reports, proposals, studies, manuals, letters, and memos. If you need to follow the standard organization typically used for documents of the type you are preparing, a template will help you both find and adhere to a selected organization.

If you know what the standard format is, you can create a template with pen and pencil using conventional methods; or you can create a template with a word processor. For example, Figure 18-9 shows a template created by a technical writer responsible for preparing use and care guides for dishwashers built by Whirlpool Corporation. Because this is the format recom-

Figure 18-9. Template for Product Manual

```
BEFORE YOU USE YOUR DISHWASHER
IMPORTANT SAFETY INSTRUCTIONS
PARTS AND FEATURES
FOR BEST RESULTS
PROPER LOADING
   Top Rack Loading
   Bottom Rack Loading
   Silverware Basket Loading
ADDING DETERGENT
BEFORE STARTING YOUR DISHWASHER
STARTING YOUR DISHWASHER
   Changing a Setting
   Selecting a Cycle
   Heat Dry or Air Dry
   To Add a Dish After Starting
ENERGY SAVING TIPS
SPECIAL TIPS
COMMON DISHWASHER PROBLEMS
IF YOU NEED SERVICE OR ASSISTANCE
   Before Calling for Assistance
   If You Need Assistance
   If You Need Service
   If You Have a Problem
```

Source: Whirlpool Corporation. (1984). *Use and care guide: Whirlpool undercounter dishwasher (model DU2900XM).* Benton Harbor, MI: Whirlpool.

mended by the company for documents of this type, one way for writers to guarantee that they follow this organization is to type the format into their word processor and save it on the disk. Now it becomes the outline, or template, when they have to prepare a use and care guide for a dishwasher.

If you regularly prepare documents of one kind, using a template will guarantee that you follow the organization consistently when you write each report. Templates can be easily created with word processing software. After you load the template in from the disk, the template becomes an organizational guide, allowing you to enter information at appropriate sections of the report. When you actually begin writing, first load the template in from your disk, position the cursor after one of the topics, and begin entering the information appropriate for that section of the report. The template will remind you of all the parts your document must have, show you the order of these parts, and help you see where each piece of information you have collected should be placed.

If the document you are writing is heavily formatted, the template almost lets you "write by the numbers." Each heading becomes a space that has to be filled in, and the heading tells you what kind of information has to be placed there. Gradually build the document by filling in the headings. A word of caution: for most writing, a template is just a guide, not the ultimate determinant of the document. You must provide the transitional words and sentences that good writers use to (a) signal that they're moving on to another point and (b) remind the reader of the organizational pattern of the document. Each organizational pattern has typical transitional words — chronologically organized documents use terms such as "first," "second," "third," "next," and "last"; spatially organized documents use such terms as "at the top," "on the outside," "to the right," "in the next layer."

When the template is very specific, writing becomes very easy. You know what has to be said and where it has to go. Other areas of the template may be more vague and leave more room for experimentation. For example, if the introduction says that you should describe the problem, it is up to you to decide how to introduce the problem. Should you use a specific example? Give a few statistics about how prevalent the problem is? Use an anecdote you found in your reading? Those decisions are up to you. The template shows you only where the material for an introduction should go.

Regardless of the organizational pattern you use, organizing information means arranging information. Because there are many ways to arrange information within a document, successful writers choose the arrangement that is the most effective for a given audience and for a specific purpose.

Activities

1. Find an example of a document organized according to one of the methods described in this chapter and identify its method of organization. List the transitional words the writer uses (a) to signal that he or she is moving on to another point and (b) to remind the reader of the organizational pattern of the document. How clear are the transitions? Rewrite the document using clearer transitions.

2. Prepare a report for your classmates describing the standard formats for documents written in your field. Your report should include the number of alternative formats that exist for similar documents, some consideration of why these formats might have developed as they did, and why they are (or are not) well-suited to their purposes.

3. Develop a template for each type of writing you have done this semester (e.g., essay, report, memo). Compare your templates with those of your classmates. How do you account for differences that occur among templates for similar documents?

4. With two or three students, find an example of a technical document organized according to one of the methods described in this chapter (chronological, spatial, order of importance, general to specific, specific to general, problem/cause/solution). Evaluate the effectiveness of the organization: consider the context of the document, the intended audience, and the purpose.

5. The memo in Figure 18-10 on the following pages appeared in chapter 15, "Technical Argumentation." The purpose of the memo was to convince Mr. Taylor, Manager of Licensing in the Nuclear Power Generation Division of Babcock and Wilcox, to change certain operating procedures at nuclear power plants operated by the Babcock and Wilcox Company. As the discussion in chapter 15 pointed out, the writer was not effective; Mr. Taylor was not convinced that operating procedures should be changed. Unfortunately, the problem Mr. Dunn had foreseen in his memo resulted in a serious accident at the Three Mile Island nuclear power plant little more than a year after the memo was written.

What is the organizational pattern in the memo? With two or three other students, analyze the memo and recommend an organizational pattern that might be more convincing to Mr. Taylor.

Figure 18-10. Letter Demonstrating Technical Argumentation

THE BABCOCK AND WILCOX COMPANY
POWER GENERATION GROUP

TO: Jim Taylor, Manager, Licensing

FROM: Bert M. Dunn, Manager
    ECCS Analysis (2138)

SUBJ: Operator Interruption of High Pressure Injection

DATE: February 9, 1978

This memo addresses a serious concern within ECCS Analysis about the potential for operator action to terminate high pressure injection following the initial stage of a LOCA [loss-of-coolant accident]. Successful ECCS operation during small breaks depends on the accumulated reactor coolant system inventory as well as the ECCS injection rate. As such, it is mandatory that full injection flow be maintained from the point of emergency safety features actuation system (ESFAS) actuation until the high pressure injection rate can fully compensate for the reactor heat load. As the injection rate depends on the reactor coolant system pressure, the time at which a compensating match-up occurs is variable and cannot be specified as a fixed number. It is quite possible, for example, that the high pressure injection may successfully match up with all heat sources at time t and that due to system pressurization be inadequate at some later time t2.

The direct concern here rose out of the recent incident at Toledo. During the accident the operator terminated high pressure injection due to an apparent system recovery indicated by high level within the pressurizer. This action would have been acceptable only after the primary system had been in a subcooled state. Analysis of the data from the transient currently indicates that the system was in a two-phase state and as such did not contain sufficient capacity to allow high pressure injection termination. This became evident at some 20 to 30 minutes following termination of injection when the pressurizer level again collapsed and injection had to be reinitiated. During the 20 to 30 minutes of noninjection flow they were continuously losing important fluid inventory even though the pressurizer indicated high level. I believe it fortunate that Toledo was at an extremely low power and extremely low burnup. Had this event occurred in a reactor at full power with other than insignificant burnup it is quite possible, perhaps probable, that core uncovery and possible fuel damage would have resulted.

Figure 18-10. Letter Demonstrting Technical Argumentation (Continued)

The incident points out that we have not supplied sufficient information to reactor operators in the area of recovery from LOCA. The following rule is based on an attempt to allow termination of high pressure
injection only at a time when the reactor coolant system is in a sub-
cooled state and the pressurizer is indicating at least a normal level
for small breaks. Such conditions guarantee full system capacity and
thus assure that during any follow on transient would be no worse than
the initial accident. I, therefore, recommend that operating procedures be written to allow for termination of high pressure injection
under the following two conditions only:

1. Low pressure injection has been actuated and is flowing at a rate
   in excess of the high pressure injection capability and that
   situation has been stable for a period of time (10 minutes).
2. System pressure has recovered to normal operating pressure
   (2200 or 2250 psig) and system temperature within the hot leg is
   less than or equal to the normal operating conditions (605 or 630
   F).

I believe this is a very serious matter and deserves our prompt
attention and correction.

BMD/lc

cc: E.W. Swanson
    D.H. Roy
    B.A. Karrasch
    H.A. Bailey
    J. Kelly
    E.R. Kane
    J.D. Agar
    R.L. Pittman
    J.D. Phinny
    T. Scott

Source: United States. (1979, October). *Staff report to the President's Commission on the accident at Three Mile Island.* The role of the managing utility and its suppliers (pp. 224–225). Washington, DC: U.S. Government Printing Office.

19

Using and Citing Technical Sources

Incorporating Source Material in Your Writing
Quotations
Paraphrase and Summary
Guidelines for Incorporating Source Material

Documenting Your Sources
In-Text Citations and Reference Lists
APA Style for In-Text Citations
CBE Style for In-Text Citations
APA Style for Reference Lists
CBE Style for Reference Lists
Number System
Guidelines for Documenting Sources

Activities

As was pointed out in chapter 16, "Collecting Information," there are three sources of information: yourself, the library, and the field. In many technical writing tasks, it is likely that you will use information obtained in the field or from library research. Field sources include interviews, letters of inquiry, questionnaires, on-site visits, and controlled experiments. Library sources include such published works as books, government documents, and periodicals.

If you write for an audience outside your organization, and especially if you write for publication, you will be required to document your sources. Technical communication within an organization often does not include either acknowledgment of sources or documentation of those sources. However, there are very good reasons for acknowledging and documenting sources for internal as well as external communications. Writers acknowledge their sources in order to increase their credibility and avoid plagiarism. Writers document their sources as a service to the reader by providing the information that allows the reader to locate the original source of the information cited.

Because technical material is often highly demanding, readers of technical writing often need to refer to original sources. If you use sources, acknowledge those sources and provide documentation for those sources. This chapter offers guidelines for incorporating sources in your writing (e.g., using quotations, paraphrases, and summaries), for preparing abstracts (a form of summary), and for documenting the sources of your information.

Incorporating Source Material in Your Writing

Reporting the results of reading is as important as the research itself. Generally there are three methods of reporting the results of your reading: quotation (using the author's exact words), paraphrase (putting the ideas into your own words), and summary (condensing the ideas into a shorter passage).

Quotations

Inexperienced writers tend to use lengthy quotations that are often difficult to understand. Experienced writers, on the other hand, use summaries and paraphrases primarily. In general, you should only use direct quotations if

1. the original language is so distinctive that you would lose something in a paraphrase or a summary,
2. you cannot say the same thing more clearly, and
3. you cannot say the same thing more concisely.

For clarity, merge quotations into the text as much as possible. While there are a number of ways to incorporate quoted material into the text, it is the writer's responsibility to relate the passage quoted to the text. In the following example, the writer uses a quotation to clarify the general description of precisely how hosts become infected by ticks:

> Steere et al. proposed that hosts become infected as the tick feeds; that is, "the spirochete may be injected into the bloodstream through the saliva of the tick or it may be deposited in fecal material on the skin" (1983).

Introduce and close quotations with double quotation marks. To indicate a quotation within a quotation, use single quotation marks:

> *The New England Journal of Medicine* reported that "Lennhoff, using a silver stain, found 'spirochete-like elements' in skin-biopsy specimens of ECM."

If, however, the quoted material is lengthy, set off the quoted material as indented blocks. There is little agreement among technical writing manuals as to exactly what "lengthy" means and just how many spaces to indent. A good rule to follow is that if the quoted material runs longer than four lines, indent the passage five spaces, single space between the lines, and do not enclose in quotation marks.

Be accurate. If you spell a source's name wrong, if you write the wrong date, if you transpose numbers, or if you inadvertently give the wrong page number in your documentation, that error might well cause a reader to question all your facts. Be careful when you copy information, and double-check when you edit.

Use ellipses to indicate omitted material. If the omission occurs within a sentence, use three periods with a space before and after each period to indicate omitted material.

Original passage:

Lyme disease, first recognized in 1975, typically begins in summer with the unique skin lesion erythema chronicum migrans (ECM).

Elliptical quotation:

Lyme disease . . . typically begins in summer with the unique skin lesion erythema chronicum migrans (ECM).

If the omission occurs between sentences, and if the sentence directly before the omission is complete, there will be four periods: one to conclude the complete sentence and three to indicate the omission.

Original passage:

Borrelia burgdorferi cells regurgitated from *I. dammini* into the host are lysed by the host's immune system components. Consequently, the toxin is released into the bloodstream causing fever, chills, and shock.

Elliptical quotation:

Borrelia burgdorferi cells regurgitated from *I. dammini* into the host are lysed by the host's immune system components. . . . into the bloodstream causing fever, chills, and shock.

If necessary, use brackets to clarify. Brackets are most often used to clarify something in the original that might not be clear to the reader. For example, because the reader might not know that in the following excerpt the pronoun "it" refers to "spirochetemia," the writer has added the reference in brackets:

Mather claims that, "Indeed, if it [spirochetemia] exists, it may be much shorter lived in catbirds than in mice."

Brackets are also used to show that an error in the text appears in the original. The most common method of indicating that an error in a quotation is not the writer's mistake is to place "sic" (Latin for "thus" or "so") within brackets:

The disease is non-discriminating, as it effects [sic] both sexes and all ages.

Paraphrase and Summary

The purposes of paraphrase and summary are basically the same: to restate the ideas of others in a more easily understood form. The decision whether to paraphrase or to summarize depends on the writer's specific purpose. Generally, a paraphrase is about the same length as the original passage, and is used when a writer wants to include many of the details that are in the original.

Original passage:

The identity of the isolate both after initial discovery and also after further cultivation was established by identical dilution end points of the direct (1:128) and indirect (1:512) immunofluorescence procedures in paired tests using the human isolate and the *I. dammini*-derived spirochete isolate as antigens.

Paraphrase:

The identity of the initial isolate was determined with the use of a specific-binding molecule that glows when exposed to ultraviolet radiation. This specific-binding molecule would also bind both the human isolate and the *I. dammini*-derived spirochete, indicating that all of the isolates belonged in the same species (Benach, 1983).

A summary is much briefer than the original, can reflect the language and tone of the original, and is used when a writer does not want to include many of the details that are in the original. Here is a summary of the passage cited above. The writer included in his summary only the information that he felt was absolutely necessary for his paper. He substituted "Lyme disease spirochete" for "the isolate" that appears in the original in order to clarify precisely which isolate was being talked about and to make sure there would be no confusion. The writer did not include specific details in his summary because he knew that if readers wanted to know specific details (such as the ratios of the direct and indirect immunofluorescence procedures), they could go to the original journal article.

> To identify the Lyme disease spirochete, *I. dammini* spirochetes were isolated using both direct and indirect immunofluorescence procedures (Benach, 1983).

Summary is the form most frequently used for reporting the results of reading. The reason for this is not to save space. Through summaries (and paraphrases) writers make complex ideas more easily understandable not only by expressing ideas in more familiar language, but also by combining information from several sources in the same summary section, by rearranging material, or by deleting material not important to the purpose of their writing.

Writing accurate, clear, and concise paraphrases and summaries requires careful reading, attention to detail, and an understanding of whom you are writing for and why you are writing. As pointed out in chapter 3, "Audience and Purpose in Technical Writing," of all the problems and choices technical writers face, none is more critical than understanding the audience and purpose for their writing. Before paraphrasing or summarizing, determine who your readers are, why they are reading, and why you are writing.

If the material you are working with is highly technical and your readers are novices, they will likely need summaries and paraphrases that they can understand. On the other hand, if your readers are experts, you will not have to define as many terms or offer as many paraphrases. If readers are concerned primarily in getting more of a general understanding of the material than in absorbing details, you will likely use more summary than paraphrase. If readers are looking for information to help them make a decision, they will likely be looking for summaries that will help them make those decisions. However, if your readers are looking for detailed information, you will likely include a lot of details in your paraphrases.

In the following example, written for nonspecialists, the writer both paraphrased and summarized the original.

Original passage:

Although spirochetes were commonly found in nymphal and adult *I. dammini* ticks, none were seen in 148 unfed larval ticks. This finding suggests that larvae may acquire the organism from animal reservoirs.

Summary/paraphrase:

The absence of spirochetes in unfed larval ticks suggested that larvae may acquire *B. burgdorferi* from animal reservoirs.

The writer summarized the first part of the passage ("Although spirochetes were commonly found in nymphal and adult *I. dammini* ticks, none were seen in 148 unfed larval ticks. This finding suggests . . .") because he determined that he could condense the original from 23 words to 10 words ("The absence of spirochetes in unfed larval ticks suggested that . . ."). For the second part of the summary ("larvae may acquire *B. burgdorferi* from animal reservoirs") he initially decided to quote directly from the original. However, the writer thought the term "the organism" might be confusing for some readers (when Steere et al. wrote their paper, the Lyme spirochete had not yet been named for Willy Burgdorfer). Thus, for clarity and precision the writer decided to substitute the term "*B. burgdorferi*" for "the organism."

Guidelines for Incorporating Source Material

1. *Be accurate.* If you are using quotations, accuracy means faithfully reproducing every detail from the original.

If you are paraphrasing, accuracy means conveying the correct sense of the original, including many of the details that are in the original, and, while you need not reproduce every detail exactly as it is in the original, not leaving out any information whose omission would significantly alter the message.

If you are summarizing, accuracy refers primarily to conveying accurately *the sense* of the original and does not include many of the details that are in the original.

2. *Be clear.* Clarity means doing all you can to make sure your readers understand exactly what you are saying. If you are quoting, this means integrating the quotation into the text so that the reader can follow the text easily. If you are paraphrasing or summarizing, clarity means determining a clear understanding of your audience so that you maintain an appropriate level of diction, and so that you determine whether terms you are using from the source need to be defined.

3. *Be concise.* Conciseness means making your sentences as economical as possible so that readers will be able to grasp your meaning quickly. For

example, to be concise while quoting means quoting only what is necessary, only what *must* be quoted. Conciseness in paraphrasing and summarizing means avoiding wordiness, needless repetition, and redundancy, which waste the reader's time and interfere with the clear communication of ideas.

Here is an excerpt from a report on Lyme disease and two summaries of that excerpt, one inaccurate and one accurate.

Original passage:

None of the 7 catbirds captured during August and September and examined by xenodiagnosis produced infected ticks, and a total of 86.4% of 22 mice examined similarly proved infective (Table II). Similarly, none of the 5 catbirds successfully sampled in May proved infective for ticks, whereas 90% of 10 mice sampled concurrently were ineffective (Table II) . . . Furthermore, using a tick xenodiagnosis to compare infectivity between all catbirds and mice assayed, we demonstrated that although the mice infected 75.8% of all ticks feeding on them, catbirds failed to infect any. This is significant in that catbirds are one of the most common birds serving as host for immature *I. dammini* in the woodland setting, and thus, this species would appear to make no contribution to the total population of spirochete-infected ticks.

Inaccurate summary:

Other animals, however, such as deer (*Odocoileus virginianus*) (Telford et al. 1988) and catbirds (*Dumetella carolinensis*) (Mather et al. 1989) are not competent transmitters of the Lyme spirochete.

Accurate summary:

Other animals, however, such as deer (*Odocoileus virginianus*) (Telford et al. 1988) and catbirds (*Dumetella carolinensis*) (Mather et al. 1989) do not appear to be competent transmitters of the Lyme spirochete.

While the two summaries at first glance might appear to be similar, the difference helps to illustrate the need for accuracy and the dangers of lack of clarity and imprecision. In the original report of the study, the tone is clearly scientific; that is, the researchers are not suggesting that they have found an absolute truth, but instead have reached a *tentative* conclusion about the infectivity of catbirds. The inaccurate summary is wrong because the words "are not" distort the original article by suggesting that the authors of the report have reached a *definite* conclusion about the infectivity of catbirds. On the other hand, the words "do not appear to be" in the accurate summary convey more precisely the probationary conclusion of the original. The authors of the article are not being wishy-washy with their use of tentative and conditional language. Because they are scientists, they know that much more experimentation needs to be done before they can ever say with certainty that catbirds "are not" competent transmitters of the Lyme spirochete.

Documenting Your Sources ————————————

It is not enough to incorporate information in your writing accurately, clearly, and concisely. You must also identify and document the sources of all ideas or expressions. (This includes all quotations, paraphrases, and summaries.) Though organizational technical writing often does not contain the meticulous and careful documentation that is demanded for professional and academic technical publications, a good rule to follow is to acknowledge all ideas not your own and not common knowledge.

Admittedly, it is often difficult to know just what is common knowledge and what is not. Generally, common knowledge refers to facts and ideas that are widely known and/or available from a number of sources. The statement that Lyme disease is named after Lyme, Connecticut, is considered common knowledge. Similarly, the statement that Lyme disease lesions may expand to 15 cm and give a "bull's-eye" appearance is common knowledge. However, because there is some disagreement over the stages during which ticks can infect humans, the statement that "Ticks at any stage can infect humans" would have to be documented.

There are three important reasons for providing thorough documentation in your technical writing:

1. Documentation gives credit to those whose information you have used. Failure to give complete credit, either intentionally or unintentionally, is **plagiarism**. Failing to acknowledge the ideas of others is unethical; publishing the ideas of others as your own is illegal. The consequences are often so dire that plagiarism is never worth the risk. If a student is caught plagiarizing in college, the student will likely fail the course. If an employee is caught plagiarizing, the employee might lose his or her job—or, worse yet, lose his or her job and wind up in court.
2. Documentation allows readers to examine your sources in greater detail. Readers who want to know more about a specific reference will have the information they need to find that reference.
3. Documentation lends authority and credibility to your work by displaying the breadth and depth of your research. It shows readers where you got your information, how recent your information is, and who the writers you have relied on are.

Though it often appears daunting to many writers, documenting your sources is one of the least demanding skills you will be required to master. Basically, there are three options: footnotes, endnotes, and in-text citations. Because in-text citation has become the standard form of documentation, it will be the only format discussed in this chapter.

In-Text Citations and Reference Lists

Unlike footnotes and endnotes, in-text citations feature a two-pronged approach: A brief notation in the text identifies the source and at the same time provides a cross-reference to the complete list of information about all the sources in the document provided at the end of the document.

The reference list is actually a directory. It provides an address for each reference in your paper. Though different disciplines often require different information, the reference list is always arranged alphabetically and includes at least the following basic publication information. For books: author(s), title, publisher, date, and place of publication. For periodicals: author(s), title of article, name of journal, issue or volume number, and inclusive pages.

If you are not sure what to include in a reference list, it is always best to include as much information as possible. It is better to be redundant than to omit the one piece of bibliographic information the reader needs. You might easily have more than one reference to the same author, in the same journal, published in the same year. The failure to list the specific reference might result in confusion at the very least. Similarly, you might have gotten different information from the same book that has been published in multiple editions. The failure to list the specific edition might send your readers on a wide goose chase looking for a fact or statement that does not exist in that edition.

Because the specific *form* of documentation you use will depend on the discipline, and because there are variations within disciplines, always check the requirements of the environment for which you are writing, and then turn to the appropriate style manual for the details and examples. Some of the most widely used manuals that describe these formats are the following:

American Chemical Society. (1978). *Handbook for authors of papers in the journals of the American Chemical Society.* Washington, DC: American Chemical Society.

American Institute of Physics, Publications Board. (1978). *Style manual for guidance in the preparation of papers* (3rd ed.). New York: American Institute of Physics.

American Mathematical Society. (1980). *A manual for authors of mathematical papers* (7th ed.). Providence: American Mathematical Society.

American Medical Association. (1976). *Stylebook/editorial manual.* Littleton, MA: Publishing Sciences Group.

American National Standards Institute. (1977). *American national standard for bibliographic references.* New York: American National Standards Institute.

American Psychological Association. (1983). *Publication manual of the American Psychological Association* (3rd ed.). Washington, DC: American Psychological Association.

University of Chicago Press. (1982). *The Chicago manual of style* (13th ed.). Chicago: University of Chicago Press.

Council of Biology Editors, *CBE Style Manual Committee. (1983). CBE style manual: A guide for authors, editors, and publishers in the biological sciences* (5th ed.). Bethesda, MD: Council of Biology Editors.

Engineers Joint Council, Committee on Engineering Society Editors. (1977). *Recommended practice for style of references in engineering publications.* New York: Engineers Joint Council.

Gibaldi, J., & Achtert, W. S. (1988). *MLA handbook for writers of research papers* (3rd ed.). New York: Modern Language Association.

United States Government Printing Office. (1984). *Style manual* (rev. ed.). Washington, DC: U.S. Government Printing Office.

The two standard methods in technical and scientific writing for citing an author's work are described in the *Publication Manual of the American Psychological Association*, 3d ed., (Washington, D.C.: American Psychological Association), 1983, and the *CBE Style Manual: A Guide for Authors, Editors, and Publishers in the Biological Sciences*, 5th ed., (Bethesda, MD.: Council of Biology Editors), 1983. These two styles will be described below.

APA Style for In-Text Citations

The American Psychological Association uses the author–year method of citation; that is, the author's name and the year of publication appear in parentheses separated by a comma:

Most compost privies have significant drawbacks (Kourik, 1990).

If you are including a direct quotation, give the page number of the source as well as the author and year. Note that each part of the reference is separated by commas, and the page reference begins with "p." followed by a single space:

Most compost privies have "significant drawbacks" (Kourik, 1990, p. 4).

If you have just mentioned the author in the text, it is sufficient to just note the year of publication (and page if applicable):

Kourik (1990) claims that most compost privies have significant drawbacks.

If you don't know the author of a work, as in a newspaper article, use an abbreviated version of the title. Use double quotation marks around the title of an article, and underline the title of a periodical or book:

Most compost privies have significant drawbacks ("Compost Privies," 1987).

If there are fewer than six authors, cite all of them the first time, and only the first author plus "et al." after that:

First occurrence:

(Telford, Mather, Moore, & Wilson, 1988)

Succeeding occurrences:

(Telford et al., 1988)

If there are more than five authors, name the first and add "et al." to indicate that there are additional authors:

(Smith et al., 1988)

If you want to include some of your source information in the main part of your sentence — that is, outside parentheses — include only the author in the sentence. *Avoid* inserting either just the year or both author and year in the structure of the sentence, as in these examples.

Awkward:

In 1990, it was claimed that most compost privies have significant drawbacks (Kourik).

In 1990, Kourik claimed that most compost privies have significant drawbacks.

Better:

Kourik claimed that most compost privies have significant drawbacks (1990).

If the author and the year are the same for two or more references, add lowercase letters (in alphabetical succession) after the year.

(Kourik, 1990a)

If the reference list includes publications by two or more authors with the same surname, include the authors' initials in all citations, even if the year is not the same.

(J. E. Kourik, 1990)

CBE Style for In-Text Citations

The Council of Biology Editors also uses the author–year method of citation but does not include a comma between author and year.

Most compost privies have significant drawbacks (Kourik 1990).

If you are including a direct quotation, do not include the page number of the source.

Most compost privies have "significant drawbacks" (Kourik 1990).

If you have already mentioned the author in the text, it is sufficient to just note the year of publication.

Kourik (1990) claims that most compost privies have significant drawbacks.

If you don't know the author of a work, as in a newspaper article, use an abbreviated version of the title. Do not use double quotation marks around the title of an article, and do not underline the title of a periodical or book. Note that in CBE style only the first word of the title of the article or book is capitalized.

Most compost privies have significant drawbacks (Compost privies 1987).

If there are two authors, cite both authors the first time and each time after.

(Kourik and Smathers 1988)

If there are three or more authors, cite only the first author plus "et al."

(Kourik et al. 1987)

Include only the author in the sentence. *Avoid* inserting either just the year or both author and year in the structure of the sentence.

If the author and the year are the same for two or more references, add lowercase letters (in alphabetical succession) after the year.

(Kourik 1990a)

If the reference list includes publications by two or more authors with the same surname, include the authors' initials in all citations, even if the year is not the same.

(J. E. Kourik 1990)

APA Style for Reference Lists

BOOK CITATIONS

The general form for book citations is:

Author.   Last name first, followed by a comma and initials for all authors. Each author is followed by a comma, with an ampersand (&) before the last author's name. List of authors is followed by a period.

Year of Publication.   In parentheses, followed by a period.

Book Title.   Underlined, with only the first word and any proper names capitalized, followed by a period.

Place of Publication.   Include only the city, unless the city is not well known or unless the city could be confused with another location. Separate place and publisher with a semicolon.

Publisher.   Give the name of the publisher in as brief a form as possible. Spell out names of associations and presses, but omit terms that are not needed to identify the publisher, such as "Co.," Inc.," and Assoc." Follow with a period.

The following examples illustrate some of the most frequently used forms for book citations:

Book with One Author.   Invert the author's name, use commas to separate surnames and initials, and finish each element with a period.

> Huth, E. J. (1982). *How to write and publish papers in the medical sciences.* Philadelphia: ISI Press.

Book with More than One Author.   Include all authors' names, regardless of the number of authors. Use commas to separate authors, and use an ampersand (&) before the last author.

> Baltimore, D., Darnell, J., & Lodish, H. (1986). *Molecular cell biology.* New York: Scientific American.

Later Edition of a Book.   Add the edition in parentheses after the title of the book.

> Frank, L. (1987). *Technical communication in an age of anxiety* (3rd ed.). San Francisco: New Age.

Edited Book.   Add "Ed." or "Eds." in parentheses after the name(s) of the editor(s).

> Glanze, W. D. (Ed.). (1985). *The Mosby medical encyclopedia.* New York: NAL.

Article or Chapter in an Edited Book.   Refer to the author(s) and title of the work you are citing first, then give information about the book in which it can be found. In the following example, note that the editor's name is given with first initials before the last name, the word "In" is added before the name of the editors, and the actual page numbers of the work cited are included in parentheses.

> Sharples, M., Goodlet, J. S., & Pemberton, L. (1989). Designing a writer's assistant. In N. Williams & P. Holt (Eds.), *Computers and writing, models and tools* (pp. 24–32). Oxford: Blackwell.

Book with No Author or Editor.   If there is no author or editor, list the book under the title. In the reference list, alphabetize books with no author or editor by the first significant word in the title.

> *Effective writing.* (1986). Lymington: Quay Software.

PERIODICAL CITATIONS

The general form for journal article citations is:

Author.   Last name first and initials for all authors involved, regardless of the number of authors, followed by a period.

Year of Publication.   Place the year of publication in parentheses, followed by a period.

Article Title.   Capitalize only the first word of the title. Do not underline or use quotation marks around the title. Follow by a period.

Journal Title.   Give the journal title in full. Underline the title, using uppercase and lowercase letters. Follow by a comma.

Volume Number.   Underline the volume number and follow by a comma. Do not use "V," "Vol.," or "Volume" before the volume number. If each issue of the periodical begins on page 1, give the issue number (in parentheses) after the volume number.

Page Numbers.   Give inclusive page numbers of the article. Do not precede the page numbers with "p." or "pp." for journal articles. Use "p." or "pp." before page numbers of newspapers and magazines. Follow the page number(s) with a period.

The following examples illustrate some of the most frequently used forms for periodical citations:

Journal Article with One Author.   Invert the author's name, use commas to separate surnames and initials, and finish each element with a period.

> Burgdorfer, W. (1984). Discovery of the Lyme disease spirochete and its relation to tick vector. *Yale Journal of Biological Medicine, 57,* 515–520.

Journal Article with More than One Author.   Include all authors' names, regardless of the number of authors. Use commas to separate authors, and use an ampersand (&) before the last author.

> Donahue, J. G., Piesman, J., & Spielman, A. (1987). Reservoir competence of white-footed mice for Lyme disease spirochetes. *American Journal of Tropical Medicine and Hygiene, 36,* 92–96.

Magazine or Newspaper Article.   For magazines and newspapers, substitute the complete date and delete the volume number. Note that the date is given with the year followed by month and day. The page numbers are preceded by "p." or "pp."

> Rogers, P. (1991, October 28). Speak nicely to your PC. *Newsweek,* p. 59.

Article with No Author.   For an article with no author, give the title of the article (with no quotation marks and only the first word capitalized), the year and date of publication (in parentheses, separated by a comma), the name of the periodical, magazine, or newspaper (underlined), and the page number(s) (preceded by "p." or "pp.").

> Researchers identify deer tick parasite. (1988, October 12). *Stevens Point Journal,* p. 3.

CBE Style for Reference Lists

BOOK CITATIONS

The general form for book citations as described in the CBE manual is

Author.   Last name first and initials for all authors involved, followed by a period.

Book Title. Not underlined, with only the first word and any proper names capitalized, followed by a period.

Place of Publication. Include only the city, unless the city is not well known or unless the city could be confused with another location. Separate place and publisher with a colon.

Publisher. Give the full name of the publisher, including such terms as "Co.," Inc.," and "Assoc." Follow with a semicolon.

Year of Publication. Not in parentheses, followed by a period.

The following examples illustrate some of the most frequently used forms for book citations.

Book with One Author. Invert the author's name, use commas to separate surnames and initials, and finish each element with a period.

> Huth, E. J. How to write and publish papers in the medical sciences. Philadelphia: ISI Press; 1982.

Book with More than One Author. Include all authors' names, regardless of the number of authors. Use semicolons to separate authors. Do not use an ampersand (&) before the last author.

> Baltimore, D.; Darnell, J.; Lodish, H. Molecular cell biology. New York: Scientific American Books; 1986.

Later Edition of a Book. Include the number of the edition directly after the title of the book, followed by a period.

> Frank, L. Technical communication in an age of anxiety. 3rd ed. San Francisco: New Age; 1987.

Edited Book. Add "editor" or "editors" after the name(s) of the editor(s).

> Glanze, W. D., editor. The Mosby Medical Encyclopedia. New York: New American Library; 1985.

Article or Chapter in an Edited Book. Refer to the author(s) and title of the work you are citing first, then give information about the book in which it can be found. In the following example, note that the word "In," followed by a colon, is added before the name of the editors, and that the editors' names are inverted. The actual page numbers of the work cited are not included in parentheses.

Sharples, M.; Goodlet, J. S.; Pemberton, L. Designing a writer's assistant. In: Williams, N.; Holt, P., eds. Computers and writing, models and tools. Oxford: Blackwell, Inc.; 1989: pp. 24–32.

Book with No Author or Editor.   If there is no author or editor, list the book under the title, not underlined. In the reference list, alphabetize books with no author or editor by the first significant word in the title.

Effective writing. Lymington: Quay Software; 1986.

PERIODICAL CITATIONS

The general form for journal article citations recommended by the CBE manual is:

Author.   Last name first and initials for all authors involved, regardless of the number of authors, followed by a period.

Article Title.   Capitalize only the first word of the title. Do not underline or use quotation marks around the title. Follow by a period.

Journal Title.   Abbreviate the journal title, using uppercase and lowercase letters. Do not underline. Follow with a period.

Volume Number.   Do not underline the volume number. Follow it by a colon. Do not use "V," "Vol.," or "Volume" before the volume number.

Page Numbers.   Give inclusive page numbers of the article. Do not precede the page numbers with "p." or "pp." Follow the page number(s) with a semicolon.

Year of Publication.   The year of publication (not in parentheses) is followed by a period.

The following examples illustrate some of the most frequently used forms for periodical citations:

Article with One Author.   Invert the author's name, use commas to separate surnames and initials, and finish each element with a period.

Burgdorfer, W. Discovery of the Lyme disease spirochete and its relation to tick vector. Yale Jour. Biol. Med. 57:515–520; 1984.

Article with More than One Author.   Include all authors' names, regardless of the number of authors. Use semicolons to separate authors. Do not use an ampersand (&) before the last author.

Donahue, J. G.; Piesman, J.; Spielman, A. Reservoir competence of white-footed mice for Lyme disease spirochetes. Am. J. of Trop. Med. 36:92–96; 1987.

Article from a Journal Paginated by Issue. If each issue of the periodical begins on page 1, insert the issue number (in parentheses) after the volume number.

Eason, K. D. Towards the experimental study of usability. Behav. and Inf. Tech. 8(2):2–7; 1984.

Article with No Author. For an article with no author, give the title of the article (with no quotation marks and only the first word capitalized), the name of the periodical, magazine, or newspaper (not underlined), the year and date of publication, and the page number(s) (not preceded by "p." or "pp."). Note that there is no punctuation between the year and month and that a semicolon separates the date from the page number.

Researchers identify deer tick parasite. Stevens Point Journal. 1988 Oct. 12;3.

Number System

Occasionally, you might be required to use a number system instead of the name-and-year system for in-text citations and the accompanying reference list. In the number system, a number is inserted parenthetically in the text instead of a name and year. The number in the text corresponds to a numbered entry in the reference list. In some numbering systems, the page is also given, separated from the number by a comma (3, 324).

Generally, there are two numbering systems. In one method, the references are numbered according to their first mention in the text. The parenthetical reference in the text refers to the number assigned to the bibliographic reference (the first reference is "1," the second reference is "2," and so on). For this method, the references are not listed alphabetically; they are listed numerically according to the order in which they appear in the text.

1. Huth, E. J. (1982). *How to write and publish papers in the medical sciences.* Philadelphia: ISI Press.

2. Donahue, J. G., Piesman, J., & Spielman, A. (1987). Reservoir competence of white-footed mice for Lyme disease spirochetes. *American Journal of Tropical Medicine and Hygiene, 36,* 92–96.

3. Glanze, W. D. (Ed.). (1985). *The Mosby medical encyclopedia.* New York: NAL.

In the second method, place a number in the text that corresponds to a specific entry in the reference list ("1" would refer to the first entry on the list, "2" to the second entry, and so forth). The reference list for this method of citation is presented alphabetically. In this system, the references will not be in the order of their first mention in the text.

1. Donahue, J. G., Piesman, J., & Spielman, A. (1987). Reservoir competence of white-footed mice for Lyme disease spirochetes. *American Journal of Tropical Medicine and Hygiene, 36,* 92–96.

2. Glanze, W. D. (Ed.). (1985). *The Mosby medical encyclopedia.* New York: NAL.

3. Huth, E. J. (1982). *How to write and publish papers in the medical sciences.* Philadelphia: ISI Press.

Guidelines for Documenting Sources

Regardless of the method of referencing you use and regardless of the format, your documentation should be clear and nonintrusive. The following guidelines will help you achieve this.

1. *Integrate the author's name in the text.* There are two major advantages to integrating the names of the authors in the text. First, this method helps the reader process information more quickly and more easily. Second, it helps establish the writer's credibility. Note that the dates always appear in parentheses.

DNA homology studies, reported by Hyde and Johnson (1984) and Schmid et al. (1984), revealed that the Lyme spirochete represented a new species of *Borrelia.*

2. *Incorporate closely related references in the same sentence.* The advantage is conciseness. Here, rather than two sentences making the same point, one about deer and the other about catbirds, both references were combined in one sentence.

Other animals, however, such as deer (Telford et al. 1988) and catbirds (Mather et al. 1989) do not appear to be competent transmitters of the Lyme spirochete.

3. *Place citations at the end of the summary.* If you feel that authorial reference might inhibit the smooth flow of the sentence, place the citation at the end of the summary. In the following example, the writer felt that the passage was so heavily laden with detail that introducing the names of the researchers would have made the passage difficult to read.

The *B. burgdorferi* is microaerophilic (low oxygen requirement), ranges in cell diameter .18–.25 micrometers and length 10–30 micrometers, contains LPS, and has a flagella called axial filaments (Beck et al. 1985 and Steere 1984).

4. *Indicate clearly where borrowed material begins.* Generally, the signals indicating where borrowed material *ends* are clear (e.g., quotation marks, parenthetical references, numbers). However, often the signals indicating where borrowed material *begins* are not as clear. Readers are often puzzled by a citation at the end of a paragraph that could refer to the entire paragraph or only to the last sentence. If you think a reader might not be able to distinguish between where your thoughts end and the thoughts of others begin, preface the quotation, paraphrase, or summary with the author's name, the title of the book or article, or a signal such as "One study indicates that. . . ."

Activities ——————————————————————————————————

1. Find a passage in a technical document that you feel uses too much quotation. Rewrite the passage with a more even balance of quotation and paraphrase.

2. For a research report you are doing on Lyme disease, prepare a summary of the following report on Lyme disease. The report, titled "Lyme Disease," is one of a series of short reports on various health-related issues published by the National Institutes of Health. The audience for your report should be educated people who are interested in but not specifically knowledgeable about medicine, biology, or health-related issues. Remember the need to be accurate, clear, and concise.

A form of chronic inflammatory arthritis, Lyme disease is caused by a tick-borne bacterium, *Borrelia burgdorferi*, which was isolated a few years ago by National Institute of Allergy and Infectious Diseases scientists. The primary persistent symptom is joint inflammation resembling rheumatoid arthritis. Early Lyme disease usually can be diagnosed by its characteristic skin lesions, rapidly changing involvement of other organ systems, and now by blood testing.

During a 5-year study of Lyme disease among the residents of Great Island, Massachusetts, researchers found 16 percent of the residents experienced symptoms of Lyme disease, and another 8 percent without symptoms had antibodies to the bacterium in their blood. Most of those with symptoms developed the primary skin lesion that is the hallmark of Lyme disease; of those who did not, most were children in whom arthritis was reported as the major symptom.

The presence of large numbers of deer on the island was determined to be the most likely source of the ticks that transmit the disease. The high prevalence of Lyme disease among those island residents and the appearance of arthritis and other symptoms years after infection in untreated persons underscores the importance of early diagnosis, prompt treatment, and surveillance and control programs.

The establishment of control programs may be more complicated than previously thought, however. Researchers have found that ticks are not the only transmitters of Lyme disease. Following reports of cases attributed to the bites of mosquitoes, deer flies, and horse flies, the researchers collected these insects in areas of Connecticut where ticks are known to carry *B. burgdorferi* and in areas where the disease is rare. In areas inhabited by ticks carrying this bacterium, up to one-fifth of the insects also carried it. Insects from areas devoid of ticks, however, did not harbor the organism, even though deer were abundant in those areas. The researchers concluded that blood-sucking insects such as mosquitoes may be a secondary source of the bacterium in areas where ticks serve as the primary source.

3. Paraphrase one paragraph from the report above on Lyme disease. Remember that the primary purpose of paraphrase is to restate the ideas of others in a more easily understood form. Compare your paraphrase with those of others in the class.

4. Occasionally, because the original language is so distinctive or you cannot say the same thing more clearly or more concisely, you will want to use a quotation rather than a summary or a paraphrase. Choose a passage from this chapter and select phrases and sentences that would be more appropriate to quote than to summarize or paraphrase.

5. With other students from your major or field of interest, prepare a brief handbook summarizing the major form of documentation used in your discipline. Distribute copies of the handbook to other members of the class.

V

TECHNICAL
WRITING TASKS

20

Literature Reviews

Situation

Form

Writing the Literature Review
   Collecting Information
   Organizing the First Draft
   Reviewing to Assure Quality
   Publishing the Literature Review

Examples of Literature Reviews

Activities

Whether it is called a literature review, a survey of the literature, a review of the literature, or a search-of-the-literature report, this document is a concise report on current research publications for a particular topic.

Situation

A literature review may be published as an independent report or as part of a larger report. Although the two types of report tend to differ in content and emphasis, the purpose of both is to provide information on current publications.

This information can be used in several ways. Executives, managers, and researchers often use literature reviews published in journals or produced by their own organizations to stay abreast of developments in their fields. A literature review can also be used as a reference work: after reading the review, a reader may turn to the works cited and described to find more information. When it is part of a larger report, the literature review serves yet another purpose: it provides a context for the rest of the report, explaining what has already been done on this topic and pointing toward the report's own research and conclusions as the logical extension.

Although there is no one audience for a literature review, the people who are most often interested in reading a summary of current thought on a particular topic are already experts in the general field. Not only are readers of literature reviews quite knowledgeable, they tend to be highly motivated: the simple fact that they are reading the review at all shows that they wish to be well informed.

Although this well-informed, highly motivated group may sound like an easy audience to write for, remember that they will have high expectations for your document. Readers often come to a literature review with specific questions in mind (What research has been done recently in my specialty? How are these research projects related? What was the most important development in the past year?), and they want to find the answers to such questions presented clearly and intelligently.

For this reason, an effective literature review must be far more than a comprehensive and accurate catalogue of sources. It must also be a clear, coherent, and persuasive analysis of the current state of the literature.

Form

Whether broad or narrow, lengthy or brief, literature reviews are highly condensed and heavily documented. If the literature review is more than several pages long, there are often internal headings indicating major categories and subcategories. Although citations are an essential part of a literature review, the document is presented as regular running text, in paragraph format. The following excerpt taken from a literature review on the causes of Lyme disease is typical of the style of literature reviews.

> In a 1983 study by Steere et al., the absence of spirochetes in unfed larval ticks suggested that larvae may acquire *B. burgdorferi* from animal reservoirs. Later studies demonstrated that animals differ in their infectivity to ticks. Some animals, such as the white-footed mouse (*Peromyscus leucopus*) (Levine et al., 1985; Anderson et al., 1987; Donahue et al., 1987) are effective transmitters of the Lyme spirochete. Other animals, however, such as deer (*Odocoileus virginianus*) (Telford et al., 1988) and catbirds (*Dumetella carolinensis*) (Mather et al., 1989), do not appear to be competent transmitters of the Lyme spirochete.

The literature review is always accompanied by a reference list of works cited. If the literature review is an independent document, the reference list usually follows immediately after the review. If the literature review is part of a longer report, the reference list either follows directly or, more usually, is incorporated into the report's reference list.

Writing the Literature Review

Writing a literature review efficiently requires a systematic approach to the task: gather information, organize the information into a first draft, review to assure quality, and publish the review.

Collecting Information

Because you will probably have a limited time to complete your assignment, you must be systematic and efficient in gathering information. Gathering information for a literature review includes locating information, recording information, and documenting information.

Research for a literature review is library research. To locate information you must develop a list of books and articles and then find out what they are about, either by reading them or by reading the abstracts.

Most of the material included in a literature review will probably come from articles in periodicals because these are generally the most recent publications. You will also want to check a library's book holdings, but books are not the most current sources. Periodical sources are also generally briefer and more focused than books, so that you can concentrate your research on directly relevant information. The most efficient way to begin your search of the literature is with a library's electronic indexes. Like print indexes, electronic indexes cover a broad (but still limited) selection of periodicals; they have the significant advantage of speeding up the selection process. Using the indexes' database features, you can use author, title, subject, and key word search strategies to quickly obtain a list of articles related to your topic in a fraction of the time it would take to locate the same material in a print index.

Once you have located the information you will have to read the articles (or their abstracts), analyze them, draw out the relevant features, and record the information in a form appropriate for a literature review. Recording information for a literature review entails primarily summarizing. Although a technical summary, as described in chapter 13, "Technical Summary," is often 200 words or more, writers of literature reviews must be able to summarize in a sentence or less. Although you will not be writing the actual text of your review as you gather information, summarizing each source now can save much work later. As you take notes, try to describe the importance of each article or book in no more than one or two sentences. When in doubt, though, be sure to note any additional information that may be relevant— you may want to use it later on.

Good documentation is absolutely essential in a literature review. Without it, a literature review becomes almost useless. Because readers will frequently want to consult the fuller reference, it is important that your documentation be complete and accurate. While conducting your library research, be sure to document every source carefully.

Organizing the First Draft

Organizing the first draft of a literature review involves filtering sources and determining an appropriate organization for the information.

FILTERING SOURCES

As you acquire more and more sources of information and investigate these sources and collect notes, you will certainly reach the point of asking, "Can I stop now? Haven't I collected enough information?" If you are using electronic indexes, the question becomes even more pressing, because these tools provide access to greater numbers of sources. Realistically, it is impossible to use every available item of information. How, then, do you determine which sources to include in a literature review?

By filtering your sources you can focus attention on those sources likely to prove most valuable to your readers. Three filters can be applied to the library research process for a literature review:

1. *Chronological.* For most projects, especially those in technical fields, the chronological filter is the most important. This filter operates on the premise that more recently published works offer the most complete and accurate information: The more recent a work is, the more likely it is that you should include this work in your literature review.
2. *Citation.* Authors and articles that are widely cited by writers in the field are likely to prove valuable resources to readers of a literature review. As you examine recent sources, take note of their reference lists. As you see the same authors' names appear again and again, be certain to examine their writings: the more frequently a work or author is cited, the more likely it is that you should include this work or author in your literature review.
3. *Location.* Finally, consider location. In any field, some publications have greater credibility than others. As you begin to filter out material, be certain to examine articles located in these respected publications: the more prestigious the journal in which a work appears, the more likely it is that you should include this work in your literature review.

ORGANIZING INFORMATION

Because the purpose of a literature review is to synthesize clearly large amounts of information and present it in a form that is easily accessible to readers, a significant difficulty often encountered by writers of literature reviews is organizing their material.

You will be expected to do more than just string together all the information you have. Readers of literature reviews don't want a string of citations; they want a coherent, comprehensive description of the literature that serves a particular purpose or answers a particular question. This is a good time to stop and think again about the purpose of your literature review: What is it that you want to show?

Making a Short List.   The quickest, easiest, and most common method of beginning to organize material is to make a short list of what you consider significant pieces of information. The primary reason for making a short list is to get at least some of your information on the page (or screen). The short list will be neither complete nor organized, but it will give you a good start toward completeness and organization.

Bob is writing a literature review on the current research into the causes of Lyme disease. He was asked to prepare the review for the rheumatology department of a large medical center. When Bob made a short list for his literature review, it looked like this:

skin lesions (bull's-eye)

ticks transmit spirochetes

life cycle of ticks

first recognized in U.S. in 1975 (Lyme, Conn.)

some animals don't transmit spirochetes to tick larvae

description of spirochete (size, shape, plasmids, etc.)

LD causes relapsing fever, A-V heart block, arthritis, etc.

Dr. Lenoff first suspected spirochetes in 1948

deer, mice, birds, etc. act as hosts for ticks

Dr. Arvid Afzelius first reported LD in 1909 (Sweden)

Bob learned from his list that, not surprisingly, most of the articles concentrated on the symptoms of Lyme disease and the agents that transmitted the spirochete. At this point Bob tried to group the information in his short list into a smaller number of topics:

Causes of Lyme disease
> etiologic agent (the spirochete)
> hosts (deer, mice, birds, etc.)
> description of Lyme disease
> vector (ticks)

Making an Outline.   Because so much information is summarized in a literature review, and because each citation must logically relate to those around it, literature reviews must be highly organized. Most literature reviews require two levels of organization: an overall structure and a method of ordering the citations within each segment. Three methods of overall organization are common: by date, by controversy, and by topic.

Literature reviews organized by date present each source in the order in which it was published. This is a good method to use if the purpose of the report is to reveal a trend. If Bob's review had been on the methods of treatment for Lyme disease, and if he had wanted to show how these have changed over time, then a chronological organization would have been a good choice.

Reviews organized by topic divide the information on the subject into topics and then present the information on each topic separately. This organization works well if the subject matter and the research about the subject are both fairly broad, or if there are obvious divisions within the subject matter. A literature review on educational reform in public schools, for example, might be divided into three topics: reform in elementary schools, reform in junior high schools, and reform in high schools. Although the order should be logical and consistent, there is no one way to order the segments of a literature review organized by topic.

Another method of organizing a literature review is to focus on the controversy that surrounds the subject. In general, this is useful if a small number of key areas seem to be the focus of current research. The information about each important controversy is presented together; common ways of ordering the information would be either least to most controversial or most to least controversial. This method worked well for Bob's review. He realized that most of the topics he came up with were really areas of controversy and that anything else could be included in an introduction. He decided to order his segments from least to most controversial.

After an overall pattern of organization has been determined, the writer of a literature review must still decide how to organize and present the information within each segment of the review. Typical patterns of organization at this level include chronological (by date of publication), topical, and narrative (in which the logical order is determined by the subject matter itself). It is not necessary to use the same pattern within each segment of the report.

This is how Bob organized the material for the body of his literature review:

Causes of Lyme Disease

Introduction:

  description of Lyme disease

  early reporting of Lyme disease

  first reporting in U.S.

Causative (etiologic) agent:

  [Since there is no disagreement among researchers that the *B. burgdorferi* spirochete is the causative agent, Bob decided to present the results of his research in this section in a primarily descriptive fashion.]

Vector:

  [Since there is some confusion among researchers as to just how hosts become infected, Bob decided to present this information chronologically by date of research.]

Hosts:

  [Although scientists know in great detail the life cycle of *I. dammini*, there appears to be much more to be learned about which animals are effective transmitters of the Lyme spirochete and which ones are not. Therefore, because research studies in this critical area clearly build upon each other, Bob presented the findings in this section chronologically by date of research.]

Conclusion:

[Bob decided to use a conclusion to include some information on the likely direction of future research.]

WRITING THE FIRST DRAFT

After you are satisfied with your outline, you will begin to write your review. Literature reviews generally adhere to the following principles:

1. The narrative style is standard for literature reviews.
2. Reviews are heavily documented. Often there are one or more citations per sentence.
3. Terms that might be unfamiliar to many readers should be defined. Typically, in a literature review, definitions are given in parentheses.
4. Every work included in the reference list is included in the review itself. The reference list is actually a list of the works cited in the review.

Here are two excerpts from the sample literature reviews presented at the end of this chapter. In both examples, which are typical for literature reviews, almost every sentence has one citation, and some sentences have two. Note that while some citations in the second example are placed parenthetically, some are included as part of the main structure of the sentence. Writers may vary the format for citations in a narrative to improve the narrative flow or to avoid a monotonous presentation.

> Such levels have been shown to cause retinal damage in albino mice (Greenman et al., 1982) and rats (Stotzer et al., 1970). Light levels of 323 lx (30 ft-candles) approximately 1.0 m (3.3 ft) above the floor appear to be sufficient for performance of routine animal care (Bellhorn, 1980).

> Strains of *B. burgdorferi* from North America and Europe differ in plasmid profiles. Of 13 spirochete isolates studied, only two had identical plasmid profiles (Barbour, 1988). Johnson et al. (1984) reported that *I. dammini* spirochetes contain two plasmids, whereas *I. ricinus* and human cerebrospinal fluid spirochetes contain single plasmids. *Borrelia burgdorferi* is an unusual bacterium because strains have both linear and circular plasmids (Hovind-Hougen, 1984). It is reported that the linear plasmids have covalently closed ends (Barbour & Garon, 1987). Only among eukaryotes and their viruses have covalently closed ends of DNA been found (Barbour & Garon, 1987).

A problem for writers of literature reviews is knowing what to document: not every point made warrants a citation. You do not need to document knowledge that is generally known to your readers or that reflects your evaluation of a wide range of reading. For example, in the review of illumination needs for laboratory animals, the writer chose not to document the following statement because it summarizes the writer's perception of the state of current research.

Precise lighting requirements for maintenance of good health and physiological stability of animals are not known.

However, the following statement was documented because it represents the result of research that readers might want to see a fuller discussion of.

Light levels of 323 lx (30 ft-candles) approximately 1.0 m (3.3 ft) above the floor appear to be sufficient for performance of routine animal care (Bellhorn, 1980).

Reviewing to Assure Quality

As with any piece of writing you do, you will want to evaluate the literature review. Make sure that you have done what you were asked to do and that your writing is clearly addressed to the readers who are your audience.

One way of evaluating the literature review is to read it over yourself. Look for obvious mistakes, make sure that you have documented your sources and that your documentation is accurate, make sure that your organization is effective and thorough, and make sure that you understand what you have written.

If the literature review is written for a nontechnical audience, and if you have the time, ask a nontechnical colleague (or a colleague who doesn't specialize in this particular technical field) to read it over. Because his literature review was written for a technical audience, Bob did not do this.

Bob did, however, ask one of the rheumatology staff members to look over the review. The staff member, a technician, suggested that because some readers might not be familiar with all the medical terms, some terms should be defined.

Publishing the Literature Review

After you have evaluated and revised your document, after you are satisfied that you have done what you were asked to do, and after you are confident that you have anticipated and met the needs of your readers, you must reproduce and distribute your document.

Two important questions must be answered: Who gets the document?, and, How should the document be printed? The first question was easy for Bob, since he was told that a copy of the literature review would be given to every member of the rheumatology department.

The second question was more difficult. Bob wasn't sure whether to photocopy his report directly from his printer copy or to send it to a professional printer. In this case the decision was easy: neither his office nor the rheumatology department was willing to support a professional printing job, so Bob printed one copy on his dot matrix printer. Although it was legible, it did not look as attractive as he wanted it to look. Because his word processor had a variety of print fonts, Bob selected a moderate size font and a low-resolution laser type to print another copy on his office's laser printer. While he liked the look of the font, the low-resolution type looked too informal. Ultimately Bob kept the font but changed to a high-resolution laser type. The finished product looked polished and attractive.

Examples of Literature Reviews

Figures 20-1 and 20-2 show two examples of literature reviews. The first review appears as part of a longer work titled *Guide for the Care and Use of Laboratory Animals.* The reference list is not included here because references to the citations in the review are incorporated into the reference list of the larger document. The second example is an independent literature review on the causes of Lyme disease and includes a reference list.

Figure 20-1. Example of a Literature Review

Illumination
Lighting should be uniformly diffused throughout animal facilities and provide sufficient illumination to aid in maintaining good housekeeping practices, adequate inspection of animals, safe working conditions for personnel, and the well-being of the animals. Precise lighting requirements for maintenance of good health and physiological stability of animals are not known. In the past, illumination levels for animal rooms of 807 to 1076 lx (75 to 100 ft-candles) have been recommended. Such levels have been shown to cause retinal damage in albino mice (Greenman et al., 1982) and rats (Stotzer et al., 1970). Light levels of 323 lx (30 ft-candles) approximately 1.0 m (3.3 ft) above the floor appear to be sufficient for performance of routine animal care (Bellhorn, 1980). This would provide the equivalent of 32 to 40 lx (3.0 to 3.7 ft-candles) to a rodent in the front of an upper cage in a cage rack. Stotzer et al. (1970) reported that these levels do not cause retinal lesions in albino rats held for up to 90 days. However, Weisse et al. (1974) found minimal retinal lesions in albino rats examined after 790 days of exposure to these levels. These observations should be considered when housing albino rats.

Source: U.S. Department of Health and Human Services. National Institute of Health. (1985). *Guide for the care and use of laboratory animals.*

Figure 20-2. Example of a Literature Review

THE CAUSES OF LYME DISEASE

INTRODUCTION
Description of Lyme Disease: Lyme disease is a tickborne bacterial disease of humans that is usually contracted during the months of May through December. The disease is nondiscriminating, affecting both sexes and all ages. Early manifestations of Lyme disease, such as skin lesions about 2 cm in diameter, are relatively mild. However, skin lesions do not always occur and they may go unnoticed. In the latter case, the lesion may expand to about 15 cm and develop a ''bull's-eye'' appearance. If Lyme disease is not treated early, more severe abnormalities can develop, including relapsing fever, atrioventricular heart block, Bell's palsy, and chronic Lyme arthritis.

Early Reporting of Lyme Disease: One of the earliest descriptions of what is now referred to as Lyme disease was presented in 1909. Dr. Arvid Afzelius of Sweden reported a migrating red rash in patients bitten by Ixodes ticks. The rash was later named erythema chronicum migrans (ECM) (Habicht et al., 1987). In 1948, Dr. C. Lenoff postulated a spirochete as the causative agent of ECM when he discovered spirochetes in the lesions of numerous dermatoses involving ECM (Burgdorfer, 1984).

First Reporting in the U.S.: Lyme disease was first recognized in the United States in Lyme, Connecticut. Since 1975 the disease has spread to at least 34 states and is widespread in Wisconsin, Minnesota, New York, northern California, and Washington. Today Lyme disease is found all around the world, including Africa, Australia, China, Japan, Europe, the Soviet Union, and the United States.

CAUSATIVE AGENT
The etiologic agent of Lyme disease has been determined to be a spirochete. To identify the Lyme disease spirochete, I. dammini spirochetes were isolated using both direct and indirect immunofluorescence procedures (Benach et al., 1983). Transmission electron microscopy results showed that the spirochetes were similar to Treponema pallidum, the etiologic agent of syphilis (Hovind-Hougen, 1984). However, like some borreliae, the Lyme spirochete is tick-transmitted and can be grown in medium (Schmid, 1984). In contrast, none of the treponemes pathogenic to humans have been grown in medium (Steere et al., 1979). DNA homology studies, reported by Hyde and Johnson (1984) and Schmid (1984), revealed that the Lyme spirochete represented a new species of Borrelia. In honor of Willy Burgdorfer, the scientist who first described the bacterium, the Lyme disease spirochete was named Borrelia burgdorferi (Beck et al., 1985).

Borrelia burgdorferi is a gram-negative helically shaped bacterium. Gram-negative bacterial cell membranes consist of an outer mem-

Figure 20-2. Example of a Literature Review (Continued)

brane, a periplasmic layer (housing a peptidoglycan layer and periplasmic space), and a plasma membrane (Baltimore et al., 1986).
Borrelia burgdorferi is microaerophilic (low oxygen requirement),
ranges in cell diameter .18-.25 μm and length 10-30 μm, contains LPS,
and has flagella called axial filaments (Beck et al., 1985; Steere,
1984). LPS is an endotoxin located within the outer membrane and is
poisonous to humans (Schmid, 1984). Borrelia burgdorferi cells regurgitated from I. dammini into the host are lysed by the host's immune
system components. Consequently, the toxin is released into the
bloodstream, causing fever, chills, and shock (Glanze, 1985). Habicht
et al. (1987) reported that LPS isolated from Lyme spirochetes is the
causative agent of ECM. Axial filaments of B. burgdorferi are struc-
turally identical to other gram-negative bacterial flagella. However, they are located in the periplasmic space between the outer membrane and the plasma membrane, not at the cell surface as seen in other
bacteria (Barbour, 1986).

Plasmid DNA from B. burgdorferi has been characterized. Plasmids
are covalently closed, circular extrachromosomal strands of DNA composed of 1,000-30,000 nucleotide pairs. Genes contained within plasmids may encode for end products, resulting in antibiotic resistance
and the organism's virulence (Baltimore et al., 1986). Barbour (1988)
reported that B. burgdorferi has a 49-kilobase linear plasmid with
covalently closed ends carrying genes for outer surface proteins OspA
and OspB. The molecular weights of these proteins, the plasmid count,
and the infectivity of B. burgdorferi change as a result of in vitro
cultivation (Barbour, 1988; Schwan et al., 1988). These phenomena may
explain the recurring fever associated with Lyme disease because they
indicate that plasmid DNA from B. burgdorferi mutates, resulting in
morphologic changes of the organism (Barbour, 1988). Consequently, B.
burgdorferi evades any immune response launched by the host functioning to destroy it. However, it is not clearly understood whether these
changes in B. burgdorferi result from certain nutrients in the medium,
Kelly's modified medium (BSK II), or whether genetic control at the
level of plasmid DNA causes these changes (Hyde and Johnson, 1984).

Strains of B. burgdorferi from North America and Europe differ in
plasmid profiles. Of 13 spirochete isolates studied, only two had
identical plasmid profiles (Barbour, 1988). Johnson et al. (1984) reported that I. dammini spirochetes contain two plasmids, whereas I.
ricinus and human cerebrospinal fluid spirochetes contain single
plasmids. Borrelia burgdorferi is an unusual bacterium because
strains have both linear and circular plasmids (Hovind-Hougen, 1984).
It is reported that the linear plasmids have covalently closed ends

Figure 20-2. Example of a Literature Review (Continued)

(Barbour & Garon, 1987). Only among eukaryotes and their viruses have covalently closed ends of DNA been found (Barbour & Garon, 1987).

VECTOR

The vector of Lyme disease is a tick. The disease is usually transmitted to a host via an infected <u>Ixodes dammini, I. ricinus,</u> or <u>I. pacificus</u> tick. Hyde and Johnson (1984) reported that these ticks are of the same species. Spirochetes were isolated from the midgut and ovary of <u>I. dammini</u> by Willy Burgdorfer, an international authority on tickborne diseases. Steere et al. proposed that hosts become infected as the tick feeds; that is, ''the spirochete may be injected into the bloodstream through the saliva of the tick or it may be deposited in fecal material on the skin'' (1983). Russell Johnson (1988) reported that the tick must feed for 8-10 hours in order for the spirochete to be transmitted to the new host.

HOSTS

The hosts of <u>I. dammini</u> include wild animals (deer, mice, migratory birds, etc.), domestic animals (dogs, horses, cattle, etc.), and humans. The two-year life cycle of <u>I. dammini</u> includes four stages: egg, larva, nymph, and adult. Eggs are usually laid in the spring and hatch into larvae in the summer, and at that stage of the life cycle they feed on blood until the fall or winter. The following spring, larvae molt into nymphs, which feed on the blood of various hosts. In the fall, nymphs molt into adults which attach to hosts (usually white-footed mice or white-tailed deer), where they mate. Males die shortly after mating, but females continue to feed to obtain protein necessary for egg development (Habicht et al., 1987). Nymph ticks most commonly attach to humans; however, ticks at any stage can infect humans (Bosler et al., 1983; Habicht et al., 1987).

In a 1983 study by Steere et al., the absence of spirochetes in unfed larval ticks suggested that larvae may acquire <u>B. burgdorferi</u> from animal reservoirs. Later studies demonstrated that animals differ in their infectivity to ticks. Some animals, such as the white-footed mouse (<u>Peromyscus leucopus</u>) (Levine et al., 1985; Anderson et al., 1987; Donahue et al., 1987), are effective transmitters of the Lyme spirochete. Other animals, however, such as deer (<u>Odocoileus virginianus</u>) (Telford et al., 1988) and catbirds (<u>Dumetella carolinensis</u>) (Mather et al., 1989), do not appear to be competent transmitters of the Lyme spirochete.

Experts feel that this disease has spread so rapidly in humans for three reasons. First, deer populations carrying <u>I. dammini</u> ticks have

Figure 20-2. Example of a Literature Review (Continued)

exploded in recent years (Anderson et al., 1987). Second, migratory birds often carry I. dammini ticks to uninfected areas (Bosler et al., 1983). Third, the tiny size of I. dammini makes them difficult for humans to detect (Hovind-Hougen, 1984). The Centers for Disease Control, in Atlanta, Georgia, has placed Lyme disease as this country's most common tickborne disease, exceeding Rocky Mountain spotted fever (Johnson, 1988). Specifically in Wisconsin, there has been a much higher rate of increase in the number of cases of the disease than in other states. State epidemiologist Dr. Jeff Davis estimated that in 1988 the number of cases would be close to 5,000, compared to 273 in 1987 (''Researchers identify deer tick parasite,'' 1988).

CONCLUSION
Currently, there are many studies being conducted on Lyme disease. One such study is focused on a more consistent method for characterizing and identifying B. burgdorferi isolates cultivated in vitro for long periods by use of DNA restriction endonuclease analysis (LeFebvre et al., 1989).

Reference List

Anderson, J. F., Johnson, R. C., Magnarelli, L. A., Hyde, F. W., & Myers, J. E. (1987). Prevalence of Borrelia burgdorferi and Babesia microti in mice on islands inhabited by white-tailed deer. Applied and Environmental Microbiology, 53, 892-894.
Baltimore, D., Darnell, J., & Lodish, H. (1986). Molecular cell biology (Vol. 2). New York: Scientific American Books.
Barbour, A. G. (1986). Biology of Borrelia species. Microbiological Reviews, 50, 381-400.
——. (1988). Plasmid analysis of Borrelia burgdorferi, the Lyme disease agent. Journal of Clinical Microbiology, 26, 475-478.
Barbour, A. G., & Garon, C. F. (1987). Linear plasmids of the bacterium Borrelia burgdorferi have covalently closed ends. Science, 237, 409-411.
Beck, G., Halbicht, G. S., Benach, J. L., & Coleman, J. L. (1985). Chemical and biological characterization of a lipopolysaccharide extracted from the Lyme disease spirochete (Borrelia burgdorferi). Journal of Infectious Diseases, 152, 108-117.
Benach, J. L., Bosler, E. M., Hanrahan, J. P., Coleman, J. L., Habicht, G. S., Bast, T. F., Cameron, D. J., Ziegler, J. L., Barbour, A. G., Burgdorfer, W., Edelman, R., & Kaslow, R. A. (1983). Spirochetes

Figure 20-2. Example of a Literature Review (Continued)

isolated from the blood of two patients with Lyme disease. New England Journal of Medicine, 308, 740-742.

Bosler, E. M., Coleman, J. L., Massey, D. A., Hanrahan, J. P., Burgdorfer, W., & Barbour, A. G. (1983). Natural distribution of Ixodes dammini tick. Science, 220, 321-322.

Burgdorfer, W. (1984). Discovery of the Lyme disease spirochete and its relation to tick vector. Yale Journal of Biology and Medicine, 57, 515-520.

Donahue, J. G., Piesman, J., & Spielman, A. (1987). Reservoir competence of white-footed mice for Lyme disease spirochetes. American Journal of Tropical Medicine and Hygiene, 36, 92-96.

Glanze, W. D. (Ed.). (1985). The Mosby medical encyclopedia. New York: New American Library.

Habicht, G. S., Beck, G., & Benach, J. L. (1987). Lyme disease. Scientific American, 257, 78-83.

Hovind-Hougen, K. (1984). Ultrastructure of spirochetes isolated from Ixodes ricinus and Ixodes dammini. Yale Journal of Biology and Medicine, 57, 543-548.

Hyde, F. W., & Johnson, R. C. (1984). Genetic relationship of Lyme disease spirochetes of Borrelia, Treponema and Leptospira. Journal of Clinical Microbiology, 20, 151-154.

Johnson, R. C. (1988). Lyme borreliosis: A disease that has come into its own. Laboratory Management, 60, 34-40.

Johnson, R. C., Hyde, F. W., & Rumpel, C. M. (1984). Taxonomy of the Lyme disease spirochetes. Yale Journal of Biology and Medicine, 57, 529-537.

LeFebvre, R. B., Perng, G. C., & Johnson, R. C. (1989). Characterization of Borrelia Burgdorferi isolates by restriction endonuclease analysis and DNA Hybridization [Mimeograph]. Department of Microbiology, University of Minnesota School of Medicine, Minneapolis [R. C. Johnson].

Levine, J. F., Wilson, M. L., & Spielman, A. (1985). Mice as reservoirs of the Lyme disease spirochete. American Journal of Tropical Medicine and Hygiene, 34, 355-360.

Mather, T. N., Telford, S. R. III, MacLachlan, A. B., & Spielman, A. (1989). Incompetence of catbirds as reservoirs for the Lyme disease spirochete. Journal of Parasitology, 75, 66-69.

Researchers identify deer tick parasite. (1988, October). Stevens Point Journal, 12 Oct. 1988:3.

Schmid, G. P. (1984). DNA characterization of the Lyme disease spirochetes. Yale Journal of Biology and Medicine 57, 539-542.

Schwan, T. G., Burgdorfer, W., & Garon, C. F. (1988). Changes in infectivity and plasmid profile of the Lyme disease spirochete, Borre-

---

Figure 20-2. Example of a Literature Review (Continued)

---

lia burgdorferi, as a result of in vitro cultivation. Infection and Immunity, 56, 1831-1836.

Steere, A. C. (1984). Conference summary. Yale Journal of Biology and Medicine, 57, 711-713.

Steere, A. C., Grodzicki, R. L., Kornblatt, A. N., Craft, J. E., Barbour, A. G., Burgdorfer, W., Schmid, G. P., Johnson, E., & Malawista, S. E. (1983). The spirochetal etiology of Lyme disease. New England Journal of Medicine, 308, 733-739.

Steere, A. C., Hardin, J. A., Ruddy, S., Mummaw, J. G., & Malawista, S. E. (1979). Lyme arthritis: Correlation of serum and cryoglobulin IgM with activity, and serum IgG with remission. Arthritis and Rheumatism, 22, 471-483.

Telford, S. R. III, Mather, T. N., Moore, S. I., Wilson, M. L., & Spielman, A. (1988). Incompetence of deer as reservoirs of the Lyme disease spirochete. American Journal of Tropical Medicine and Hygiene, 39, 105-109.

Activities

1. Prepare an analysis of the two example literature reviews in Figures 20-1 and 20-2, focusing on the following questions: Which of the two reviews do you find easier to read? What specifically makes that review easier to read? Can you tell who the audience is for each of the reviews? If not, why can't you tell? If so, how can you tell?

2. Write a summary of each of the two example literature reviews at the end of this chapter.

3. Working with two or three students, locate a literature review. Prepare a brief report on the group's understanding of the audience for and the context in which the literature review was written. Include in the report what in the review led to the group's understanding of the audience and context.

4. In technical journals, find several literature reviews organized in different ways. Prepare an oral report for the class on the organization of the reviews. In addition to describing how the reviews are organized, discuss why the writer might have chosen the particular form of development.

5. Some literature reviews arrive at a conclusion about the subject being reviewed. Using an index such as the *Readers' Guide to Periodical Literature*, find a literature review that arrives at a conclusion. Write a one- to two-page report in which you analyze the reasonableness of the conclusion. Is it clear

how the author arrived at the conclusion? Is it possible to arrive at a different conclusion? What might be an alternative conclusion?

6. Prepare a three- to five-page literature review on a topic of current controversy. If you are in an agricultural field, the topic might be the debate over the use of pesticides, or the topic might be something outside your field that interests you, such as whether pornography is harmful or why so few officers from Southern states resigned their commissions and joined the Confederate Navy. The review should be written for your instructor, who may or may not be knowledgeable about the subject. The review should not be a list of sources on a particular topic, but rather should reflect your point of view on the topic after having read a number of reviews. After thoroughly reviewing and revising your literature review, publish it by producing a well-designed document and submitting it to your instructor.

Research Reports

21

Situation

Form
   Title and Author
   Abstract
   Objective
   Literature Review
   Procedures
   Results
   Conclusions
   Appendixes

Writing the Research Report
   Collecting Information
   Organizing the First Draft
   Reviewing to Assure Quality
   Publishing the Research Report

Example of a Research Report

Activities

A report of research or observation describes the author's research and the results of that research. Both the research report and the literature review describe research. But while a literature review uses publications to survey research done by others, a research report provides an in-depth, first-hand description of a single research project.

Situation

There are two basic reasons for writing research reports. One is the need to report on basic research. Such research can take many forms: field studies, surveys, and laboratory research are common examples. Once a researcher has studied animal behavior in the wild, surveyed employee preferences, or made a study of chemical reactions in the laboratory, the research report is the means by which the researcher communicates his or her findings to the world.

Research reports may also be the result of a specific need for information — information often required before a decision can be made or action can be taken. For instance, as part of an environmental impact statement a writer might include information about the habits of an endangered species that inhabits a site near proposed construction. As part of a feasibility report, a writer might describe observations he or she made of a group of employees struggling to meet deadlines with outmoded equipment. As part of a report on construction delays a writer might describe tests that were run on soil samples. In such cases the research report becomes part of a larger document, providing background information that will allow readers to make informed decisions.

The important point to make about both kinds of research report is that sooner or later, people will be taking an action as a result of the report. This makes research reports very different from the laboratory reports that students typically write for their college courses. The point of the latter report is to prove that the student has done the work and has come up with the expected results. It is not very likely that the professor reading the report will be surprised by the results or feel challenged by them (unless they were terribly wrong!). The only consequence of this type of report is generally a passing grade.

In a professional context, however, it is highly unlikely that readers would know the results of your research until you report it. After all, if they already knew the answer, why would they read your report? Not only might they not know your results, they might not like your results. Your results might differ from their expectations, or threaten their position, or cost them money.

For example, a report on an endangered species might show it is being threatened by new construction. This will obviously affect decisions about whether or not to proceed with the planned construction. The report will be read very thoroughly by those who favor that construction. Even a simple observation report concluding that new equipment must be purchased will be examined carefully by those who are responsible for watching company resources. Research done professionally has consequences.

In addition to providing the basis for decisions, research can affect people's actions by becoming the basis for other studies. Those who are interested in your field will use your experiment in their own research. They may repeat all or part of your study, or they may build on your techniques and results. In either case, a good study enters into the ongoing professional discussion of its subject. The point is that research done professionally has consequences; it never exists in isolation. For this reason research reports must be thorough and accurate.

Those who read research reports do so for one of three reasons: they wish to learn more about a field, they are already doing research in the area and want to learn more to help their own research effort, or they are looking for evidence to help shape a policy.

Reports of basic research are rarely read from beginning to end by more than a few other specialists in that field. On occasion a published piece of medical research may make the six o'clock news, but most studies are not seen by the general public or intended for them. Even studies commissioned to resolve policy matters are rarely presented directly to the public or to management. They are usually included as appendixes to larger reports, available to those who wish to see them, but are not placed where a layperson might encounter them accidentally. At most, nonspecialists may read the report's abstract or its conclusions, the parts that describe the essence of the report without requiring the reader to follow the actual research process.

Because research reports are read primarily by people already knowledgeable in the field, technical terms or concepts can be freely used, generally without explanation. Similarly, generally known and accepted assumptions, theories, methodologies, or studies can be referred to with little or no explanation (although sources may have to be cited). As always, having an expert audience also means that your readers have enough knowledge and experience to be suspicious of any information that seems inaccurate or obsolete, or of any argument that seems flawed. Expert audiences are highly critical audiences.

The use of research reports as bases for decisions is another reason why they must be absolutely accurate, thorough, and up-to-date. The conclusions derived from your research will have many effects, and some people will be made uncomfortable by those effects. How might they respond? The best way for them to challenge your conclusions is to challenge the data upon which the conclusions are based. If your tests were conducted using nonstandard equipment or in a manner that is not normally accepted, then those who disagree with your results are in a position to dismiss your conclusions.

Whether reporting on basic research or on research destined to help answer a specific policy question, a research report must do two things to be effective. First, a research report must clearly and persuasively explain what the results and consequences of a study are. These are the portions of research studies that are read most frequently, and are often the only portions that nonspecialists read.

Second, a research report must explain how research was conducted and why it was done that way. This is partly so that future researchers will be able to use the same procedures in their own research. But it is also so that you will be able to demonstrate that your research is valid. Remember that some readers may disagree with the conclusions or dislike the implications of a study. If they wish to debate your conclusions, they will have to dispute the findings of your research, and the simplest way is to find a weakness in the procedures used. For this reason research reports include exhaustive descriptions of subjects, methods, materials, statistical approaches, and other detail that is often difficult reading. Without this background, readers are unable to determine the validity of any claims a researcher is making.

Form

The need for a research report to provide information and to stand up to rigorous scrutiny is reflected in the form such a report takes. A report on research or observations normally follows this outline:

> Title and Author(s)
> Abstract (optional)
> Objective
> Literature Review (optional)
> Procedures
> Results
> Conclusions
> Appendixes (optional)

The order and contents of this outline make sense, given the way such reports are presented and used. The report is broken into sections, each of which is fairly independent. A reader who is interested in the quantitative results of a study, for example, may go directly to that section and need not read the writer's interpretation of the results. The sections are also quite standardized. Almost every report published in a professional journal follows this format or one close to it. This is because a research report, as a contribution to the ongoing professional discussion of a topic, is likely to be read by a variety of people from a variety of language and technical backgrounds. Communication is much easier if a universal format is followed.

The abstract, placed at the beginning, summarizes the most important facts about the report so that readers can decide whether or not they wish to read the whole report. The appendixes place at the end any related or lengthy material that is really not necessary to an understanding of the report. The rest of the report format follows the idealized order in which researchers themselves go about conducting research:

1. What do I want to know? (Objective)
2. What is already known on the subject? (Literature Review)
3. By what methods do I find out what I need to find out? (Procedures)
4. What did I find out when I did this? (Results)
5. What does this mean? (Conclusion)

Title and Author

The titles of research reports are often long, reflecting the need to precisely delineate the nature of what was studied. Long, precise titles also aid the electronic storage and retrieval of material, based on a search for key words in the title. For instance, a database program might select all references with "pollution" or "quality" in the titles, followed by a second, more selective key word search ("pollution, water") to limit the number of titles accessed.

The problem with a computer access strategy limited to key words found in a title is that if a title does not contain the expected key words, the report will be missed by the computer. Because electronic accessing of titles in databases is common, writers carefully compose the titles of their reports. The title below is typical:

Cuticular water loss unlikely to explain treeline in Scotland (Grace, 1990).

The title contains four terms that would help a computer search pick up the report: cuticular, water loss, treeline, Scotland. Each would help identify the article as one of interest to researchers in these areas.

By contrast, the title could have been written as "Research into water loss." The first word, "research," is of no value since all reports involve research. "Water loss" will be picked up, so someone interested in that subject would see the article listed by the computer, but readers interested in trees or Scotland or cuticles might never know the article even existed.

Publishers will also often ask authors to supply additional key words to help with computer searches, but the title is crucial in helping with the search process and helping readers quickly determine whether an article identified in a computer search contains relevant information.

There are two reasons for identifying what author or group conducted the research. One is to give readers a source for more information or clarification if they need it—they know whom to call with questions. The second reason is authority. When senior managers see on the cover of a report the name of a trusted employee who has been with the company many years, they read the report very differently than they would if it had been written by someone new. This is true for basic research as well: reports written by an established investigator in the field will be read differently from reports by lesser known researchers. The credibility of research results should be based on science, not on the reputation of the author, but as a practical matter, trust and experience still play a role in decision-making. For that reason, if you are working on a report written by a team, you may want to put the most senior member's name first.

Abstract

Not all research reports contain abstracts, but an increasing number do. Good abstracts facilitate the reading of a report and help the reader decide whether to read the report in its entirety. The type of abstract used in research reports is the *informative abstract*. It provides information about each major section of the report: what the study objectives were, which methods were used, and what the conclusions were. This information should be clear and complete enough so that the casual or busy reader can, if necessary, read the abstract rather than the report itself (see chapter 13, "Technical Summary").

Objective

The reason for stating the objectives of a study is for clarification: readers are told the intentions and scope of the research.

> Objective: Determine ability of the secondary O-ring to seal with ice in joints under dynamic conditions. (Presidential Commission, 1986.)

This objective statement, from a report on a study conducted after the crash of the space shuttle *Challenger*, makes it clear what is being tested and under what conditions. Readers know what to expect from the study and what not to expect: the report will not consider the performance of O-rings under extremely hot conditions. The limits of the study are now clear.

The "Objective" section of a report would also contain the formal statement of the research hypotheses, if any. These hypotheses should be listed consecutively and stated positively:

> 1. Predictable stages of growth on important features of language development will emerge (Loban, 1976).

Literature Review

In longer reports writers may include a review of studies already conducted on the topic. A literature review in a research report serves three purposes. First, the review is an overview of other studies that a reader might find useful. Providing it is a service. Second, the review is an opportunity to provide the background for the study. It helps explain why the study took the form it did. Third, the literature review begins to establish the quality of the study. By showing that the study is directly connected to accepted research done previously, the literature review can build credibility for the research report (see chapter 20, "Literature Reviews").

Procedures

The "Procedures" section of a report describes how the research was conducted; it may also offer brief explanations of why research was done in this manner and not in some other way. Conventionally, a "Procedures" section covers three topics: subjects (e.g., people, animals, or sites studied), methods (the ways in which data were collected), and materials (the tools or substances used in the research). This section can be quite long and often is divided into subsections, typically one each for subjects, methods, and materials.

The point of this section is simple: it is here that readers determine whether they can trust your conclusions. They look at your procedures and ask, are they reasonable? Appropriate? Standard? If the research procedures used are common in the field, this section can be fairly short. But if the study involved novel techniques or if it arrived at controversial conclusions, many readers, especially those who disagree with the results, will scrutinize this section carefully.

The "Procedures" section will also be read carefully by investigators who wish to use the techniques in their own research. They may replicate the procedures exactly or adapt the methods to their own purposes. In either case, they will want exact details.

For these reasons, "Procedures" sections are often quite lengthy. Descriptions of human subjects, for instance, might include such information as age, ability, ethnic background, socioeconomic level, prior training, and so on. Descriptions of laboratory animals typically include species information, age, and general condition. Anything that might eventually prove relevant should be included here.

Sections on methods can be even more detailed, showing exactly how information was obtained and how it was classified. In the example below, the researcher describes how teacher ratings of students were obtained and classified.

In every year of the study each subject's teacher rated him or her on a specified series of language factors, with each factor scored on a five-point scale. Throughout the course of the research, the following features of oral language, each defined for the teacher, were included:

1. amount of language
2. quality of vocabulary
3. skill in communication
4. organization, purpose, and control of language
5. wealth of ideas
6. quality of listening

In addition, beginning in grade four, the teacher was also asked to rate the subject on quality of writing and on skill and proficiency in reading. Inasmuch as a cumulative average of teacher's rating was the basis by which the investigator selected the subgroups for special study, the scale merits particular attention. A sample of the teacher's rating scale may be found in Appendix A. Thus, we have at least thirteen teacher's ratings per child. These ratings were averaged in order to select the thirty-five more proficient and the thirty-five least proficient subjects in language (Loban, 1976).

This level of detail makes the method of collecting data quite clear, and it is this level of detail which generally characterizes research reports. The one exception arises when the study uses methods that are standard within the field and known to most readers. In the example below, an entire series of tests is simply referred to in one sentence:

The tests were based on the results of the O-ring Dynamic Blowby Test series. (Presidential Commission, 1986)

Since the authors were using standard tests, they needed no additional explanation.

Results

The purpose of the "Results" section is to describe the results of the research clearly. This is not always easy since there is often a large quantity of information that comes from a study. In addition, the results may take many forms: raw data, compiled data, or statistical interpretations of the data, for example. Consequently, the "Results" section often makes up a significant portion of the total report, and, like any other technical description, it relies heavily on a logical and rigorous organization. The "Results" section also typically makes good use of tables, charts, or other graphics. The table in Figure 21-1 and the bar graph in Figure 21-2 demonstrate one report's presentation of results.

Figure 21-1. Table in "Results" Section of a Research Report

PROCESSOR, MEMORY & DISK

	AT&T	COMPAQ	IBM	TANDY	GATEWAY
80386 INSTRUCTION MIX	4.57	4.48	4.46	4.38	4.12
FLOATING POINT CALCULATION	8.45	9.67	8.46	8.02	7.19
CONVENTIONAL MEMORY	0.83	0.83	0.83	0.74	0.55
DOS FILE ACCESS (SMALL)	67.20	64.04	76.59	68.22	72.18
DOS FILE ACCESS (LARGE)	5.67	7.31	7.79	9.21	8.75
BIOS DISK SEEK	21.49	24.77	26.78	26.82	29.70

Of course, tables and graphs do not stand alone; their content is also summarized in the text of the "Results" section. The written description helps readers understand what a table or chart represents and highlights particularly significant results that might be difficult to see in a visual display of data.

Figure 21-2. Bar Graph in "Results" Section of a Research Report

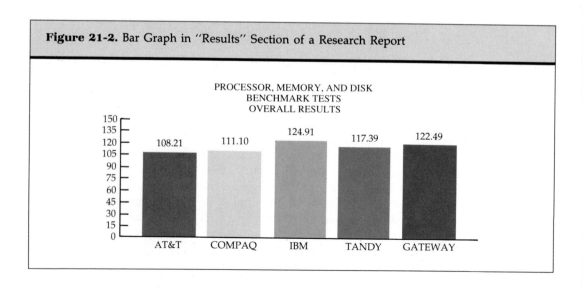

PROCESSOR, MEMORY, AND DISK
BENCHMARK TESTS
OVERALL RESULTS

The "Results" section does not interpret the findings presented; that occurs in the next section of the report. The objective of the "Results" section is to present the information discovered in the study in a way that is clear to readers. The example below comes from a study group examining fragments of the *Challenger* (Presidential Commission, 1986).

Right hand
    sooted appearance
    propellant quantity from burnback charts
    ocean location south of right track

A piece of the wreckage is described as having a "sooted appearance." No conclusions are drawn about how the piece was burned. That is left for a later portion of the report. For now, the section simply states what was found and where it was found.

This neutrality about the results extends to decisions about what information should be included. Writers should be careful to present all results, not just those that support a particular interpretation or prior expectation. If they are uncomfortable with some results because they believe the results were accidental or flawed, they can say so and explain why they feel the results are marred. But it is always the writer's responsibility to be complete.

Conclusions

The "Conclusions" section of a report interprets the results and discusses their implications. For that reason, this section is sometimes labeled "Discussion." This is the writer's chance to state what conclusions should be drawn from the study and to comment on the implications of these conclusions for future studies or for the field at large. If the research was conducted and the report written to form the basis for a larger decision, this is the section in which recommendations might be given.

Be aware, however, that other readers will develop different interpretations based on your results. They may see a pattern you missed, or they may have additional experience with this kind of research, or they may have an axe to grind. Once you present your results they become public information and can be used by anyone. Readers may come to the same conclusions about your data as you do, but don't be surprised if they form alternative conclusions. Disagreement is part of the scientific process.

A report's conclusions should be directly related to the results of the investigation and should not be more elaborate than the results warrant. Remember the scope of the research you stated in the objective section. If your conclusions start drifting from that initial objective, it is time to bring some discipline to your writing.

Appendixes

The appendixes in a research report are generally used to include information that would be too lengthy or interruptive if placed in the body of the report. For example, the "Procedures" section of the educational research report quoted above described the rating scale researchers used to collect data from teachers, but the scale itself was not presented in the report's "Procedures" section, it was included in an appendix. Appendixes are commonly used to show the actual questionnaires, surveys, or other data-collection documents used in the research. They may also be used to present the raw data generated by the research, if the data are too lengthy to include in the "Results" section but might be of interest to some readers.

Writing the Research Report ————————————

Collecting Information

One of the hardest problems of writing research reports is remembering to take good notes while doing the study. Whether they are doing a study in the laboratory, conducting a survey, or doing field observations, it may be days or even weeks before writers have an opportunity to create the report. By then they may have great difficulty remembering the details of how they conducted a study. They may even begin confusing some of the results.

An easy way to improve notetaking is to create a special form for each aspect of a study. For instance, if observing human subjects, you could quickly use a word processor to create a form with headings such as this:

Subject Name:

Date and Time of Test:

Subject Description:

Test Description:

Test Materials:

Sequence of Behaviors Observed:
    Time   Behavior

Such a form will help ensure that you collect the same information for each of your observations and that none of the details will be lost.

Remember that readers who disagree with a study will not only carefully read the results, they will also examine the details of the procedures. If the procedures are uncertain or incomplete, the study will be dismissed. There is also the possibility of legal action if the results of a study are used to choose a certain policy and errors are subsequently found in the report.

One way to be certain that information about procedures is not lost is to write as much of the report as possible even before doing the study. Information about such areas as subjects and expected research methods are usually known before a study begins. These sections can often be written in advance of the study. As the study is completed, the sections can then be changed if changes were made in the procedures. This is often much easier than trying to write the procedures from memory.

Organizing the First Draft

Since reports on research are presented in a fairly standard way, organizing the first draft of the paper will consist mostly of putting information where it belongs in the outline.

Unfortunately, not every report fits the pattern perfectly. Short reports, for instance, often merge the results and conclusions. Longer reports may have subsections to explain procedures, such as subsections devoted to subjects, methods, and materials. Even conclusions can be organized around groups of results or tied to specific issues of concern. Ultimately, organization of a research report should reflect the nature of the material and the needs of readers. This is also the point at which you should decide whether you need to include drawings, photographs, or other visual evidence.

While organizing the first draft may be simple in general, there are two portions of the draft that may take special effort—"Results" and "Conclusions." The "Results" section must present every relevant piece of data you collected in a manner that is organized and meaningful but not biased. Tables of results can help keep details organized and make them more accessible to readers. Always consider using a table when you have a large number of discrete pieces of information to present in this section. It is critical that the basis of the analysis be discussed in great detail. Notice the level of detail used in the example conclusion below.

> Reliability based on the Kuder-Richardson formula pooled for the 438 subjects yielded coefficients of .76, .55, and .70 respectively for the twenty-four-item instrument, the twelve-item H subtest, and the twelve-item S/C subtest (Fagan, 1985).

The author is careful to include the details of the results, and the name of the formula used to determine reliability. Such detail is critical in reporting research results.

The "Conclusions" section must be a logical analysis of the data. The strength of this analysis will determine in large measure how seriously the research report will be taken. (See chapter 17, "Analysis.")

This section can also be quite large. One way to organize it is to have it reflect the original objectives or hypotheses. Each of these original objectives could be listed in turn, followed by the conclusions and a discussion of relevant results. A way to begin each of these subsections would be to state whether the hypothesis or objective was proved. If so, why does this seem to be the case? If not, why might the objective not have been achieved? If there are multiple conclusions to be made for each objective, each conclusion should be listed and discussed in turn. Such an organization was used in the example below:

> *Comparison 5.* This comparison was made on grades obtained in the spring semester. Again, as in comparison 1, the improvement of the revision grades over the original grades for the two teachers using computer-assisted composition was compared with the teacher whose class did not use computer-assisted composition. This contrast was significant, $F (1,96) = 8.08$, $p < .01$. Inspection of the means in Table 2 reveals that there was greater improvement of the originals in the two classes using computer assisted composition than in the class not using computer-assisted composition (McAllister, 1988).

Reviewing to Assure Quality

The two sections of any research report that should be reviewed most thoroughly are "Procedures" and "Conclusions."

For "Procedures," start by putting yourself in the position of your readers. Reread the section. Can readers repeat your research by following what you have said? Are there any sections that report unusual procedures? Are those sections clearly stated? Are your techniques justified? Do you use any acronyms or terms that not all of your readers would know at first glance?

For the "Conclusions" section, envision someone who completely disagrees with you. List the arguments this reader might make to refute each of your conclusions. Now return to "Results." Is each of your conclusions justified by these results? Is any other interpretation possible? Is yours the best? Why? This is the time to drop any conclusions that were not fully supported by the results and to consider the possibility of other interpretations of the data. You may even want to describe other interpretations in your report; if there are good reasons why you think these other interpretations are less accurate than yours, you can explain them.

Publishing the Research Report

There are two aspects of research reports that are particular problems when putting the final draft of the report together. One has already been mentioned—the fairly common need to include large numbers of results.

These results are often placed in tables or otherwise grouped so that they do not overwhelm the report. Tabulated data and appendixes should be labeled carefully and explained in the report as necessary.

A second problem arises from the need to include drawings, photographs, or other visual evidence. Most of these items should be placed at the end of the report as appendixes. However, if a particular photograph or illustration is especially important in explaining results or drawing conclusions, it should be included and explained in the relevant section of the report. Be wary of including too many such illustrations, however, since they break up the report. When in doubt about an illustration, put it in an appendix.

Example of a Research Report

The following short research report in Figure 21-3 is drawn from the studies that were done after the crash of the space shuttle *Challenger*. It was used to determine to what extent ice might have contributed to the leaks that caused the explosion.

Activities

1. List words or phrases from the example research report in Figure 21-3 that demonstrate that the report was intended for a technical audience. Discuss substitutes you would use if this report were intended for members of Congress.

2. What does the example research report do to ensure that its conclusions will measure up to careful scrutiny? Point out any places that most clearly make this effort.

3. In a small group, write an abstract for the example research report. Check to be sure that it summarizes each portion of the original report.

4. Review a recently published research study in your area of interest. Carefully read the procedures section. Could a professional in the field repeat this study? Identify sections that might be clearer.

5. Find a recently published study that contains a lengthy "Results" section. How is this section presented? How much of the information is presented as text, how much in tables, and how much in graphs?

Figure 21-3. Example of a Research Report

Test No. MTI 114   Test Report Date: TBD
Title: Ice in Joint Tests

Objective:
Determine ability of the secondary O-ring to seal with ice in joints under dynamic conditions.

Test Description:
The test fixture is shown in Figure C.30. Following assembly and successful leak testing, a pressurization fixture is used to apply pressure to the downstream side of the second O-ring. This operation is to reposition the O-ring to the upstream side of the groove, as indicated by tests at MSFC.
Distilled water is then introduced into the gap using a hypodermic syringe and needle. The fixture is then conditioned to 250F and tested.

Test Limitations:

- Subscale
- Cold gas

Test Variables:

- Extrusion gap

Results/Conclusions:
Six tests were conducted in this series. The tests were based on the results of the O-ring Dynamic Blowby Test series. That series of tests showed the following important results:
1. At 250F with an initial gap of 0.004 inch, the primary O-ring consistently leaks at a chamber pressure between 200 and 400 psi. There is a significant amount of blow-by past both seals generally followed (4 out of 5 times) by seating of the secondary O-ring with no further indication of leakage.
2. At 250F with an initial gap of 0.020 inch, the primary O-ring seals with little or no leakage or blow-by. The secondary O-ring in this case is not forced to seat.
   Because of the pronounced difference in behavior at these two sets of conditions and in order to assess the effect of ice on the secondary O-ring sealing capability, different approaches had to be taken for the two sets of test conditions.
   With an initial gap of 0.004 inch, it was simply a matter of repeating the earlier tests but with ice introduced into the secondary O-ring groove. The results of these tests were then compared to the earlier tests to assess the effect of ice. The comparison shows that ice does seriously impair the sealing capability of the secondary O-ring with this set of test conditions.
   For an initial gap of 0.020 inch, it was necessary to create a situation in which the secondary O-ring was forced to function. Therefore, these tests were run with the primary O-ring removed. In order to provide a basis for evaluation, the first two tests were run without ice. When the comparison was made of the tests with and without ice, it was again

Figure 21-3. Example of a Research Report (Continued)

obvious that ice had a serious detrimental effect on the capability of the secondary O-ring to seal.

The results to these tests indicate that ice in the secondary O-ring groove will impair the sealing characteristics of the secondary O-ring.

Source: Presidential Commission. (1986). *Report to the President on the space shuttle Challenger accident.* (Vol. 2, p. L-235). Washington, DC: U. S. Government Printing Office.

6. Conduct a small research study and write a report on it. Some suggested research topics include

a. A survey of student opinion on marketing

b. An observation of student interaction during meals

c. A study of memory or visual perception

The study should be six to eight pages long, and written to inform your classmates. You may want to do the research in small groups and share the task of writing the report. After thoroughly reviewing and revising your research report, publish it by producing a well-designed document and submitting it to your class.

Proposals

Situation
Solicited Proposals
Unsolicited Proposals
Internal Proposals
Audience
Common Constraints

Form
Summary
Introduction
Description of the Solution
Organizational Qualifications
Evaluation
Budget
Appendixes

Writing the Proposal
Setting a Schedule
Collecting Information
Organizing the First Draft
Reviewing to Assure Quality
Publishing the Proposal

Example of a Proposal

Activities

A proposal is an offer to perform one of three services: to furnish goods, to research a subject, or to provide a solution to a problem. Proposals are among the most common documents in technical writing. They range from one-page memos to thousand-page "books," from formal to informal, from extremely technical to almost nontechnical. Nonetheless, all proposals have one thing in common: they are trying to persuade the reader to follow a certain course of action.

Even though proposal writing is common, for many writers the proposal is the most difficult writing task they are asked to do. It demands all of a writer's skills: thorough research, persuasive argument, effective organization, and the skillful writing of definitions, descriptions, analyses, and summaries.

Situation

Many proposals offer to solve a problem by correcting a situation. For example, the city of Stevens Point, Wisconsin, may have discovered that the drinking water systems of three local communities had unacceptably high nitrate levels. Proposals written in response to this situation would offer to solve a clear problem: potentially unsafe drinking water. On the other hand, some proposals are written when no actual problem exists. These proposals simply offer to provide a possible benefit that, while not necessary, is highly desirable. For example, a privately funded fine arts center might sponsor a competition for a ceiling-mounted design that would enhance the courtyard of the building. Proposals written in this situation would not be solving a problem—the lack of such a design doesn't cause any hardship—but offering to improve an already acceptable condition.

Solicited Proposals

Most proposals are prepared in response to a notice calling for a proposal. Such a notice is most often referred to as a request for a proposal (RFP). An organization issues an RFP when it has identified a need for goods or services but doesn't know who the best supplier would be. By using the RFP to describe what it wants, the organization creates an efficient, competitive situation: every interested supplier is challenged to come up with a plan for meeting these needs at the lowest possible cost and to submit a proposal describing this plan.

RFPs are used by many different types of organizations and in many different situations. A local government, for example, might request proposals providing solutions to the problem of high nitrate levels in drinking water. A corporation might request proposals for a new underground sprinkler system. A grant agency might request proposals for establishing teacher-training programs. Or a private organization, such as a fine arts center, might request proposals for a new ceiling-mounted design to enhance its building.

Because there is no one source for RFPs, it is often difficult for organizations to locate them. Most RFPs issued by private businesses and industries appear in newspapers, trade magazines, and business journals. The major source of RFPs issued by the U.S. Government is *Commerce Business Daily*, published by the U.S. Department of Commerce. (It is available from the Superintendent of Documents, Washington, DC 20402.)

RFPs are usually very detailed, containing specific requirements for both content and format that acceptable proposals must meet. Following is a partial list of the specifications included in the fine arts center's design competition for a ceiling-mounted design:

> Each submission must include complete dimensional orthographic drawings of the proposed design (scale $1/8'' = 1'0''$)—plan view and four elevation views.
>
> Each submission must include one full-color perspective rendering to scale of the proposed design as it would appear installed in the courtyard.
>
> Drawings and rendering are to be mounted on foamcore boards not to exceed $24'' \times 30''$ (overall dimensions of boards).
>
> Each design proposal must include a cost estimate for materials, labor, and installation of the design.

The major question to be answered in a solicited proposal is, why us? The proposal that most closely meets the RFP's specifications—and usually price is an important factor—will most likely win the contract. Proposal writers must persuade the readers that their company and plan is the best choice.

Unsolicited Proposals

An unsolicited proposal is born when an individual or a company determines it has a product or service that another company can use and benefit from. The major question to be answered here is, why act? That is, the proposal writers must convince the proposal readers that they should act on the proposal. In most cases this means that the proposal will first have to convince readers that they have a problem, and then that the individual or company represented by the proposal can help them solve it.

For example, a manufacturer that has developed a new type of filter for improving the quality of drinking water in municipal water systems might submit proposals for installing its new filter to one or more local governments. The manufacturer's proposal must first persuade the local governments that unfiltered water is a potential hazard and then persuade them that the new filter is the best way to solve this problem. As with solicited proposals, though, some unsolicited problems attempt to provide a benefit rather than solve a problem. For example, a food services company may want to encourage its customers (cafeterias in several office buildings) to offer a new line of desserts on their menus. The proposal describing this option would try to persuade the readers that it would be a worthwhile improvement, but it would not need to identify an actual problem. In both cases, the proposal must first convince the readers to act and then convince them that the proposal offers the best possible course of action.

A proposal typically not written specifically in response to an RFP is a grant proposal. One good source of information about all aspects of grants, including where to find foundations and grant information, how to research previously funded grants, and how to prepare grant proposals, is the Grantsmanship Center (1015 West Olympic Blvd., Los Angeles, CA 90015). A nonprofit educational institution, the center conducts workshops, employs a research staff, maintains a library, and publishes *The Grantsmanship Center News*. Another excellent source of grant information is the *Illinois Researcher Information System* (IRIS), an on-line computerized file containing descriptions of over 4,000 funding opportunities from federal agencies, private and corporate foundations, and other not-for-profit organizations.

Internal Proposals

Most proposals are external, that is, they are directed from one company or organization to another. However, some proposals are internal: they are directed from one person or department to another within the same organization. Like external proposals, internal proposals may be solicited or unsolicited. The communications office within an organization might solicit proposals for improving mail distribution; proposals written by various individuals or committees would thus be solicited internal proposals. If an employee realized that communication among branches of his company would be improved if each branch used the same word processing software, he might write a brief unsolicited proposal recommending this improvement to his supervisor.

Internal proposals can be extremely brief and informal (for example, mentioning orally to the office manager that purchasing a few more inexpensive staplers would improve productivity), or they can be as lengthy and detailed as any formal external proposal. The length and level of formality will largely depend on how much time and money are involved. In any case, careful presentation and a well-developed persuasive strategy are just as important for most internal proposals as for external ones.

Audience

Proposal readers are decision-makers. Initial screeners decide whether your proposal should be considered at all; ultimately, readers will decide whether your proposal should be accepted. Because an effective proposal must elicit a positive decision at each step along the way, knowing the readers' interests and needs is extremely important.

Like other decision makers, proposal readers are much more favorably disposed toward a proposal if they can quickly and easily find all the information they need to make a well-informed decision. No one wants to spend half an hour flipping through a 500-page document trying to find an important calculation or to wade through a lengthy description when a concise one would have been enough. Proposal readers also tend to be highly critical. They almost always have several options available — in the case of solicited proposals, they usually have more than one proposal to choose from; in the case of unsolicited proposals, they can decide either to act or not to act. So they can afford to scrutinize each proposal carefully and to be skeptical of every claim.

Find out as much as possible about your readers. If you are writing an internal proposal, this should be somewhat easier: you probably already know something about your readers, either by reputation or from first-hand experience. But try to find out more. Above all, try to evaluate the situation from your readers' point of view, keeping their interests in mind. If you are preparing an internal proposal recommending that all branches of your company use the same word processing software, you will need to recognize that managers will primarily be interested in whether the proposal will save money or improve productivity.

An organization may have interests or needs other than those specified in an RFP; it is up to you to uncover this information and use it to your advantage. For example, would a fine arts center be more interested in a conservative, classical design or in a bold, innovative statement? How important is it to the city of Stevens Point that the solution to their water supply problem be environmentally friendly?

Although the ultimate decision to accept or reject your proposal will most likely be made by executives, your proposal may have many readers, each of whom may take part in the decision-making process. In various parts of the proposal you will need to address and convince busy executives who need to see all the pertinent facts at a glance, skeptical experts who will pore over every fact and assertion, assistants who will check to make sure that your proposal fits the required format, and various other managers and employees who may be primarily interested in how the proposal, if accepted, would affect them and their jobs.

An effective proposal must convince these readers of three things: that your organization understands the readers' needs, that the proposed plan is feasible, and that your organization is able to implement the plan. In most cases there will be many proposals, but only one will be selected. Therefore, all aspects of the proposal must be persuasive.

Common Constraints

Most solicited proposals are written in a highly competitive atmosphere. In almost every case, proposal writers can be sure that someone else at a different company is working just as hard to make a persuasive case for the acceptance of *that* proposal. Even unsolicited proposals are often competitive. The proposed plan may have to compete with a product or service currently in use, or it may be trying to preempt a product or service that has not yet come on the market.

Time constraints, always present in technical writing, become critical in the case of proposals. Solicited proposals have deadlines. Although it may be possible to turn in one of a series of progress reports a day late, a proposal that does not meet the deadline specified in the RFP usually will not be considered. Timeliness is often a factor for unsolicited proposals, too. Because every other supplier is a potential competitor, it is an advantage to have your proposal be the first one considered, before anyone else has been able to submit one.

Because proposals must meet deadlines and because they are often both lengthy and complex, they are frequently written by more than one person. Teams of writers collaborate to develop a persuasive strategy, conduct and report on basic research, assemble background information, and prepare the final report. There may even be parts of the proposal that are borrowed from other proposals—not all of the writers may be sitting in the same conference room. (See chapter 6 for more information on effectively managing group projects.)

Form

Regardless of the level of formality or the length of the proposal, proposals have certain characteristics in common. Typically, proposals do the following:

1. Describe the problem (if necessary, arguing that there is one)
2. Describe the solution
3. Describe how the solution can be implemented
4. Argue that the proposer is the best person or organization for the job

The actual form of a proposal is more variable than the form of a research report; it depends heavily on the audience, the purpose, length and formality of the proposal, and on whether the proposal is solicited or unsolicited. Additionally, the RFP for a solicited proposal may include many specifications as to the organization and format of the proposal. However, in order to meet the objectives above, many formal proposals include these elements:

Summary
Introduction
Description of the Solution
Organizational Qualifications
Evaluation
Budget
Appendixes

Summary

The summary is an informative abstract of the proposal that briefly outlines the most important points. Summaries are usually included in both solicited and unsolicited proposals, although short, informal proposals may omit a summary. The summary of a proposal is important for several reasons:

- The summary provides an overview of the proposal, serves as an advance organizer, and sets the tone.
- Readers who have many proposals to look at, and who are screening proposals for other readers, will often pass along or reject a proposal on the basis of the summary alone.
- Because the summary may be the first or only part of a proposal that is read by executives, it can play a large role in the final decision.

A summary for a proposal should be organized to help with decision-making. The writer has to ask, what information will readers need in order to act? Because the summary is usually less than one page long, writers must be concise.

Below is the summary of a proposal submitted to the Wisconsin Department of Natural Resources recommending a solution to the problem of the threat of pollutants. This summary includes several common elements: a statement of the problem, a description of the solution, and a description of the benefits of the proposed plan.

SUMMARY

Problem
Pollutants are saturating our atmosphere. Regionally we have the effects of acid rain. Globally we have two similar pollutant threats which are capable of causing

catastrophic changes to Earth's climate: chlorofluorocarbons and the greenhouse effect.

The first pollutant threat is caused by the release of chlorofluorocarbons, manmade chemical compounds, into the atmosphere. As they accumulate high above Earth they consume ozone, Earth's protective layer. Ozone shields life on Earth from ultraviolet radiation. Increased levels of ultraviolet radiation causes skin cancer, eye ailments, and other serious health problems, as well as causing environmental damage to crops and marine ecosystems.

The greenhouse effect, the second threat, is caused by continuous buildup of carbon dioxide, nitrous oxide, and trace gases, including CFCs (chlorofluorocarbons) in the atmosphere. Human activities have increased the concentrations of these gases since the Industrial Revolution. Rising into the atmosphere, they create a blanket-like layer. Visible light reaching Earth is converted into infrared light (heat), which ordinarily escapes back into space. The buildup of greenhouse gases can trap the heat and ultimately results in rising temperatures.

Solution

Dupont, the largest manufacturer of CFCs, has a simple solution. On March 25, 1988, Dupont called for a worldwide phaseout of CFCs by the year 2000. This will cost hundreds of millions of dollars in research and plant construction. This statement followed a report by the Ozone Trends Panel of the National Aeronautics and Space Administration which indicted manmade chemicals as culprits in the destruction of the stratospheric ozone layer.

Benefit

Successful reduction of worldwide CFC levels within the next six to eight years will affect all living human beings and every ecosystem. The atmosphere will be able to maintain a constant chemical climate of oxygen, nitrogen, and water vapor. A constant chemical climate will maintain the radiation balance by blocking out excess ultraviolet radiation. In other words, our grandchildren and great grandchildren may experience life on Earth as we have.

Introduction

The introduction is probably the second most important part of the proposal. After the summary, busy readers typically look to the introduction for more detailed information about the proposal. If the introduction is lengthy or confusing or vague, readers are not likely to read further. Therefore, be brief, clear, and specific.

A successful introduction must accomplish several things. First, the introduction provides background information on the situation. Lengthy or excessively technical information is often confined to appendixes, not presented in introductions. The introduction, however, may include a general discussion of how a situation has developed and why it is significant. Second, the introduction identifies the problem to be solved. Here you must be specific and clear: What is the problem, why is it significant, and what are the immediate objectives? Third, if the proposal is unsolicited, the introduction argues that the problem needs to be solved. Fourth, the introduction summarizes the solution to the problem. While the discussion should not be too detailed here, the reader will want at least a brief description of the solution. Finally, the introduction helps to establish your credibility. While most, if not all, of your readers will already know much of the background information, you must demonstrate to them that you understand the scope and complexity of the problem or situation and have a thoughtful solution to recommend.

Following is the introduction to the proposal presenting a solution to the problem of high nitrate levels in drinking water. The writers provide background information on the problem, describe the problem, and offer a solution.

> Nitrates are salts of nitric acid that can cause a disease called methemoglobinemia (blue baby syndrome) in infants. Excessive consumption of nitrates can cause hemoglobin problems in adults.
>
> Nitrates are filtering through the ground and contaminating local wells and drinking water systems. Sources of nitrates include nitrogen fertilizers (commercial and residential), breakdown of manure, and human wastes in septic effluent.
>
> The U.S. Environmental Protection Agency (EPA) has set a mandatory standard level of 10 parts per million (ppm) under the Safe Water Drinking Act passed by Congress. Presently, several area wells have been shut down or are very near to exceeding EPA limits. Consequently, the surrounding communities of Whiting and Plover are left with the problem of finding a new adequate water supply.
>
> After examining three alternative solutions, we feel that the most economical solution to this serious health problem, the solution that would benefit both the city of Stevens Point and the villages of Whiting and Plover, would be for the city to annex the villages.

Description of the Solution

Whether a proposal is offering to furnish goods, research a subject, or provide a solution to a project, proposal readers will want to know what the plan is and why it is a good idea. In other words, this section of the proposal must serve both as a technical description and as a persuasive argument.

First you must clearly and precisely describe your plan: what, exactly, do you intend to do? Readers are not likely to be persuaded if you simply say that your company "will install an adequate computer network." Without overwhelming the reader with details, you must describe which computers, the kind of network, and the method of installation.

Next, you must argue that your plan is the best one: why is installing a computer network the best solution to the problem? Why these computers and not some others? As part of your argument you may want to describe any additional benefits that would result. For example, the new computers might significantly increase the company's graphics capabilities. You might also identify precedents for your plan—another company in a similar situation that has successfully adopted this solution, for example. A further strategy is to anticipate and counter possible objections: if the new computers seem too expensive, you could show why they are a wise investment.

Organizational Qualifications

In addition to describing what you will do, you will have to persuade the readers that you or your company are capable of carrying out the task you have described. This is often accomplished by presenting the names, titles, and qualifications of the primary members involved in the project. Most government proposals require that proposers list not only key personnel involved in a project but also those who are peripherally involved. The organization's qualifications as a whole may also be described—its reputation, history, experience, and resources are all relevant here. Additionally, many proposals cite previous successes in this field or on similar projects.

Evaluation

Although many client organizations prefer to evaluate the results of a project themselves or to hire an independent outside consultant to evaluate the work, many RFPs require that those submitting proposals include an evaluation section. In this section of the proposal, the writer must describe the method to be used for evaluating the work done, how long the evaluation will take, and when a formal report can be delivered. Even if you are not required to do so, consider including an evaluation section with your proposal: most readers will be interested to know how the success of the project will be measured and what would be involved in such an evaluation.

Budget

If your proposal involves money—and most proposals do—this is where you indicate the cost. If the proposal is solicited, readers often rely heavily on this section to compare one proposal with another: if every proposal presents a good solution, then the least expensive option is usually the best one. If the proposal is unsolicited, this section helps the reader decide if implementing the proposed action is worth the expenditure.

The best guideline is to be as careful as possible. Make sure you understand and include every relevant budget item, including fringe benefits for personnel (health insurance), travel, overhead expenses (rent, lighting, heat, maintenance), and materials and supplies. In addition, many budgets include estimates of the future funding that would be needed to continue a project or to maintain equipment.

The biggest danger here is not, as many novice proposal writers believe, estimating too high, but just the opposite — estimating too low. This statement taken from a Department of Health and Human services RFP explicitly emphasizes the importance of providing a reasonable and realistic budget:

> Proposals which are unrealistically low or do not reflect a reasonable relationship of compensation to the professional job categories so as to impair your ability to recruit and retain competent professional employees, may be viewed as reflecting a failure to comprehend the complexity of the contract requirements. The Government is concerned with the quality and stability of the work force to be employed on this project. The compensation data required will be used in evaluation of your understanding of the project requirements (U.S. Dept. of Health and Human Services, 1985).

The budget below is in the format required for proposals submitted to the Fund for the Improvement of Secondary Education (FIPSE). It uses standard headings for budgets. (Note that FIPSE requires that budgets include "indirect costs," a fixed percentage of the request, as part of the total amount requested.)

Budget Item

A. Direct Costs:
1. Salaries and Wages	
a. Professional	$27,800
b. Clerical	1,750
2. Employee Benefits	5,600
3. Consultant	4,000
4. Travel	0
5. Materials and Supplies	0
6. Equipment (Purchase or Rental)	0
7. Production (Printing, Reproduction)	0
8. Other	0

B. Indirect Costs	3,132

TOTAL Requested from the Fund	$42,282

Appendixes

An appendix contains supplementary material that is too lengthy to include in the main body of the proposal or that is of interest to only a few readers. For example, although the most important qualifications of the team that is to carry out a project might be discussed under "Organizational Qualifications," complete résumés for each member of the team might be included in an appendix. Similarly, although the water-supply proposal discussed earlier includes a general description of the problem in its introduction, independent reports on the danger of nitrates or on the current state of the water supply might be included in appendixes.

Writing the Proposal

If you are preparing an unsolicited proposal, know exactly what it is that you want to do. If you are preparing a solicited proposal, know exactly what you are being asked to do. Read the RFP again carefully, looking for any additional information about the project. For example, the "Statement of the Problem" that begins the ceiling design RFP simply calls for "the installation of a ceiling-mounted design." However, in a later section, titled "Structural Considerations," the RFP specifically describes the materials that currently exist in the area where the ceiling design will go:

> The predominant materials used on the interior of the fine arts center are unpainted concrete, neutral-colored brick, and glass. Small amounts of natural oak are used for window trim, mullions, and doors. There is little color in the courtyard area, except for the floor treatment (red/orange carpet) in the balcony area—the carpet should not be considered as a design constraint.

This description allows the writer to coordinate the new design with the existing materials and colors.

Setting a Schedule

More than almost any other writing task, proposals demand that you set a schedule. You will need information from various sources, and you will likely be working with other people to write the final document. You must set strict deadlines for yourself and for others.

Collecting Information

Because each idea and fact in a proposal must stand up under scrutiny, and because the proposal could become the basis for a decision, proposals must be exhaustively researched.

It is not likely that you or your group will be the sole source of information for your proposal. You will probably need technical and financial information from other people and sources both inside and outside your own organization. But don't underestimate how much you *do* know. Regardless of the proposal's scope or subject matter, spending a few minutes brainstorming about what might be included in each section of the proposal or about what a good argumentative strategy could be will probably prove quite helpful.

Gathering information for a proposal usually means getting information from others. You might have to interview a design engineer to find out alternative solutions to the method of packaging software that you are proposing, you might have to write letters of inquiry to city governments to find precedents for your proposed solution to waste-water disposal, you might have to contact the personnel department of your organization or agency for a list of personnel qualifications, or you might need to determine some of the details of the solution and its implementations, such as contacting suppliers for budget figures. It is important to begin collecting information from other people as soon as possible. Be sure to tell them exactly what information you need and when you need it.

The writers of the proposal offering a solution to the problem of high nitrate levels in drinking water had to interview a county attorney about the process of annexation, and, in order to provide information about the status of area wells, the writers needed property tax information from three municipalities.

Libraries are invaluable sources of information. For example, the writers of the proposal offering a solution to the problem of high nitrate levels thought it would strengthen their argument for annexation if they could show precedents for their solution. They conducted a library search and found other cases in which annexation had solved the problem of water quality.

For an unsolicited proposal, the library can often supply data on the scope and the problem. Census and other demographic data, for example, are essential for documenting many human resource problems.

Organizing the First Draft

Because a proposal must be persuasive, it is critical that the writer anticipate competitors' arguments in order to present information in a way that makes the reader understand and agree with that information.

ANTICIPATING THE COMPETITION

If you are writing a solicited proposal, you should expect that other proposals will be submitted. Anticipating your competitors' arguments allows you to strengthen your own arguments. You might use your knowledge about the competition in a general way, as a guide to the kind of information to include, or more aggressively. In an aggressive strategy you detail the arguments you expect your competitors to present and counter them with arguments showing why your proposed plan of action is better. For example, in the proposal offering a solution to the problem of high nitrate levels in drinking water, the writers discussed and rejected several alternative solutions they expected their competitors to present.

SELECTING DETAILS

In gathering information for a proposal, you will almost always accumulate more details than you can use. The successful proposal writer figures out what information the reader wants to know, not what information the writer wants to give. If your audience is highly knowledgeable, you might not have to provide as much background information. However, if you do not expect your readers to have a firm grasp of the subject or the situation, you should provide sufficient background information that your readers feel qualified to read the proposal. Think of the proposed project or solution from your readers' point of view. Their questions or concerns may be very different from yours.

In the proposal offering a solution to the problem of high nitrate levels in the water supply, for instance, the writers had to determine what the readers, who were city officials, wanted to know. Since they were recommending the annexation of three communities as a way to assure quality drinking water, they expected their readers would want to know the economic ramifications of the solution. Therefore, they emphasized the low cost of the solution in their proposal.

DETERMINING THE PROPER FORMAT

If the RFP indicates a format, *follow that format.* You don't want to lose the contract because you didn't follow the required format. Below are three required formats for proposals, representing widely varying levels of specificity.

The first example is the format required by the Department of Health, Education and Welfare for proposals submitted to the Fund for the Improvement of Postsecondary Education. The guidelines for this format are quite broad, indicating only the barest outline for a proposal.

Proposal Title
Brief Abstract of Proposal
The Problem
Objectives
Description of Program
Implementation
Evaluation
Budget

The second format, required by a private foundation, is more detailed than the first (Fig. 22-1). The writer has a clear idea of what information is

Figure 22-1. Proposal Format Specified in an RFP

I. Title of Proposal
II. Purpose of Proposed Project
   A. Target population, community type and size, and high school size (if appropriate)
   B. Name and other identification of theory being applied
III. Description of Proposed Program
   A. Overview of how the program will work
   B. Explicit explanation of how theory is incorporated into the program to produce the desired results
   C. Degree of success expected at the end of one year of program operation
IV. Plan for Implementation of the Proposed Program
   A. Selection of program staff
   B. Selection/recruitment of program participants
   C. Starting, modifying, and continuing the proposed program for at least one year
   D. Specific resources needed
      1. Personnel
      2. Institutional supports
         a. Endorsements, licenses, approvals
         b. Contributions of staff time
         c. Contribution of funds
         d. Contribution of services and supplies
         e. Contributions in the form of access to facilities
      3. Physical plant and equipment
   E. Budget for one year's operation
      1. Personnel
         a. Wages, salary
         b. Fringe benefits, insurance
      2. Materials
         a. Capital purchases ($500 and above)
         b. Other materials and consumable supplies
      3. Services
         a. Liability insurance as needed
         b. Utilities (telephone, light, heat, air conditioning, etc.)
         c. Transportation rentals, mileage expenses
         d. Equipment leases, maintenance contracts
         e. Professional services, testing and evaluation
V. Program Evaluation—Attainment of Goals
   A. How will goal attainment be demonstrated?
   B. How will success of the program be judged?
   C. How often will the program be evaluated (monthly, quarterly, annually)?
   D. What standards will be used in deciding to reduce or expand the scope of the program?
   E. What is the smallest practicable unit of program increase or decrease which can be used without causing severe program disruption?
   F. Who is empowered to make decisions concerning each of these program decisions?

Figure 22-2. Proposal Format Specified in an RFP

    4.6.4. Environmental Requirements
        Provide a complete description of the environmental requirements for the proposed system and its peripherals for full and initial capacity indicating capacity to provide additional requirements (e.g., cards, shelf, cabinet, nodes). The following areas should be described:
  4.6.4.1. Space requirements (square footage, room dimensions, floor plan) indicating cabinet sizes, numbers and weights
  4.6.4.2. Floor loading
  4.6.4.3. Electrical requirements
  4.6.4.4. Air conditioning (in tons)
  4.6.4.5. Heat dissipation (in BTU/hour)
  4.6.4.6. Humidity controls
  4.6.4.7. Room treatment
  4.6.4.8. Wall space required
  4.6.4.9. Power consumption (must be 3 phase)
  4.6.4.10. Special fire protection/preventative devices
  4.6.4.11. Minimum and maximum temperature requirements
  4.6.4.12. The power consumption of the PBX at the equipped line size under normal traffic conditions
  4.6.4.13. The power consumption for the air conditioning system for the equipped line size.

Source: Reprinted, with permission of the editors, from *Request for proposal: Telecommunications system.* (1984). Stevens Point, WI: University of Wisconsin.

required and in what order. The disadvantage of a highly detailed format is that it is less flexible.

The third example, a brief excerpt from a highly detailed RFP issued by a telecommunications system, offers even more specific guidelines for the preparation of the proposal (Fig. 22-2). Like many RFPs, this one specifies that "[p]roposals must be submitted in the format and structure specified in this section. Please do not restate the RFP language, but do use the section numbering and titling provided" (*Request for proposal: Telecommunications system*, 1984).

If there is no required or recommended format, remember that almost all proposals have certain characteristics in common and that most proposals contain some or all of the following elements:

Summary
Introduction

Description of the Solution
Organizational Qualifications
Evaluation
Budget
Appendixes

Whether or not there is a required format, compartmentalize the proposal and use clear captioning techniques. You want to make it as easy as possible for your readers to find their way around.

Because there was no required or recommended format for the proposal offering a solution to the problem of high nitrate levels in the water supply, the authors of that proposal chose their format for the following reasons.

1. They included a summary because they had written proposals before and knew that readers expect a summary. Although the summary traditionally is placed first, it is almost always written last.
2. Because the proposal was solicited specifically to help solve a problem caused by unacceptably high levels of nitrate in the drinking water of area wells, the writers knew they would not have to first convince the readers that there was, in fact, a problem. However, the writers did want to convey that they understood the problem fully, so they included a section titled "Nitrates: The Problem."
3. Because the writers anticipated competing solutions, they next discussed what they felt were the three most likely alternative solutions to the problem and why those solutions were unsatisfactory.
4. The writers ended by recommending their solution because it would provide clean drinking water to the most people for the least cost over the longest period of time.

The final outline for their proposal looked like this:

Summary
Nitrates: The Problem
Alternative Solution 1
Alternative Solution 2
Alternative Solution 3
Recommended Solution

Reviewing to Assure Quality

Although the ultimate measure of a proposal's quality is whether or not it is accepted, you will want to evaluate the proposal as best you can before sending it out.

When reading over your own document, pay special attention to the style and vocabulary. As far as possible, the style and vocabulary of the proposal should follow that of the RFP. A particular hazard of proposals is the use of stuffy, overblown language. Federal government writers have a special fondness for abstruse expressions. Inflated language is difficult to read and hampers communication. Reviewers of proposals want clear, factual language. Rewrite any parts of the proposal that do not meet this criterion.

If there is no format specified by the RFP, make sure that the section headings of your proposal are clear and logical, which will make the proposal easier to read.

Colleagues can often spot problems a writer misses. Ask a colleague to look for the following:

- Is the presentation clear? Are the design and layout of the proposal easy to follow? Are the captions and graphics easy to understand?
- Is the proposal coherent? Do all the paragraphs and points clearly relate to one another?
- Is the argument convincing? Is there anything that would make it more convincing?

Although reader-based testing is not as common for proposals as for instructions and computer manuals, occasionally it is possible to have a potential reader look at your document before it is submitted. Sometimes there are formal procedures for this; the document submitted in this case is called a pre-proposal, generally only a three- to five-page description of the complete proposal. These procedures are most commonly used by foundations and government agencies; FIPSE, for example, always uses pre-proposals.

If there is no pre-proposal procedure, and if you cannot find a potential reader to look at your proposal, ask someone who is *like* a typical reader for your proposal to read it, or ask a colleague to look for weak spots in the proposal.

Publishing the Proposal

The last step is to reproduce and distribute the proposal. Because proposals entail many hours of work and can result in substantial business contracts or funding, this step is crucial in the writing process. Unless you are submitting an informal, internal proposal, use good paper and good printing. On the other hand, avoid costly packaging, which may jeopardize your standing with nonprofit granting agencies by suggesting that too much of the budget goes for appearances and too little for direct services to clients. The proposal should look professional but not flamboyant.

Example of a Proposal

The document shown in Figure 22-3 is the solicited proposal written in response to the problem of contamination of the water supply by nitrates in two Wisconsin communities.

Activities

1. The example proposal in Figure 22-3, *Proposal for Ground Water Problem*, was written for the Stevens Point City Council. What evidence indicates that the writers had council members in mind when they prepared their proposal? Prepare a brief memo to the writers of the proposal suggesting recommendations that would strengthen the proposal.

2. Select a different audience for the example *Proposal for Ground Water Problem*. Rewrite the summary to reflect the new audience's needs and expectations.

3. Find a request for proposal. (One of the best places to find an RFP is in the government documents section of the library.) Prepare a written analysis of the proposal: What can you tell about the audience(s) to be addressed? What is the purpose of the proposal? What features of the RFP make the audience and purpose clear? What information would you need in order to prepare the proposal? What kinds of sources might you use in order to gather information for the proposal?

4. Prepare a proposal for your instructor outlining a research project. Include a description of the intended audience, the purpose, the objectives, and the rationale for the project.

5. Write a proposal recommending a change in one aspect of the class (e.g., assignments, grading, texts, syllabus). The proposal should describe the problem and explain how the change will improve instruction. Address the proposal to your instructor.

6. With two or three other students, prepare an unsolicited proposal offering a solution to a problem on your campus that needs to be corrected (e.g., inadequate computer facilities, long lines at registration, insufficient parking for commuters). After thoroughly reviewing and revising your proposal and developing an effective design, publish it by submitting it to the appropriate individual or office.

Figure 22-3. Example of a Proposal

Proposal for Ground Water Problem

May 9, 1988

Prepared For: Stevens Point City Council

Submitted By: Frank Gagnon and Associates
1776 Main St.
Plover, WI 54489

SUMMARY

The contamination of area wells by nitrates has forced some communities to find a new
source of water. Annexation into Stevens Point will provide Whiting and Plover with clean
drinking water, will result in a growing community, and will provide extra revenue to help
maintain one quality water supply.

NITRATES: THE PROBLEM

Nitrates are salts of nitric acid that can cause a disease called methemoglobinemia (blue baby
syndrome) in infants. Excessive consumption of nitrates can cause hemoglobin problems in
adults.

Nitrates are filtering through the ground and contaminating local wells and drinking
water systems. Sources of nitrates include nitrogen fertilizers (commercial and residential),
breakdown of manure, and human wastes in septic effluent.

The U.S. Environmental Protection Agency (EPA) has set a mandatory standard level
of 10 parts per million (ppm) under the Safe Water Drinking Act passed by Congress.
Presently, several area wells have been shut down or are very near to exceeding EPA limits.
Consequently, the surrounding communities of Whiting and Plover are left with the problem of finding a new adequate water supply.

After examining three alternative solutions, we feel that the most economical solution
to this serious health problem, the solution that would benefit both the city of Stevens Point
and the villages of Whiting and Plover, would be for the city to annex the villages.

ALTERNATIVE SOLUTION 1

Edward Lubar, the mayor of Stevens Point, has proposed to sell the surrounding communities water at a rate of $1.00 per $1000 assessment value. For example, if you have property
worth $60,000 you would pay $60 in water taxes to Stevens Point. Presently Stevens Point
supplies Whiting with water, while Plover and Whiting have several private wells that are
contaminated by nitrates. This proposal would bring Stevens Point approximately $150,000
a year from Plover and an additional $50,000 a year from Whiting. The advantages of this
would be that one well would be used, and the extra monies could be funneled toward the

---

Figure 22-3. Example of a Proposal (Continued)

---

upkeep and quality control of the city's water supply. Worries are that without this extra money for research in the study of ground water, Stevens Point's well might develop nitrate problems because of excessive use. If this happened, nobody will have an adequate water supply.

ALTERNATIVE SOLUTION 2

Each of the outlying communities can purchase land and build their own wells.

Recently, the water nitrate level in Plover was tested to be very close to the mandatory national standard set at 10 parts per million (ppm) by the Safe Drinking Act. If Plover decides to build a new well, it may well decide to share with Whiting. This would mean, along with Plover constructing a new well, a pump station, and a connecting water main, the two villages would have to pump and transport the water from the new well to Whiting's existing municipal well. There they intend to treat and blend the two water supplies before distributing to customers in both villages. Even if Plover chooses to spend millions of dollars on a well, there is no guarantee that the nitrate level will stay below the national standard.

The entire cost of the project has not been fully totaled. It will cost Whiting $13,000 to $15,000 to begin pumping from its own well that was shut down in 1979 by the EPA. In the past year, nitrate levels have been down below the 10 ppm, at close to 8.3 ppm. If Whiting decides to use its own water supply, it will have to install fluoridation and chlorination equipment, along with other improvements.

Whiting may also build another well system which could cost about $500,000. This would include the cost of two new wells and pump stations. The village of Whiting has a couple of problems in mind. One, they only have a one-year contract with the city of Stevens Point for water; two, the village doesn't know if it will be able to connect with Plover's new well if it is of high quality water.

ALTERNATIVE SOLUTION 3

To control the nitrate saturation and clean up the wells would solve everyone's problems, but there are some problems with this proposal.

- The Blue Top Ranch, the main contributor to this problem, refuses to stop pumping their manure into the ground.
- Local potato farmers are still using nitrogen fertilizers on their fields.
- Septic runoff is not being controlled, mainly because property owners are not aware of the contamination happening.
- It could take several years from the time we start to clean up to when the well reaches a safe drinking level.

Steps should be taken, regardless of what action is pursued, to clean up the ground water. Things that could be done are: put a stop to the methods used by the Blue Top Ranch to dispose of their manure, place restrictions on the types and amounts of fertilizers farmers can use, and control the septic runoff from area homeowners who use septic tanks.

Figure 22-3. Example of a Proposal (Continued)

Something has to be done to control the present problem or the Plover aquifer could move and contaminate other wells, resulting in a more serious problem than we are facing right now.

RECOMMENDED SOLUTION

We recommend that the city of Stevens Point annex the villages of Whiting and Plover. Residents of the city of Stevens Point presently pay $10.27 in property taxes per $1000 property assessment. The outlying communities of Whiting and Plover presently pay $0.50 and $0.87 respectively. After annexation is put into effect, both communities will pay the same as Stevens Point. This would mean an additional $9.77 for Whiting and $9.40 for Plover. This amount may seem staggering at first to the residents of Whiting and Plover, but this increase includes the use of all city parks, city sanitation and disposal, police and fire protection, sewage, and, most of all, clean drinking water.

The advantages to this proposal are many:

1) City taxes for everyone may decrease overall because of the additional revenue brought to the city;
2) Some of the additional revenue can be put back into the city to improve conditions;
3) The city will be allowed to grow and prosper because people will no longer be drawn out of the city limits in search of lower property taxes.

In addition to all of these benefits, those who now live outside of Stevens Point and take advantage of the excellent parks and recreational facilities will be paying for these services. If adopted, the plan would call for mandatory annexation of any surrounding communities who wish to use the city's services, including those who desire to share our clean water supply.

There have been similar cases such as this in the past. For example, Brown County, Madison, Chippewa Falls, and Fond du Lac all had similar problems and found annexation to be the solution. The city of Eau Claire took the village of Eau Claire to the state supreme court, winning their case and forcing the village to annex into the city.

Each village discussed here is in search of clean water, so why should each village build their own water system, not knowing if the new wells will be any better than the present wells? The municipalities should combine all available resources, enabling them to preserve one existing water system and supply clean drinking water to surrounding cities.

Feasibility Studies

What Is Feasibility?

Situation

Form
Recommendations
Original Situation
Background of the Investigation
Comparison of Alternatives
Conclusion
Appendixes

Writing the Feasibility Study
Collecting Information
Organizing the First Draft
Reviewing to Assure Quality
Publishing the Feasibility Study

Example of a Feasibility Study

Activities

A feasibility study assesses whether a plan — or each of several plans — is practical, or feasible. If several feasible plans are available, a feasibility study may also determine which option is best. Some common situations involving feasibility studies are described below.

- A small company has a standing contract with a computer firm for computer services to handle payroll and accounting records. A computer salesperson from a different company offers a "really good deal" on a microcomputer and software that will do payroll and accounting in-house. Is owning a computer and doing the job in-house a good idea?
- The local waste disposal site is filling fast. One local group is pressing for more recycling. Another wants to start burning much of the trash. A third wants to expand the local disposal site. Which is the best option?
- Federal legislation requires that smokestack emissions be lowered at an industrial plant. A furnace company arrives with a description of a new boiler that can meet these new emission controls, but at a high cost. A coal supplier suggests moving to a higher grade of coal that is much more expensive per ton. Which is the better solution?
- A manager is convinced that her company would benefit if it dropped its current telephone arrangement and set up its own private branch exchange (PBX), a locally owned telephone switching system. Is such a move practical?

In each of these situations a team of employees would write a feasibility study analyzing the alternatives and making a recommendation.

What Is Feasibility?

Feasibility can be defined in three ways:

1. *Operational feasibility*: Will the solution meet our needs?
2. *Technical feasibility*: Will it work?
3. *Economic feasibility*: Does the solution make economic sense?

Although economic feasibility is often considered the bottom line for decision-making, a solution that is cheap but that does not perform adequately is no solution. Therefore, we will look at the other two kinds of feasibility first.

OPERATIONAL FEASIBILITY

A solution is operationally feasible if it will accomplish what you need it to do. For instance, an operationally feasible solution for each of the examples above would have to adequately perform payroll and accounting duties, dispose of waste, meet clean air standards, or handle all telephone traffic, respectively.

Operational feasibility also requires that a solution perform *well*. For instance, in the computer payroll example, the system should perform not only accurately but efficiently (no one likes a late paycheck!) and should have built-in security. Determining what is operationally feasible mandates a detailed examination of the operation's requirements.

TECHNICAL FEASIBILITY

Many solutions that are operationally feasible are not technically feasible. To be technically feasible, a solution not only must work in general, it must work in the specific environment of proposed use. Take the waste disposal example. Burning waste may work quite well in some places, but in a community with air pollution problems, burning may not be an option. It is not technically feasible. Similarly, changing to a new kind of coal might not be possible if a company is too distant from reliable sources.

An often overlooked area of technical feasibility is human resources. In the telephone system example, the technology may be available to buy and install, but if no one in the company has the expertise needed to operate or maintain the equipment, that solution is not feasible. Determining technical feasibility requires examining the proposed solution in terms of the local environment.

ECONOMIC FEASIBILITY

How can we prove that an alternative is a wise expenditure? Two common techniques are

1. payback analysis, and
2. return-on-investment analysis.

In each case the costs of one approach are compared with the costs of alternative approaches, or with the cost of doing nothing at all. The two techniques use a very different process of computing feasibility.

In a payback analysis one is interested in knowing how long it will take for an investment to pay for itself. In the clean air example, there might be a large initial expenditure to buy a new furnace, but if the new furnace could burn cheaper coal enough savings might be realized on coal costs over the long run to pay for the furnace. A payback analysis computes the length of time this would take; a decision would be based on whether the payback would come fairly quickly or would take much longer. Companies usually have standard payback rates they expect for investments, typically in the range of three to five years.

Return-on-investment analysis calculates an annual rate of return from an investment, that is, how much money the investment is earning each year. Return on investment is comparable to the interest rate for money in a bank: if you deposited money in a passbook savings account, the rate of return (the interest rate) would be around 5%. Calculating the return on investment allows you to see whether the money invested in a project would be better spent elsewhere.

We will use the example of the computer payroll system to compare the two techniques. If the new system costs $50,000 but saves the $5,000 per year that is now spent on the outside company, payback analysis shows that the investment will pay for itself in ten years. This is a long payback period and would generally be rejected.

A potential purchaser would now calculate the annual return on investment. Return-on-investment analysis considers the average savings per year over the lifetime of the equipment. If the computerized payroll system is expected to last for fifteen years, and pays for itself by the end of ten years, an additional $25,000 ($5,000 for each of years 11 through 15) would be saved. A $25,000 return on a $50,000 investment represents a total return of 50%, but that return must be averaged over fifteen years. The annual return is actually less than 4%. In fact, it is a 3.33% annual return (50% return ÷ 15 = 3.3%). Would you invest your money for a 3.33% return? Wouldn't the company do far better just to put its money in a savings account that earns 5% annual interest? Return-on-investment analysis allows decisions about economic feasibility to be made quickly.

Although payback analysis and return-on-investment analysis work in principle, both calculations are complicated by the time value of money and by opportunity costs—essentially, money now has more value than money later. These factors tend to reduce the value of projects that have lengthy payback periods or substantial initial costs. Therefore, the calculations shown above are very optimistic in a real-life setting.

Situation

Feasibility studies are decision documents. They originate with some proposal for change, involve an often lengthy investigation, and result in a document that will be used to select among alternatives. Sometimes the choice is between many alternatives, for example selecting which of many suppliers to use. Sometimes the choice is between continuing with current practices or changing to a new procedure. In every case there are at least two alternatives presented; often there are many more.

Feasibility studies originate in several ways. They may be written after one or more proposals have been submitted by an external supplier. In this case the feasibility study would summarize all of the proposals, presenting the most important information in an accessible form. Often a manager will recognize a problem and ask an employee or work group to investigate the problem and present alternative solutions. On other occasions an employee may initiate a feasibility study. The employee might be aware of special opportunities and might want to convince management to try them, or might be aware of a special problem and has an idea for improving the situation.

In any case the feasibility study presents alternatives that will be used to make a selection. Like proposals, feasibility studies provide information about a particular problem and recommend a solution. But feasibility studies differ from proposals in several important ways. First, they are always internal documents: the interests of the writer and the interests of the reader are one and the same. Second, feasibility studies are more analytic than persuasive. Third, they provide information on a range of options rather than focusing on a single alternative.

Because a feasibility study is used to make decisions, it will be read by those who have the power to make decisions—usually a manager or an executive, but in some instances a work group. In either case the audience consists of knowledgeable individuals in a position to make decisions. At the initial levels these will be people familiar with the situation, but as the study moves up the decision-making ladder, readers may be less and less aware of the situation the study is addressing. These readers may also be less knowledgeable concerning the specific technologies involved. A successful feasibility study provides enough background information that an executive not familiar with the specific project can judge whether the study's recommendation is warranted, and can make a decision on this basis. If decision-makers must look to other resources before reaching a conclusion, the feasibility study is incomplete.

Form

The body of a feasibility study report usually has these main sections:

Recommendations
Original Situation
Background of the Investigation
Comparison of Alternatives
Conclusion
Appendixes

Recommendations

This section briefly states the course of action the writer thinks should be taken. Because the rest of the report may not be read, the writer should also use this section to summarize briefly the original situation and the reasoning underlying the recommendation.

Original Situation

Although most people reading a feasibility study will be familiar with the original situation, as the study moves through decision-making circles it may reach others who are less knowledgeable. In this section the writer presents an organized and complete review of the reasons why the report was necessary in the first place.

The impetus for a feasibility report need not be a problem. A special opportunity may have arisen, or there may just be routine changes in the works. You need not dramatize your report by intimating that a "problem" exists that needs resolution. The people currently doing the job may be embarrassed to find out there is a problem when they thought they were doing their jobs.

Background of the Investigation

This section describes how you went about investigating the situation. Readers will want to know which departments were contacted, what documents were read, what sites visited, which vendors interviewed, and what systems tested. In this section they find out whether or not you did your homework; based on this, they will decide whether they can trust your conclusions.

Comparison of Alternatives

The section comparing the alternatives presents the facts of the case. This section may have as many as three subsections:

1. Criteria for evaluation
2. Individual descriptions and evaluations
3. Chart of differences

The criteria are necessary so that readers will know the basis on which the final judgment was formed. Some of these criteria will be implied in the original problem statement, but they should be listed so that everyone recognizes them. Most criteria will come from the investigation. In this part of your report you specify each of the criteria, describe them, and give some indication of their relative importance. Here is an example:

Operational Expenses
These include regular maintenance costs on the computer, costs for paper and other supplies, the salary of an employee to operate the computer, and the costs of additional security measures. It is important that these costs be less than $2,000 per month.

After the criteria are established, each of the alternatives is evaluated on this basis. This section can be quite long and detailed. If there are many alternatives and a long list of criteria, one way to simplify comparison is a summary chart that serves as a "score card."

Conclusion

By this point in the report everything that really matters has already been said, so it isn't necessary to belabor the point. The "Conclusion" briefly summarizes the investigation and recommendations. This is not the place to introduce new materials or ideas, since it is the section of the report most likely to be skipped by hurried readers.

Appendixes

Typical contents include copies of vendor material, summaries of interviews, copies of suggestions, and results of tests performed. In general, examples or transcripts of the actual material on which the recommendation was based should be included as an appendix.

Writing the Feasibility Study ⎯⎯⎯⎯⎯⎯⎯⎯⎯⎯

The process of writing a feasibility report will be dissected in an extended example. Your company has been contracting with an outside vendor for computer services to handle payroll and accounting records. A computer salesperson from a competing vendor suggests an alternative, a microcomputer and software system that will do payroll and accounting in-house, and offers a good price. Your manager asks you to investigate this offer and prepare a feasibility report that she can take to management within two weeks. What process do you use to create such a report?

Collecting Information

The first step is to write down alternatives. You already know two: continuing with present practice or buying a new computer system. At this point you should ask colleagues if they are aware of other alternatives. If there were time, you might issue an RFP (request for proposal) to other vendors to see if they have products you should consider. Professional publications might have articles describing other alternatives as well. Begin by listing as many alternatives as you can, and add other alternatives as they become known to you in the course of your investigation. The worst thing that could happen to you is for you to present a recommendation to a superior only to be asked, "Yes, but did you ever consider . . . ?"

One way to generate this list of alternatives is to review each of the three kinds of feasibility. During this review you will often discover new approaches, and you will be able to determine the standards that will later be used to decide which solution best meets your needs.

OPERATIONAL FEASIBILITY

An operational feasibility study considers whether the solution will perform the functions required. The first step is to list all the functions the system must perform. Usually this is done with the help of people who know the problem best. In our example, it would mean talking with the people in payroll and personnel as well as with any area managers. You want to know at least the following:

1. *What is done currently?* What are the strengths and weaknesses of the current approach? Often there is a loyalty to current processes, but there may be a list of problems available from sources such as employee suggestion boxes and repair records.
2. *Who are all the people involved?* Include both people directly involved and those who would be indirectly affected. For example, a payroll system affects not just the payroll clerks, but all employees who receive a paycheck. A pollution abatement system affects people who live around your plant as well as your employees.
3. *What are the minimum requirements for the operation?* These would include such things as legal requirements, and requirements based on stated company policy. For instance, air quality laws may set a limit of 5 parts per billion for some pollutant. That is the law. If you review a solution that releases 6 parts per billion, that solution is not operationally feasible.
4. *What are some valuable additions?* Following the law or company policy is just the minimum. Look beyond that. Items to include here might be an enhanced image in the community, better service to customers, or a healthier relationship with employees.

In the case of our computer payroll example, questions about operational feasibility might yield the following notes:

> Current System: Works quite well in general, but there are some problems just before April 15 each year because the company that does the payroll also does taxes and is very busy then. Otherwise no complaints.

> Staff: Susan Martinez collects all the hourly records from division managers and checks them for completeness. She then makes one copy and sends it to the computer company. Within three to five days the payroll is completed and returned to her. She verifies the accuracy of all checks and stores them until payday. Other people involved include division managers who collect hourly records.

> Minimum Requirements: Original hourly reports must be retained in case of disputes. Payroll calculations have to be made to include FICA, state and federal tax, and health insurance coverage. All records must be kept at least seven years. Payroll information is personal and must be kept private.

> Additional Features: Several employees want automatic deductions for savings bonds.

TECHNICAL FEASIBILITY

Now that you know what a system has to do, you can investigate whether the system can perform in your environment. This is the nuts and bolts part of your investigation, but it also involves personnel. You want to know what technology is involved, whether you have it, and if you have staff who can run it. Here are some of the questions you might ask:

1. *Will this system actually work in our company?* For instance, if you were considering installing a telephone PBX, you would want to know if the building could be easily rewired. Is there room for new conduits? In the case of the new payroll system, is there room for a new computer? Is there an adequate power supply? Security? Temperature control?
2. *What system support is available?* This includes repairs, utilities, and consumables. Utilities are often overlooked, but are crucial. In the earlier landfill example, one solution might be a new landfill, but there may be few access roads to the selected site, or the roads may go through residential neighborhoods.
3. *Do we have the people for this alternative?* Too often it is the people who are forgotten. In the smokestack emission example, a new control system might be perfect, but if no one in the company has training in the use of real-time systems, you would need to hire such a person. If you bring in a computer, is there anyone who can operate it?

In the case of the payroll system you might have gathered these notes:

Performance: The new computer system can do a payroll in less time than before, and has built-in security/password protection. The company selling the system is new, though, so it is hard to gauge reliability.

Support: The supplier guarantees it will repair the computer if it breaks. There is room for the computer in the payroll office, and there is sufficient wiring and air conditioning to keep the computer running.

Operations: Susan Martinez has taken a night-school course on computer operations and could handle the payroll system. Several managers, though, voiced anxiety about using computers. They would need training. Who would do the training?

ECONOMIC FEASIBILITY

We have already examined how to calculate payback periods and returns on investment. Although the calculations are relatively simple, they are often done incorrectly because not all costs are included. To be sure you are complete, ask yourself these questions:

1. *What are the original purchase prices?* This is the easy part, but it still requires that you make a complete list.
2. *What are start-up costs?* Any new venture you select will have a number of start-up costs you might not expect. These costs might include training workers, buying new furniture, or installing special wiring or air conditioning. In buying a telephone PBX, for instance, you have to have a place to put it and a power supply. This means the cost of a desk, and the cost of installing new wiring.
3. *What are personnel costs?* People will be paid to operate any new equipment. Even if you don't hire new people, your current employees will be spending time on the new system that they used to spend on other things.
4. *What are operating expenses?* Any system has running costs. Computers use printer paper, and most equipment uses electricity. All equipment requires repair and maintenance.

While looking for all costs, you should also look for all benefits. There might be indirect financial gains from a project. You may not be able to quantify these gains, but training people for one new project may make it easier to train them for the next. Similarly, by learning how to do a project in one area, you may be more able to do another project in that area. In effect, you plan to recoup some of your expenses later while keeping the company up to date and competitive. These benefits, never easy to quantify, should nevertheless be noted.

If you had carefully examined the economic feasibility of your new payroll system, you might have these notes:

Purchase Costs: Computer—$2,400, printer—$940, software—$4,200, tape backup system—$580.
Start-up Costs: Training for Susan—$300. Training for managers—$800. Desk and other office furniture—$1,000. Disk, paper, and other supplies—$300.
Personnel Costs: Susan Martinez will need an extra 15 hours per week to enter employee data and run the program. Either she must be paid overtime, or an assistant for her must be hired. Costs would approach $150 per week.
Operating Costs: Check forms—$30 per month. Disks—$10 per month. Maintenance contract—$40 per month.
Growth Gains: All of the managers would see what microcomputers can do and would learn about them. This might pay off in future automation efforts, and would reduce the cost of future training, but it would be hard to estimate just how much would be saved.

As you complete this investigation, you should begin to see not only what alternatives are available, but also what your detailed requirements are. These requirements should be listed and will become the criteria by which you determine which alternative is acceptable.

For instance, your review of operational feasibility may lead you to list these requirements:

Operational Requirements
    Able to maintain records on 4,000 inventory items
    Able to produce a payroll report by department
    Able to produce a quarterly tax statement
    Simple facility for adding new employees

This list, and the others you produce after considering technical and economic issues, will be critical in later evaluations of alternatives and in justifying your final selection. Any substantial requirements that you miss will invalidate your decision.

Organizing the First Draft

There are two common ways of arranging the segments of a feasibility study:

Recommendations	Original Situation
Original Situation	Background
Background	Comparison of Alternatives
Comparison of Alternatives	Recommendations
Conclusion	Conclusion

The first organization has one major advantage: it places recommendations at the beginning, where they are easier to find. Most readers will look first for this section anyway. The second format organizes the report in chronological order. This format begins with what started the process, then describes what investigators did, what alternatives were available and how they compared, gives recommendations, and closes with conclusions. The difference between the two formats is in the placement of recommendations. The first makes recommendations immediately. The second presents recommendations after readers have read all the background information and fully understand the choices.

Which order should you choose? Putting recommendations first is far more common. It eases the reader's burden. A controversial recommendation may be rejected immediately, however, and the rest of the report remain unread. In such cases it may be better to use the second format and present the controversial recommendation only after taking readers through all the other alternatives.

The rest of the outline follows from the decision to place your recommendations first or toward the end of the report. In either case, each section of the report should be sufficiently clear in itself that readers can make decisions based on the contents.

Begin your report on the situation with a dispassionate description of the original situation. This will include the following:

- A general description of the operation involved, so that people new to that area of the company can understand how things work. (This section should not be overly long, since most of your readers will be familiar with the situation.)
- A review of the current situation, including strengths and weaknesses.
- A discussion of special circumstances that motivated the report.

Materials for this section will come from the notes you took earlier. Your review can come from interviews you conducted, logs you examined, and memos you read.

In the telephone system example, you might begin by describing how telephones are currently handled by the company. It is unlikely that even higher management knows how they get a dial tone when they pick up the phone. How technical should your description be? That depends on your readers. Remember that reports generally circulate farther than you originally expect, so you may in fact be writing for people different from your original audience. When in doubt, be less technical, or allocate technical information to appendixes.

You also want to show what works well with the current phone system, and where things are less successful. Again, rely on your interviews for data. People may be happy with their phone service or angry that it takes a long time to install new lines. Workers may be satisfied with service, but managers might be concerned by costs.

In any case, make every effort to present accurately information you have found and to give sufficient detail so that managers can understand the scope of the problem. If there are complaints, quantify them: "In the last two months there have been four written complaints from customers concerning their inability to reach us by phone."

At this point you also want to describe the motivation for the report. "In an effort to discover if cost savings were possible, . . . " or "Due to recent customer complaints, . . . " or "Given new competitive realities . . . " all are common beginnings for this section.

How you state your motivation will largely determine the expectations of your readers and contribute to their willingness to accept your recommendations. If you begin with the statement, "Due to constant problems in shipping," everyone in shipping and receiving will be on the defensive and looking for unsupportable statements in your report. An alternative and better-worded statement is, "In an effort to further improve customer satisfaction with deliveries. . . . " Your motivation is now recorded in positive language and wounds no more people than absolutely necessary.

The next section describes the background of the investigation. Rather than present a diary ("On Tuesday I spent an hour and a half as Maria Escobar described each step of the quality assurance process"), give precise facts about your actions and present lists of activities. For example, you could state, "This investigation covered the week of June 23d. During that time I interviewed 14 people, including Maria Escobar, Wally Kowalski, and Samuel Brown of Quality Assurance, Stephan Gomez of Customer Relations. . . . " If this is a lengthy list, you might organize it by activity such as interviews, site visits, vendor demonstrations, etc.

For the point-by-point comparison of the alternatives, individual descriptions of alternatives can be arranged in one of two ways. In the first, one alternative at a time is described and taken through each of the evaluation criteria:

Alternative 1: Current phone system
   Cost
   Performance
   Flexibility
Alternative 2: Company PBX
   Cost
   Performance
   Flexibility

In the second, one criterion at a time is described and each alternative is evaluated according to this criterion:

Criteria 1: Cost
   Current System
   Company PBX
Criteria 2: Performance
   Current System
   Company PBX
Criteria 3: Flexibility
   Current System
   Company PBX

The first organization emphasizes each choice, the second emphasizes the criteria for evaluation. The first organization allows readers to focus exclusively on one alternative that might be of special interest to them, but it also means that they could easily skip a description of an alternative you favor. The second organization pits each alternative against the others point by point and so makes direct comparisons easier.

A summary chart may improve the presentation of this material if there are many alternatives and a long list of criteria. You list the alternatives down the side and the criteria across the top, and put an x in each place where an alternative meets a criterion. The smokestack emission example described at the beginning of this chapter might produce a summary sheet like the one shown in Figure 23-1.

A summary chart makes it visually clear why some alternatives are feasible and others are not. However, it is not a substitute for narrative descriptions, since it is the descriptions that explain why an alternative meets or does not meet each requirement.

Figure 23-1. Table Used to Compare Alternatives in a Feasibility Study

	Cost < $3 mill	Reliable Fuels	Proven Technology	Maintenance Contract
Acme	x	x		
HorseHair Ltd	x	x	x	x
Stellar MFG	x	x		x
Stacks R Us		x	x	x

Reviewing to Assure Quality

After you complete the first draft of your report, you will want to revise it using all the techniques described earlier in this book. Two aspects of feasibility reports require special attention.

First, since you are basing much of your investigation on interviews and observations, you will need to verify your quotations and observations. The credibility of your report will be destroyed if one person you quote says "No, that isn't what I said at all. What I really said was. . . ." A good policy is to verify every quotation you use. It just takes a minute to go back to a person and say, "I have you quoted as saying. . . . Is that accurate?"

Second, your draft should be read by others to ensure that you are not trivializing or aggrandizing a problem or misunderstanding the situation; this is standard procedure for technical writing. But a feasibility study poses a special problem when it comes to selecting readers. If your report makes recommendations about personnel (and most do), the report should remain confidential until it has been approved by management. The reason for confidentiality is to avoid upsetting people until it is certain that the course of action you are recommending will be approved. Choose your readers carefully (or with the approval of management), or work alone. Your study may also consider a major new direction for the company: Should the company open a new office? Introduce a new product line? Again, finding readers involves a special sensitivity to the nature of the study being done.

Publishing the Feasibility Study

A feasibility study is published much as any other business report. The only aspect of the study that might be somewhat unusual is the likelihood of substantial appendixes. Appending such information to the main report can be difficult. One solution is to photocopy selected portions of such material. This creates standard-sized copy that can easily be bound with the rest of the report. And of course, you will need to be selective in what you include. The point of the appendixes is to provide material that a few readers might want to form their own conclusions or to verify your conclusions. There is no point to placing every conceivable brochure or record in an appendix.

Example of a Feasibility Study

Figure 23-2 is a sample feasibility report analyzing solutions to the computer payroll situation described earlier. In this case, the payroll manager has asked a systems analyst in data processing to study the possibility of the new computer system.

Figure 23-2. Example of a Feasibility Study

TO: Grace Owens, Director, Payroll Department

FROM: Elaine Brown, Systems Analyst, Data Processing

DATE: July 12, 1990

SUBJECT: Proposed Computer Payroll System

Recommendations

The sales call last month by ABC Computing crystallized an issue that had been discussed in our department for the last several years: should we continue to contract out for payroll services, or acquire our own system? I recommend that we purchase our own payroll system, and that we select Phoenix Microsystems as our source. The Phoenix system is slightly more expensive than competing products, but it gives an economic advantage over our current practice (payback = 2.7 years). Their system also contains all the features we require: security, backup, ease of use, and a maintenance contract. Our payroll people are more than competent enough to operate this system, and in fact are eager to use it.

Original Situation

In 1982 Local Accounting began supplying us with payroll services. They also provided us with tax services and did account auditing. Payroll services involved receiving wage reports from us every two weeks, calculating appropriate payroll information, and printing checks charged to our account. We are currently billed $900 per month plus $5 per employee, and are required to keep a minimum balance of $75,000 in an account they maintain for us. Total related expenses last year were $47,568.35. In June a representative of ABC Computing called on the payroll department and demonstrated a microcomputer-based payroll system costing $22,600. Since this seemed to represent a substantial potential savings over current practice, I was asked to investigate the feasibility of this alternative.

Background of the Investigation

To determine whether such an alternative was feasible, I first reviewed the current payroll system. I spent one day interviewing payroll personnel, including Nancy Alvarez, Jaime Gonzalez, and Suzanne Hartwig. I also spent an afternoon observing the current process at Local Accounting. Next I asked for a demonstration by ABC Computing. After that demonstration I contacted three other vendors of business computing products, and phoned four data processing colleagues I know at other firms through our professional association. Following these discussions I created a list of requirements any payroll system would have to meet. I verified this list with two people in payroll, and then further explored four computer systems which were in common use. Two of these systems (High Tech Data Wares and Fred's Checks) appeared to fall so short of our needs that I dropped them from further considerations. I bench-tested the remaining two systems, ABC Computing and Phoenix Systems, and phoned for references for these systems.

Figure 23-2. Example of a Feasibility Study (Continued)

Evaluation of Alternatives

After discussions with several people in the payroll department, and after speaking with senior management, I determined that any payroll system would have to meet a number of standards. I have grouped these standards by importance.

Critical Criteria:
Cost. Any new system could not cost more than $200,000 or have a payback period in excess of 4 years.
Security. Employee privacy must be protected.
Accuracy. Underpayments and overpayments must not occur.
Updates. It should be easy to add new employees, and to make changes in pay rates. The system should also accommodate changes in federal and state tax laws.
Backup. Records must be kept for 6 years. No records should be destroyed by hardware problems.

Secondary Criteria:
Maintenance. 24-hour repair service should be available for all computer hardware.
Training. Training should be available for computer operators and for department heads.
Ease of Use. It should not take more than one week for new operators to learn to use the system.

Supplemental Criteria:
Employee Information. Detailed information about deductions should be clearly presented to employees.
Management Reporting. Reports describing year-to-date payroll expenditures by department, and giving managers year-to-year comparisons, should be created.

Alternative 1: Local Accounting is the vendor we have been using for nearly twenty years. They have done a good job, but there are areas for improvement, as noted below.

Critical Criteria:
Cost. Annual costs are currently $47,568.35. Costs are increasing at roughly 7% annually.

Security. No security lapses have occurred.

Accuracy. Payroll reports are an increasing problem here. Four checks last month were in error.

Updates. There are complaints that new employees sometimes are not fully entered into the system for five to six weeks.

Backup. I have verified that they keep complete records.

Secondary Criteria:
Maintenance. They maintain their own system.

Figure 23-2. Example of a Feasibility Study (Continued)

Training. Little required, since they do all operations.

Ease of Use. Not applicable.

Supplemental Criteria:
Employee Information. Adequate but not extensive.

Management Reporting. Not available. Could be done, but additional charges would result.

As you can see, there are few major failings with the current system. There are a few complaints, but generally Local Accounting has done an adequate job for us. The one weakness in the current arrangement is cost.

Alternative 2: ABC Computers

Critical Criteria:
Cost. Initial cost would be $22,600, but there would be additional charges for a maintenance contract and training, plus operating costs. Payback period would be 2.6 years (see Appendix G for computations).

Security. Password system.

Accuracy. Few range checks done. A check for $1 million would go undetected by the system!

Updates. Could be done in roughly thirty minutes per employee.

Backup. Tape backup system included, but system is not highly rated by users in published evaluations (*PC Week*).

Secondary Criteria:
Maintenance. Contract is available, but work is done in California, so the system would have to be packed and shipped for repairs.

Training. Three days included in purchase price.

Ease of Use. Reference sites complained that screens were cluttered and that new employees needed weeks to understand the system fully.

Supplemental Criteria:
Employee Information. All basic information is included. Output seems clear.

Management Reporting. Not available now. They are working on such reports as a future product.

Although the ABC Computing System is slightly cheaper than the alternatives, it fails several important tests. System accuracy is a problem, and the backup system is unimpressive. A larger concern is maintenance. Shipping the computer to California for repairs

Figure 23-2. Example of a Feasibility Study (Continued)

would leave us dependent on shipping services and on repair facilities we do not know. The system is also hard to use and confusing to operators. The savings do not appear to warrant the risks of this system.

Alternative 3: Phoenix Systems

Critical Criteria:
Cost. Initial cost would be $28,500, plus additional charges for a maintenance contract and training. Operating costs would also have to be added, but would be less than for ABC System, for reasons cited below. Payback period would be 2.7 years (see Appendix G for computations).

Security. Password protection.

Accuracy. Range checks eliminate major errors.

Updates. Can be accomplished in five to ten minutes through a simple menu system.

Backup. Uses optical disk technology. Very reliable.

Secondary Criteria:
Maintenance. Local servicing within 24 hours.

Training. Three days supplied with system.

Ease of Use. Our payroll people and reference sites were universally impressed with the ease of operation.

Supplemental Criteria:
Employee Information. Standard information is included.

Management Reporting. Year-to-date reporting is included. Year-to-year comparisons are not available.

The Phoenix System has a slightly higher initial price than ABC Computing, but because it is simpler to use, operating costs would be lower, giving the system comparable costs. It has a very attractive payback period. It meets all the primary and secondary criteria and one of the supplemental criteria. People who have used the system universally praise it, and our payroll personnel are excited about the possibility of using it. It would seem a wise investment.

Conclusion

While our current contract for payroll services seems to be working adequately, newer microcomputer based systems offer an attractive alternative. After determining our company's needs and reviewing several possible products, I recommend we choose the payroll system developed by Phoenix Systems. We have an opportunity to save substantial costs in the payroll department, and to upgrade our payroll information for managers.

Figure 23-2. Example of a Feasibility Study (Continued)

Appendixes:
A. List of Requirements
B. Contract with Local Accounting
C. Proposed Contract from Phoenix Systems
D. Sales Literature from ABC Computing
E. Sales Literature from Phoenix Systems
F. Payback Calculations

Activities

1. As a class, critique each section of the example feasibility report in Figure 23-2.
   a. Was the recommendation convincing? If so, highlight places that made it so. If not, how might the author have improved it?
   b. Is the description of the original situation complete? What more could have been said?
   c. Does the background show a careful investigation? What actions seem impressive? What more could have been done?
   d. The comparison of alternatives was presented in text form. Rewrite them in the form of a table. Which is easier to read?
   e. What information could be added to the appendixes?
2. Assume your college library is considering going on-line and will replace the current card catalog with a computer system. In small groups, do the following:
   a. List some of the capabilities this new system must have for it to be operationally feasible.
   b. Create a list of questions you would ask before you would know if such a system were technically feasible. Don't forget local human resources. What questions would you ask about librarians, professors, and students?
   c. Determine whether such a system would ever have a payback period or return on investment. If not, could it ever be economically feasible? How would you justify the costs to your fellow students, who might have to pay for the system with higher tuition? What is the payback for them?
3. The quality assurance department in your paper mill has run up a large bill for overtime while tracking down an equipment problem that is

causing much of the paper to have a poor finish. Four customers have complained, and one is threatening to sue. In writing a feasibility report recommending a totally new paper machine, one of your colleagues began the "Background" section with, "Due to a total breakdown in quality. . . ." In small groups, write a better draft of the "Background" section. Compare how each group handled the problem of explaining the problem without creating adverse reactions from other employees.

4. As a city manager you are confronted by a citizens' group that wants you to stop buying water from BigCity and dig wells locally. Their research says this would save the city $70,000 per year. A local well field would cost $500,000. Is this move economically feasible? Calculate the payback analysis and return on investment, assuming that a new well field could operate twenty years without major additional expense. Write a two-page report explaining your analyses to the citizens' group.

5. Under what circumstances might the "Recommendations" section of a study be placed at the end of the report? How would you react to such a placement if you were a manager?

6. As a class, discuss new projects being considered on your campus (building additions, dorm remodeling, new majors, various automation projects). Assign one group to consider the feasibility of each project and write a report. Members of the group should meet with campus decision-makers to discover how the evaluation process usually occurs. How are projects determined to be feasible? After gathering information about operational, technical, and economic aspects of the project, write your own feasibility studies. Remember to review and revise the studies to make them as persuasive as possible. When they are complete, publish the feasibility studies by presenting them to the head of your college.

Progress Reports

Situation

Form
  Introduction
  Summary
  Body
  Conclusion
  Appendixes

Writing the Progress Report
  Collecting Information
  Organizing the First Draft
  Reviewing to Assure Quality
  Publishing the Progress Report

Example of a Progress Report

Activities

What is a progress report? When you see some friends you have not talked to recently and explain to them how well you are doing in school and what your plans after graduation are, you are making a progress report. When your parents telephone you to complain about how slowly the addition on their house is coming along, this is also a progress report. The progress reports discussed in this chapter are similar to the progress reports you have been giving and receiving most of your life. Progress reports, also called "activity reports" or "status reports," provide information about the status of an ongoing project: how much work has been accomplished and whether that work is on schedule, within budget, or otherwise on target.

Situation

Issuing progress reports is part of the standard operational procedure of many companies, especially for large projects. Progress reports can be made on any continuing project, regardless of the size or length of the project. For example, for the monthlong preparation of an employee manual, a brief progress report might be prepared after two weeks had elapsed. At the other extreme, a ten-year project to develop a manned space vehicle would generate an extensive and detailed series of progress reports. In general, if a project is short-term, entails few people, and requires a simple series of steps, progress reports tend to be few, brief, and informal. If the project is lengthy, involves many stages over a period of months or years, and represents a large investment of money and resources, there tends to be a regular series of progress reports that are carefully planned and written.

The audience for a progress report may be internal, external, or both. For an internal project, one that a company or organization is performing for its own benefit, the progress report is primarily addressed to an internal audience. The most important readers are usually the supervisors of the project, other people who are working on the project, and any executives who might have to make decisions about the project. Other significant readers might include other employees who are affected by the progress of the project. For example, the customer service department of a computer software company might be interested in the ongoing development of a new computer manual.

Many progress reports are submitted by a contracting company to a client company. The primary external readers for these reports include the people directly responsible for the project within the client company and any executives who make decisions about the project. In some cases separate progress reports may be issued for those inside the company and those outside, but in most cases a single report addresses both audiences.

The schedule of a series of progress reports is usually determined by the supervisor (in the case of an internal project) or by the client organization (in the case of an external project). In general, progress reports should be issued frequently enough to keep readers sufficiently informed but not so frequently that the reports become burdensome or redundant.

An important reason for writing progress reports is to identify problems before they become too serious. A yearlong construction project that loses one day each week due to transportation difficulties would eventually finish more than two months behind schedule. A good progress report created early in the project would allow supervisors to decide whether buying or leasing additional trucks was necessary. Supervisors also use progress reports to plan their allocations of resources, including people, equipment, and money. For example, a complex computer installation project that usually requires one third of the department might require the entire technical staff in week 13; if the project is two weeks behind (or ahead), a good progress report will allow supervisors to schedule staffing without unnecessarily taking these employees away from other projects.

Readers in the client company will want to be assured that the project is proceeding more or less as planned. It is their money that is involved, and they are the ones who will suffer most if a project is poorly executed. The client may also use a progress report to make important decisions about the project (for example, whether to continue or cancel it, whether to request changes) or about related matters within its own organization (when to schedule the hiring and training of new employees).

Finally, a progress report serves as a valuable record after the project is completed: the information presented may be used to answer questions about the project or to plan other projects.

Form

The form of a progress report varies greatly according to the circumstances. Depending on the purpose and audience, the report might be in the form of a letter, a memo, a formal report, or an oral presentation. Furthermore, each company, organization, or agency usually develops its own format.

However, most progress reports must do these things: (1) identify itself and the project, thereby providing a context for the report, (2) describe the status of the project, (3) comment on this status, (4) if necessary, provide further information or documentation to substantiate the report. So, in one form or another, most progress reports have these parts:

Introduction
Summary or Abstract (optional)
Body
Conclusion
Appendixes

If the progress report is in the form of a brief memo, each part might represent a sentence or two. If the report is presented as a one- or two-page letter, these parts of the report might appear as separate paragraphs. If the report is presented as a formal, multipage report, each part of the report will likely appear as a distinct section.

Introduction

The "Introduction" to an initial progress report or to the first in a series of reports must be more complete than the "Introduction" to subsequent progress reports. The following example illustrates most of the information that should be included in an initial introduction:

> Progress Report
> U.S. Department of Energy
> Subject: Report No. FE-1207-42
>
> This report presents progress achieved during January 1978 on the general program, "Gas Generator Research and Development," being conducted by Bituminous Coal Research, Inc., for the U.S. Department of Energy and the American Gas Association. The overall program was initiated under Contract 14-01-0001-324, December 20, 1963, and was subsequently transferred to Contract 14-32-0001-1207 on August 19, 1971, Contract No. E(49-18)-1207 on March 31, 1975, and Contract No. EF-77-C-01-1207 on March 22, 1977 (Committee on Foreign Relations, 1978).

A TITLE OR A REFERENCE NUMBER

The introduction provides a context for the report. Because every project may have several reports, and because progress reports can look very similar to one another, it is important to identify the report clearly at the beginning. The title of a progress report is usually the same as the title of the proposal that spawned the project. If a progress report has a reference number, as is often the case with state and U.S. government progress reports, that number is usually assigned by the agency responsible for the project and progress reports.

IDENTIFICATION OF THE DOCUMENT AS A PROGRESS REPORT

Although this might seem obvious to the writer, busy readers will appreciate knowing what it is they are reading.

PROJECT IDENTIFICATION

There might be more than one project going on at the same time, so the writer must be specific about which project is being described.

BACKGROUND TO THE REPORT

This section (a sentence or two) describes the purpose of the project and the methods being used to accomplish this purpose. This information is especially important for readers who are not familiar with the project or who might be reading a number of similar reports. Following is an example of an expanded description of the background to a report. This excerpt appears in a lengthy, three-paragraph introduction to a progress report.

> The objective of the combined bench-scale and PEDU studies is to develop a fluidized-bed methanation process for use in upgrading a coal-derived synthesis gas. At the bench-scale level, catalysts are being investigated under conditions imposed by the BI-GAS process. These conditions include high carbon monoxide concentrations, high pressure, and a 3:1 hydrogen to carbon monoxide ratio. The PEDU, which is a 6-inch diameter reactor designed for 6000 scfh, will generate process data applicable to BI-GAS pilot plant operations (U.S. Department of Energy, 1976).

The introduction in subsequent progress reports need not be as complete as in the initial progress report:

> Progress Report No. E(49-18)-2461
> This report presents progress achieved in the bench-scale and methanation PEDU gas-processing programs during January 1978 (U.S. Department of Energy, 1978).

Summary

If the body of the report is lengthy, a summary or abstract is often presented before the body (or as the first part of the body). Following is an abstract that appears at the beginning of a seven-page progress report titled "Bear Population Indices (Survey and Inventory)." It evaluates the results of the study and presents the recommendations made in the progress report.

> ABSTRACT
> Analyses of 5,764 bear ages obtained since 1973 showed a decrease in the average age of bears harvested and an increase in their average annual mortality rate in recent years. Visitation rates on the 1982 and 1983 bait station transects were significantly lower than from 1979 through 1980. These data suggested that the bear population was being overexploited, and recommendations were made to reduce the harvest to an acceptable level (Wisconsin Department of Natural Resources, 1984).

Body

This is the most important part of a progress report: it describes the current status of the project. This normally includes describing work completed, work in progress, and work still to be done. You can think of this part of the report as a schedule: each of the tasks is listed chronologically and then an explanation is given as to whether or not it has been completed yet. First, the work completed is described, usually in chronological order. Second, the work in progress is presented. Third, after the work completed and the work in progress have been described, the description of work to be done is presented.

This section also tells the reader whether the project is behind schedule, on schedule, or ahead of schedule. If the project is behind or ahead of schedule, explain the reasons. Sometimes the current status is presented in narrative form; sometimes it is presented as a log of events. Regardless of whether the project is behind, on, or ahead of schedule, this part usually includes a projection describing how work is expected to proceed in the future.

Conclusion

The primary function of the "Conclusion" is to provide an overall evaluation of progress to date. Generally, the conclusion is either that work is progressing better than expected, work is progressing well, or work is not progressing well at all. The difference between the overall evaluation and the schedule status described above is that the overall evaluation is more qualitative; it presents a broader and more judgmental assessment of progress to date.

Here is the conclusion to the progress report on gas generator research described above. The conclusion offers a balanced evaluation; it tells the reader what area of work is progressing well and what area is not progressing as well as expected.

The major objective of PEDU Test No. 27 was to obtain data with the Homer City catalyst under conditions favoring combined shift and methanation (i.e., a low hydrogen/carbon monoxide ratios). These conditions were readily achieved by blending supplemental carbon monoxide into the feed gas from a tube trailer. However, throughout the test, catalyst activity was lower than expected based on results of earlier PEDU tests. The reasons for this are not yet clear. Carbon deposition was originally suspected, but the results of subsequent reactivity tests on a "regenerated" catalyst sample have cast doubt upon this hypothesis. It is hoped that further tests, to be undertaken by the catalyst supplier, will shed some light on the cause for this unexpected change in catalyst behavior (Committee on Foreign Relations, 1978).

Appendixes

Use an appendix to include material that clarifies or supplements the report but is either too technical or too lengthy to be included in the report. Materials often found in progress report appendixes include case histories, charts, computer printouts, glossaries, maps, questionnaires, references for further reading, tables, and texts of interviews.

Writing the Progress Report

The most important thing to keep in mind while writing a progress report is that you are writing for a very specific purpose, to a very specific audience, and (usually) with a specific format. The most common obstacle to writing a good progress report is forgetting why and for whom you are writing.

Collecting Information

Collecting information for a progress report usually involves other people and departments. You might be asked to prepare a progress report for a project you know little or nothing about. If you are working for a large organization, you might have to contact managers and technicians from several different departments. If you are not involved in the project, the first person you would normally contact would be the project manager. The project manager can often give you the answers to the following questions:

- What has been done so far?
- Is the project on schedule?
- Is the project within budget?
- What is left to do?

If the project manager cannot answer all your questions, you might have to contact budget personnel or even conduct on-site visits.

The sooner you know what information you need, the better off you will be, because gathering information from others is time-consuming, and progress reports usually have deadlines.

Organizing the First Draft

If you are assigned to prepare a progress report, find out what format is preferred or required. If no format is specified or required, find existing progress reports and use them as models. If you cannot find any progress reports that have been prepared in your area, consider your audience, purpose, and general situation when deciding how to use and arrange the common components described above.

Because it is an essential part of the progress report, the schedule update requires special care. Be as accurate, complete, and clear as possible — important decisions may be based on this section. It is usually presented either in narrative form or as a log of events. The schedule update below is a narrative reporting on a project that is behind schedule. The narrative tells the reader how many days behind schedule the project is, why the schedule is behind, and offers an estimate of when the project will be back on schedule.

> The project is eleven days behind schedule. Efforts to conduct Test G-4 were to have begun on January 13th but actually began on January 24th. The test was hampered by numerous problems and delays, most of which were caused by cold weather. We expect to be back on schedule soon (U.S. Department of Energy, 1978).

Because schedules indicate when a project is behind or ahead of schedule, the absence of such a signal usually is evidence that a project is on schedule. The second example, presented as a log of events in outline form, reports on a project that is on schedule. Because no indication to the contrary is given, it is understood that each of the events has been completed on time. (The time span indicated after some of the events is the predicted amount of time needed for completion.)

INTERMEDIATE EVENT LOG

Event No.	Description
1.1.1. A	PEDU Test No. 27 (5 days)
1.1.1. B	Life Test, Catalysts No. 3896, 3897
1.1.1. C	Life Test, Catalysts No. 3898, 3899
1.1.1. D	PEDU Test No. 28 (5 days)
1.1.2. A	Interim Report received from BYU
1.1.2. B	Termination of current BYU program; Final report due
1.1.3. A	PEDU Test No. 27 (5 days)
1.1.3. B	PEDU Test No. 28 (5 days)
1.2.2. A	Estimated completion of AAI Phase II hookup
1.2.4. A	Completion of Revised Computer Simulation Model
1.3.3. A	Ambient sampling complete
1.3.4. A	Completion of current MIT Kinetic Study (Chart)
1.3.4. B	Completion of JAYCOR Gasifier Modeling Program

(Committee on Foreign Relations, 1978)

The description of work to be done can also be presented as a narrative or as a numbered list, although for ease of reading, numbered lists are often preferred. Following is a work forecast presented as a numbered list. It is clear, concise, and easy to follow.

WORK FORECAST
1. Achieve extended gasifier operation with steady slag tapping.
2. Achieve flow control of char through the three gasifier char feed legs as designed.
3. Achieve flow control of coal through the two gasifier coal feed legs as designed.
4. Collect data necessary to perform heat and material balances around the gasifier.
5. Continue the coal grinding program aimed at reducing the quantity of fine coal that is generated.

(Committee on Foreign Relations, 1978)

Reviewing to Assure Quality

Whether the progress report is one page or ten, it is essential that you evaluate the report. As with any piece of writing, regardless of its audience, purpose, or size, you will want to do what you can to assure its quality.

The first step in evaluating your own progress report is to look it over yourself. In addition to proofreading for spelling, punctuation, and grammar errors, check for the following:

- Are your facts accurate?
- Do *you* understand what you have written?
- Is the report clearly identified as a progress report?
- Have you included all the parts of the report?
- Are the parts clearly labeled?
- Have you anticipated your readers' needs?

If you are writing for a broad audience, ask a colleague who is not a specialist in the area you are writing about to read the report. However, remember to respect your colleague's time—don't ask for your document to be proofread, and don't ask overly broad questions. Tell your colleague what the purpose of the report is and to whom it is being written, then ask specific questions:

- Does the report need more details?
- Are the graphics labeled clearly?
- Are there any terms that need to be defined?
- Should any material now in the report be moved to an appendix?

If you are preparing an external progress report, it is unlikely you will be able to ask an intended reader to read it before it is submitted. However, if you are preparing an internal report or a report that is both internal and external, you can usually find an intended reader within your organization to look over the report. You might want to ask your reader these questions:

- Is everything clear?
- Is there anything you expected to find out but didn't?
- Is there anything in the report that should be left out?
- Is there anything not in the report that should be there?

Publishing the Progress Report

After you have evaluated and revised your report and after you are confident that you have anticipated and met the needs of your readers, you must reproduce and distribute the report.

Two basic questions to be considered are, who gets the report? (this will determine the number of copies to be made), and, how should the report be printed? It is often difficult to know who should get a copy of a progress report. Normally, there are two ways of finding out this information. One way is to find a recent progress report in your area and check for a distribution list; the other way is to ask your supervisor or manager for a distribution list.

How the report should be printed can be a difficult question. Before the advent of computers and desktop publishing, writers were limited to the typewriter unless they took the report to a print shop. However, a large range of options is now available to even the smallest office. Except for very large reports for very important projects, however, it is unusual for a company to spend as much money on producing a progress report as it would on a proposal, for example. Progress reports are primarily a working tool, not a selling tool: you don't have to sell your ideas or company any more, just maintain confidence in the project. The best advice is to make the report as professional looking as you can, within budget limitations. Though an attractive report will not hide flaws in the writing, an unattractive presentation will detract from the report.

Example of a Progress Report

The progress report in Figure 24-1 was written in the early stages of a program being conducted by Georgetown University for the Energy Research and Development Administration.

Figure 24-1. Example of a Progress Report

INDUSTRIAL APPLICATIONS
OF
FLUIDIZED-BED COMBUSTION

GEORGETOWN UNIVERSITY
Washington, D.C.
Date Published—December 1976
PREPARED FOR THE UNITED STATES
ENERGY RESEARCH AND DEVELOPMENT ADMINISTRATION
Under Contract No. E(49-18)-2461

INTRODUCTION

General

As a result of evaluations over the past decade by the Federal Government, it has been determined that the fluidized-bed combustion process can efficiently convert the energy of coal to usable power in an environmentally acceptable manner. In the fluidized-bed combustion process, crushed coal is burned in a bed of granular particles which are maintained in mobile suspension by an upward directed stream of air. Energy from this hot fluid mass is transferred to another fluid which passes through tubes located in and around the bed of hot granules. The overall heat transfer is greater than that obtainable in conventional combustion processes. In addition to this, if limestone is used as the granular bed material, the sulphur contained in the coal combines with the available calcium and forms $CaSO_4$. This eliminates the need for flue gas desulphurization which is required when fuels containing sulphur are burned.

Objectives

The objectives of this program are:

a. To design, construct and operate an atmospheric fluidized-bed boiler burning high sulphur coals in an environmentally acceptable manner in an urban institutional complex.
b. To obtain sufficient data from the prototype operation so that industry can move directly to the design and construction of commercially warranted industrial size fluidized-bed boiler units.

Scope of Work

ERDA Contract E(49-18)-2461, which was awarded to Georgetown University on June 30, 1976, provides for the engineering, design, construction, testing, operation, and evaluation of an atmospheric fluidized-bed boiler using high-sulphur coal functioning as a source of steam. The boiler, which will be located on the campus of Georgetown University, will have the capacity of generating 100,000 pounds per hour of saturated steam 250 psig and be able to meet the changing load conditions of a University complex. The work will be performed in two distinct phases.

Figure 24-1. Example of a Progress Report (Continued)

a. Phase I—A nine month effort for the development, engineering, and design of the demonstration unit.
b. Phase II—A fifty-one month effort encompassing the construction and operation of the demonstration unit.

Participants

Work under the contract will be accomplished through a teaming arrangement with Georgetown University as the prime contractor and the Fluidized Combustion Company (FCC) as a subcontractor. FCC is a joint venture of wholly-owned subsidiaries of Pope, Evans, and Robbins Incorporated (PER), an engineering consulting firm, and Foster Wheeler Corporation (FWC), a supplier of equipment to the steam power industry.

Georgetown University will provide overall program management as the prime contractor, in addition to providing environment assessment, scientific testing, fluidized bed performance evaluation, and facility operation. PER and FWC, through FCC, will engineer and design the demonstration unit, award the construction and boiler subcontracts, provide engineering and consulting services during construction and start-up, and oversee the operation of the demonstration unit, including training of operating personnel.

ABSTRACT

In line with a determination by the Federal Government that a fluidized-bed combustion process can efficiently convert the energy of coal to usable power in an environmentally acceptable manner, Georgetown University was awarded contract E(49-18)-2461 by the Energy Research and Development Administration to construct and operate a demonstration plant. This report, which covers the initial progress under the contract, reflects that work is proceeding as planned with no significant problems identified.

SUMMARY OF PROGRESS TO DATE

Upon receipt of Letter Contract E(49-13)-2461 dated June 30, 1976, Georgetown University gave Fluidized Combustion Company (FCC) authorization to proceed with subcontract effort. Work is proceeding as planned with no significant problems identified. FCC concentrated major effort during this period on preliminary engineering (Phase I, Task 3 of the Statement of Work) and the preparation of the 30 percent design review package. Action was initiated to obtain clearance of the project through the various agencies designated by the D. C. Government as required for issuance of the various required permits.

DETAILED DESCRIPTION OF TECHNICAL PROGRESS

Phase I. Development Engineering and Design
Task 1.   Program Plan

> **Figure 24-1.** Example of a Progress Report (Continued)

A. Work accomplished
1. Post-award orientation meeting held at ERDA on July 19, 1976. Representatives from ERDA, McKee, GU, FCC, FWC, and PER in attendance. Report requirements, program plan, statement of work, and other technical/contractual matters discussed.
2. Program plan prepared and submitted. (FCC)
B. Work forecast—this task is considered to be completed.

Task 2. Preliminary Engineering
A. Work accomplished
1. Initiated action to obtain clearance of the project through the various agencies designated by the D.C. Government as required for issuance of the required permits. Several meetings held with various D.C. agency officials to provide them with advance knowledge of the project and to gain a full understanding of the documentation and filing requirements of the various agencies. Preliminary architectural plans, elevations and plots provided. (GU and FCC)
2. Prepared and submitted for ERDA approval an environmental impact assessment for the installation. (GU and FCC)
3. Submitted revised cost estimate. (GU and FCC)
4. Investigated possible sources of supply for coal and limestone. Requested and received authorization to utilize G.S.A. as a source of supply for coal. (GU)
5. For the Steam generator initial effort by FCC included work on the heat and material balances, combustion calculations and unit general arrangement. Performance engineering on the steam generator is proceeding with the main effort now directed toward the analysis of the circulation characteristics of the water walls and horizontal banks.
6. Solids material handling systems were reviewed with experts in this field, i.e. the firm of Janicke & Johnson, Inc. Pertinent comments are being embodied in the 30 percent review package. (FCC)
7. Several technical and economic evaluations were initiated and/or completed by FCC during this period. These include:
   a. Steam Generator: A detailed comparison was made between partial shop assembly and complete shop assembly. This study has been completed and due to the excessive costs associated with the shipment, off and on loading, and trucking of a completely shop assembled steam generator, FCC made the recommendation that the unit be partially shop assembled with the completion of assembly being done in the field. This approach has an economic advantage of approximately $100,000 for the Georgetown location.

   It is important to note that as units of this nature are sold on a commercial basis, that evaluations similar to the one made for Georgetown be incorporated in that it is obvious from review that there are various locations where the shop assembly and barging would be more economical than the partial shop assembly and field work.

Figure 24-1. Example of a Progress Report (Continued)

  b. Precipitator: Several parties have shown interest in the possible substitution of a baghouse filter for the precipitator. A cost comparison was made between the use of an electrostatic precipitator and a baghouse filter for particulate removal. This comparison indicated that on a first cost basis a baghouse filter is approximately $270,000 more economical than an electrostatic precipitator of the same efficiency. In addition the annual operating costs including electrical requirements, maintenance, etc. favor the baghouse filter at an annual savings of approximately $15,000.

  c. Coal Drying: The GU Pon response document was based upon coal being received dry to the required surface moisture at the job site. Due to uncertainties associated with obtaining coal adequately dried to meet materials handling equipment requirements and considering the near future operation of Rivesville which will provide valuable information as to the need of coal drying, FCC has recommended that a coal dryer be included in the present day system design parameters with a bypass arrangement. If after further information is received on the coal to be supplied and feedback information from Rivesville indicates that dryers are not necessary, the dryer will be excluded from the plant material list. FCC would expect this decision to be reached before it is necessary to purchase the coal dryer. FCC feels that this approach is the most conservative at this point and allows for evaluation based upon operation. The costs of such a dryer would certainly be substantially less than the cost of the equipment associated with the bed recycle system which has presently been deleted and hence the effect of the addition of a coal dryer if determined necessary would not require a cost increase to the overall projected program estimates.

  d. Electrical Support Systems: There have been various discussions between parties pertaining to the electrical support system which will be required for this power plant addition. Philosophy of this system has now been agreed to and the additional costs associated with the same have been incorporated in the 30 percent review estimate.

  e. Bed Recycle system (Elimination): Upon further analysis it was determined that this system was unnecessary. It was economically favorable, while maintaining operational reliability, to remove the system. No cost is included in the 30 percent estimate for bed recycle.

 8. Major effort was made by FCC in the preparation and submittal of the 30 percent review package for comments. The package includes a 30 percent budget estimate, outline specifications for prepurchased equipment, preliminary bid documents and outline technical specifications for underground structures, preliminary bid documents and outline technical specifications for general contractor, boiler proposal, and at least 35 drawings.

 9. Work initiated by FCC to develop vendor specifications for steam generator appurtenant systems.

10. Evaluation initiated by FCC of potential means of simplifying fluidized combustion boiler supporting system.

Figure 24-1. Example of a Progress Report (Continued)

> 11. Continued to refine scope of effort envisioned by the scientific program. Submitted for ERDA approval a list of long-lead items required for the chemistry department. (GU)
> B. Work forecast—Submit 30 percent review package which will complete effort under this task.

Task 5.   Documentation and Reporting
A. Work accomplished—monthly progress reports; accumulation of data and verbal reports to ERDA on costs incurred. (GU and FCC)

Phase II.   Construction and Operation of Demonstration Plant
Task 2.   Construction
A. Work accomplished—None.
B. Work forecast—Solicit bids for prepurchased equipment and award first increment purchase orders. (FCC)

Task 3.   Design of Demonstration Facility
A. Work accomplished—None.
B. Work forecast
> 1. Prepare for and participate in 30 percent design review. (GU and FCC)
> 2. Initiate preparation of final prepurchased specifications and submission of same for ERDA approval prior to soliciting bids. (FCC)
> 3. Continued development and refinement of design drawings for purposes of optimizing arrangement and assuring of cost effectiveness. (FCC)
> 4. Initiate design drafting effort on the steam generator. (FCC)
> 5. Completion of performance engineering effort on the steam generator and its appurtenances. (FCC)
> 6. Completion of in-house engineering schedules and the integration of same between PER and FWC. (FCC)
> 7. Preparation of Failure Modes and Effects Analysis. (FCC)
> 8. Prepare for and participate in 50 percent On Board Review. (GU and FCC)
> 9. Preparation of Failure Modes and Effects Analysis. (FCC)
> 10. Complete drawings and bid package for substructure subcontracts. (FCC)
> 11. Continue the process of clearing the project with the agencies designated by the D.C. Government. (GU and FCC)
> 12. Submit test plan recommendations. (FCC)
> 13. Continue efforts on identifying source of supply for limestone. (GU)
> 14. Continue efforts on identifying source of supply for limestone. (GU)

Task 4.   Program Management
A. Work accomplished
> 1. Gave authorization to FCC to proceed on FBB project. (GU)
> 2. Continuing effort to arrive at a mutually acceptable subcontract agreement. (GU and FCC)
> 3. Continuing administrative and management functions. (GU and FCC)

Figure 24-1. Example of a Progress Report (Continued)

B. Work forecast
1. Continue effort to arrive at a mutually acceptable subcontract. (GU and FCC)
2. Provide additional answers and material to ERDA in an effort to formalize the contract between ERDA and the University. (GU)
3. Continuing administrative and management functions. (GU and FCC)

CONCLUSIONS

Work is proceeding as planned with no significant problems identified.

Source: U.S. Dept. of Energy. (1976, December). *Industrial applications of fluidized-bed combustion.* Quarterly progress report. Washington, DC: U.S. Government Printing Office.

Activities

    1. With two or three classmates, write a report for the rest of the class in which you analyze the example progress report in Figure 24-1. Your report should cover the following:

        a. Who is the intended audience: Experts? Novices? Executives? Technicians? Combined? How much do you know about the audience? What evidence in the report indicates the intended audience?

        b. What is the purpose of the report? How do you know what the purpose is? Has the purpose been fulfilled? If not, why? If so, how?

        c. Describe the development of the report. Is the organization logical?

    2. You are the author of the example progress report and you plan to ask a nonspecialist colleague to help evaluate the report. From what you learned in chapter 5, "Quality Reviews," and in this chapter, what criteria would you ask your colleague to use in evaluating the report?

    3. Using an index that includes technical documents, such as the *Applied Science and Technology Index*, the *Biological and Agricultural Index*, the *Current Technology Index*, the *Engineering Index*, the *General Science Index*, the *Index to U.S. Government Periodicals*, or the *Social Sciences Index*, locate a progress report. Prepare a summary of the report to be presented orally to the class.

    4. Prepare a personal academic progress report. The report should include at least the following: number of credits earned toward graduation, number of credits left, courses needed to fulfill the university's general degree requirements, courses needed to fulfill your major (and minor) requirements, grade point average in general courses and in your major. Choose a specific purpose and an appropriate audience (parents, grandparents, spouse).

    5. Write a one-page report to your classmates on the progress of one of your ongoing projects.

    6. Write a two- to three-page report to your instructor in which you describe your progress in the course so far. Your report should include a discussion of your expectations for the course, your instructor's expectations for the course, what problems you have encountered, how you solved or failed to solve those problems, and what you feel you need to do for the remainder of the course. After reviewing and revising, publish your progress report by submitting it to your instructor.

Instructions

Situation

Form
  Parts List
  Warnings
  General Background
  Preparatory Steps
  Sequential Instructions
  Trouble-Shooting Guide

Writing Instructions
  Collecting Information
  Organizing the First Draft
  Reviewing to Assure Quality
  Publishing the Instructions

Examples of Instructions

Activities

Instructions tell readers how to perform a task. Writing instructions is not easy. Every year around Christmas the newspapers carry cartoons showing harried adults surrounded by wrapping paper and half-assembled toys. The caption is usually something to the effect, "It says the instructions are simple enough for an eight-year-old. Next year Susy puts together her own toys." Consider these situations:

- Your company manufactures a small propane stove intended for backyard use. To keep shipping costs down, you sell these stoves unassembled. They arrive in a brown corrugated box with three pages of instructions. Hooking up the propane tank is not difficult if consumers are careful and follow every step in the instructions exactly. If they don't, they might injure themselves. How can you be absolutely certain they will understand and follow your instructions?
- You send out a new maintenance announcement on a $4 million paper machine. A bushing must be carefully removed each time a particular assembly is lubricated, or the machine will overheat. How do you describe exactly which bushing you mean? How do you explain the subsequent lubrication process? Millions of dollars depend on the clarity of your instructions.

Instructions may determine how consumers will feel about a product they have purchased, whether a product is safe to use, and whether huge losses can be avoided. Instructions are part of the technical world. Anyone entering a technical career must learn how to produce them effectively.

Situation

As the examples above suggest, many instructions are associated with consumer products. But instructions are also common in the workplace: they may tell an accountant how to fill out a new form, show a technician how to set up and operate a new piece of equipment, or instruct a surgeon in the proper use and care of a new tool.

It might seem that readers of instructions would all be novices, but that is only sometimes the case. Readers of instructions for consumer products are often novices, since it is unlikely that anyone assembles a gas grill more than once, or assembles a specific child's toy more than once, or sets up any new product more than one time. These readers never have sufficient practice to develop expertise.

But there are cases where some readers will be highly knowledgeable about a product or procedure. Photocopier repair people, for example, are repeatedly called upon to work on new and different models. However, they begin their work already possessing a great deal of knowledge about similar machines. These readers will want and need a different kind of information than pure novices.

This is not to say that there are no commonalities among readers. Two are actually quite important. First, most people resist reading instructions. Novice or expert, readers feel they already know all they need to know, or that they can figure it out for themselves, or that reading will slow them down. As a result, readers may omit reading instructions. Second, readers form much of their impression of a product based on the instructions they are given. If the instructions are confusing, readers will assume there is something wrong with the product.

The principal purpose of instructions is to help readers perform some task. But there are three additional purposes to consider when writing instructions.

First, instruction writers should be interested in the safety of the people following the instructions and in the safety of the machines or materials they are using. No one wants to be responsible for injuries to customers, clients, or employees. Writers also want to protect those products. If instructions are unclear and consumers misassemble part of a product, the product may be ruined. Ensuring safety should be an important part of instructions.

Second, it may be possible to teach users about a product while they are working on it. The illustration in Figure 25-1 is part of assembly instructions for a gas grill. Notice that each part is labeled. These labels not only help while the product is being put together, but later when it is being used.

Third, instructions can help improve the image of the company or product. A *Consumer Reports* reviewer had this to say after following the manufacturer's instructions for assembling a gas grill: "Good points: I didn't spend my own money to buy this grill" ("Gas Grills," 1991). People who have become confused, frustrated, and anxious will hardly be the best advertising.

As important as instructions are, it is hard to imagine a worse situation than is typical of their use.

1. People read instructions as they are about to carry out a task. There is little time between reading and acting. Therefore, if what they read is confusing, or if they misread, the damage will be done almost immediately. Each and every instruction has to be clear and correct.
2. People alternate between reading instructions and carrying them out. As a result, they are constantly leaving a set of instructions and going back to them. They must be able to find exactly where they were before, or they may skip steps.
3. The physical act of reading instructions can become complicated. We seldom read instructions while seated in the library or behind a desk. More normally it occurs while we are surrounded by equipment, sprawled out on the floor or on a work bench, reaching over to the instructions while holding a wrench or part of an assembly. Readers will be distracted, tired, often sweaty or dirty.

Figure 25-1. Partial Instructions

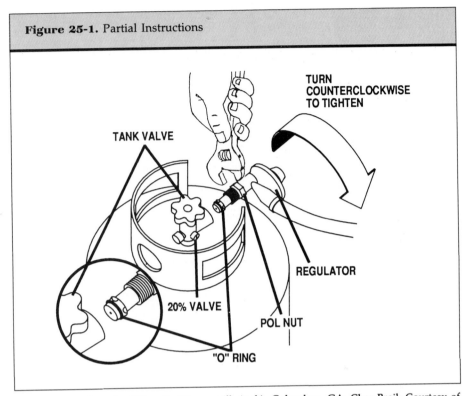

TURN
COUNTERCLOCKWISE
TO TIGHTEN

TANK VALVE

REGULATOR

20% VALVE

POL NUT

"O" RING

Source: Instructions for "Char-Broil" gas grill. (n.d.). Columbus, GA: Char-Broil. Courtesy of Char-Broil.

Clearly, a great deal of care and preparation must go into the writing of good instructions.

Form

Instructions vary widely in their length and in their readership, so there is no set organization for them. Nevertheless, instructions frequently contain the following components:

> Parts list (optional)
> Warnings (optional)
> General background (optional)
> Preparatory steps (optional)
> Sequential instructions
> Trouble-shooting guide (optional)

Parts List

Before readers can assemble a product or carry out a procedure, they need to know the names of the parts they will be using. This list is often given using names, descriptions, and, increasingly, illustrations of the parts. A good example is shown in Figure 25-2.

Figure 25-2. Parts List from Instructions

PARTS LIST

Part No.	Quantity	Description
CL60	1	Battery holder
CN38	1	10 μF electrolytic capacitor
CN102	1	0.01 μF ceramic disc capacitor
HA900	2'	Solder
IG102	1	555 timer IC
LP17	1	Light-emitting diode
PO158	1	10,000 ohm potentiometer
RE36	1	100,000 ohm 1/2 watt resistor
RE161	1	470 ohm 1/2 watt resistor
RE164	1	1,000 ohm 1/2 watt resistor
SO116	1	Breadboard socket
SP20	1	2" speaker
WR901	5'	Red hookup wire

Source: *Contemporary electronic circuits and components.* (1983). Washington, D.C.: McGraw-Hill Continuing Education Center, p. 52. Courtesy of McGraw-Hill Continuing Education Center.

Warnings

Once readers become immersed in a project, they may become so busy or bothered they could miss warnings about possible dangers. A clear statement at the start of a set of instructions might save problems later. Unfortunately, people are often so anxious to get started that they skip introductory material. For that reason, warnings must be displayed prominently.

General Background

If instructions about a difficult or complex process are being written for novices, background information about the product or procedure may be welcomed by some readers — although others may find it patronizing. If instructions are being written for technicians or experts, background information justifying the steps being taken may be essential. This would be especially true if your recommendations differed from previous practice: experts will want to know *why* they should do what you're telling them to do.

Preparatory Steps

Some instructions indicate the tools that will be necessary for a project or suggest site preparation work. The reasoning is that once readers start on a project, it may be difficult for them to stop work to locate a particular tool. They should be told what tools they need before they start, especially if a special tool or site preparation is necessary and if damage could result if readers attempted to continue without it. Figure 25-3 shows what such a section might look like.

Sequential Instructions

The major part of any instructional material is the sequence of steps readers will need to follow. This portion of the instructions is generally the most heavily illustrated.

Trouble-Shooting Guide

Problems should be anticipated. The machine doesn't operate, the results don't seem right, there are parts left over. Many products now come with a toll-free number for help, but it is better if the instructions themselves review

Figure 25-3. Description of Preparatory Steps within Instructions

Hardware stores and other stores sell some solder with an acid core. Acid flux in any form should never be used in electronics. The acid will corrode the leads and terminals and may damage some of the parts. Use only rosin-core solder in electronics work. Never use liquid or paste flux.

PREPARING THE PARTS

Your first step in getting ready to perform the lab projects in this module is to prepare a series of connecting wires.

Get the red hook-up wire from the parts you received with this module. Using a side cutter or wire stripper, cut wires of the following lengths:

3 wires — 1-1/4"
4 wires — 3"
4 wires — 6"

Now, using a wire stripper or knife, remove about 1/4" of insulation from each end of the wires as shown in Figure 2. You will use these wires as leads for your battery holder and speaker, as well as for connections on your breadboard.

Your next step is to solder two of the 6" leads to the terminals of the speaker. Lay the speaker face down on your work surface. On the rear of the speaker you will see two tiny terminal lugs with holes in them. To each terminal, you will connect a piece of 6" hookup wire.

Bend the bare end of one piece of hookup wire into a hook as illustrated in Figure 3. Slip the hook onto one of the speaker terminals and crimp the

Figure 2. Remove approximately 1/4" of insulation from the end of each hookup wire.

Figure 3. Make a hook in the end of a piece of 6" hookup wire.

Source: *Contemporary electronic circuits and components.* (1983). Washington, D.C.: McGraw-Hill Continuing Education Center, p. 53. Courtesy of McGraw-Hill Continuing Education Center.

solutions to common problems. A trouble-shooting guide listing standard situations may be enough for people either to find a solution or to realize that the problem is too serious to resolve without help.

For example, a popular VCR manual lists fifteen common problems with a suggested "correction" for each. Here is one such entry:

> TV program can't be recorded . . .   Check the connections between the VCR, the external antenna, and your TV.
>
>   Make sure the record tab on the cassette is still intact.
>
>   (Quasar VHS Owner's Manual)

Such suggestions not only save possible problems with the equipment, but they give the person following the instructions more confidence.

Writing Instructions ————————————————————

Writing instructions follows the same four steps used in other forms of technical writing: collecting information, organizing the first draft, reviewing to assure quality, and publishing.

Collecting Information

If asked to write about equipment or processes that are new to you, begin by learning all you can. Typically you will be asked to describe a process you know well, such as maintenance for a piece of equipment you often use. It is tempting just to sit down and write the manual. No matter how completely you understand a process, however, there is still much to learn before you are ready to write your instructions. Several of the most important areas to research are described below.

LABELING

Naming parts of a machine or naming parts of an operation would seem easy, but different people may use different names for the same object. Before you label any diagrams, you want to be sure that your readers use the same names for parts. Do not rely on a single source—either a book or a person—for labels. Language practice changes, and your source may not give the most common or most correct name for an operation. When in doubt, remember that an important element of your instructions is likely to be a parts list. It is here that you can not only list the most common name for an object but also give popular alternatives.

PRIMARY SEQUENCES

Some steps must be assumed or abridged. Your job is to identify which steps are essential to the task.

Interviewing people as they perform a task is one way to gauge the relative importance of operations; another way is to compare the way different people do the job. If each person is careful to do one task first, you know you have identified an important task. It is these tasks that you need to identify and fully describe.

Similarly, if there are steps in a process that several people seem able to skip, and if you think a novice could also safely skip these steps, drop the steps out. You will save your readers time and avoid confusion.

HUMAN VARIANTS

Few processes are done exactly the same way by all people. Physical stamina, age, and sex will lead people to address a task differently. A major difference comes from the experience level: once people have done a process many times, they can often modify a process significantly, yet still do it safely.

Your investigation should lead you to view a range of people performing the operations you are describing. You will not write a different set of instructions for tall people or older people, but you will be able to tell what approach is most typical, summarize it, and note any major alternatives.

VISUAL NEEDS

As you watch people perform a task, be aware of steps where they take extra time, seem somewhat confused, or are particularly careful. These are the steps that will be hardest for the novice to learn or where the novice is at most risk of injury. These steps should be illustrated as well as described.

CONDITIONS AFFECTING THE PROCESS

A procedure may be performed differently under different conditions. For example, vehicle maintenance in a heated shop is quite different from roadside repair. For this reason, observe operations being done under a range of conditions so that you can identify dangerous conditions or give hints for solutions to problems that may arise under certain conditions. If you are uncertain about the range of conditions that is possible, ask.

DANGERS

Some dangers are obvious, and you will have no trouble listing and describing them. But even if you are well acquainted with a piece of machinery, there may be additional dangers unknown to you. A good source to check is safety reports. The Occupational Safety and Health Administration (OSHA) requires all companies to keep records of injuries sustained on the job. If you can get access to these records, they may help you understand additional dangers from equipment or procedures.

Besides trying to learn about all possible dangers, you will want to study ways of avoiding those dangers. Are there ways, for example, for a person to spot early signs of exposure to pesticides or chemicals? What forms of prevention work best? What special clothing or special equipment should be found before starting a task? What safeguards are built into the equipment? What help is available should a problem occur?

If the danger is not to the person but to the equipment, is there a way to spot early signs? Can the damage be easily rectified, or should the user be warned that no repair is possible? Electrical circuits are an example of equipment that can be easily damaged. If a circuit is incorrectly installed, once a power source has been turned on, the damage is instantaneous. If you become aware of such risks, you can include them in the instructions.

Organizing the First Draft

As you put together the first draft of your instructions, there are a number of things you can do to address the special concerns for instructions.

1. *Create an outline that suits the situation.* Instructions may include a parts list, warnings, general background, preparations steps, sequential instructions, and a trouble-shooting guide. You must decide which of these sections to include in your instructions, and in which order to use them. That decision should be based on the procedure you are describing and on the readers' needs. For instance, a procedure that is dangerous would probably begin with a substantial section of warnings. Instructions written for technicians would likely begin with general background information so that they can compare these procedures to those they already know. Having done a thorough investigation, you should be able to determine who the readers are likely to be and which possible components of the instructions are essential in this case.

2. *Use visual aids.* Graphics take up space and add significantly to production costs for instructions, but it is hard to argue with their effectiveness. The reader can grasp spatial relationships quickly from a diagram or photograph.

As an example, to repair the carburetor on your car, you need to clean a particular part. Which would make it easier for you to find and remove the part—a paragraph-long description of its appearance and location, or a diagram with an arrow pointing to the part? A picture may not be worth a thousand words, but it is usually better than a paragraph or two.

3. *Make your descriptions as clear as possible.* Many instructions leave out necessary detail. A photograph is valuable, but it is much clearer to readers if particular reference points are called out. A diagram should show all the important parts and be clearly labeled. Written descriptions need to be checked for accuracy and completeness. What seems clear to you may be incomprehensible to a novice.

4. *Carefully sequence and group steps.* Your instructions give readers a step-by-step sequence to follow. Be sure that the steps are in the correct order, and that there are no unnecessary digressions in the instructions.

One technique to use to organize steps is to begin by making a complete list of every step that needs to occur. It may take some time, but make an effort to include every possible step. Next, mark the most important steps. Some will be important because they involve some danger if missed. Others are major points in a process. Third, try to group steps. The advantage of grouping will become clear very quickly if you look at a set of instructions with dozens of steps. After the first few steps they all blend together and it is hard for readers to remember whether they are on step 23 or 24, or what any of the steps have to do with the others. Grouping breaks a maze of steps into a few manageable steps and thereby makes the process more understandable to the reader.

Often these groups can be predicted. For instance, the first few steps of any process typically involve gathering and preparing materials. Surfaces may need to be sanded or painted, parts may need to be laid out or cleaned, components may need to be tested. Rather than present these as the first five or six steps of a thirty-step process, it is clearer if they are grouped into a single step—*Preparation*. The reader now knows the purpose of the smaller steps and has a better understanding of the entire process. If the other steps can be grouped in a similar way, a thirty-step process is reduced to five or six major steps, each of which is clearly labeled for the reader.

5. *Use a page design that adds clarity.* The layout of words on a page is especially important in instructions because it is assumed that the reader will be both reading and carrying out instructions at the same time. The reader of instructions must be able to find a spot in the text quickly. For this reason, multiple instructions should not be lumped together in the same paragraph. Notice the example below:

> (1) Wrap tissue specimens in gauze or paper toweling, place into plastic bag and saturate with ten percent neutral formalin. Pour off excess fluid, remove as much air as possible, then heat seal the plastic bag. Insert into second plastic bag along with identifying information, remove as much air as possible and seal the second plastic bag (Armed Forces, 1976).

Buried in this paragraph are ten specific steps that need to be followed by the reader. Someone trying to follow the instructions would do one of the steps, then go back to the text and try to find the next step. Nothing in the way the steps are presented would make this easy: each step is placed next to the other, often merged into the same sentence. No visual cues are given to help a reader distinguish steps. The paragraph could, however, be broken into a series of steps that were much more clearly labeled and formatted. Here is one possible way:

Transmittal of Tissue

Preparation

1. Wrap tissue specimens. Use either gauze or paper toweling.

Inner Bag
1. Place specimens in a plastic bag.
2. Saturate specimens. Use ten percent neutral formalin.
3. Pour off excess fluid.
4. Remove as much air as possible.
5. Heat-seal the plastic bag.

Outer Bag
1. Insert into second plastic bag.
2. Include identifying information.
3. Remove as much air as possible.
4. Heat-seal the second plastic bag.

With this description readers can read an instruction, perform the instruction, and then return to the text for the next instruction. The placement of the instructions on the page makes it easier for readers to keep their place, and less likely that steps will be missed.

6. *Include progress and test points.* One of the concerns of readers, especially when performing complicated or delicate procedures, is that they are making some mistake that they will not recognize until it is too late. To help readers check on their own progress, and to reassure them that everything is proceeding as planned, look for places where you might give them a short test. A likely place is at the end of a group of steps. Here is an example from the instructions above:

Inner bag

1. Place specimens in a plastic bag.
2. Saturate specimens. Use ten percent neutral formalin.
3. Pour off excess fluid.
4. Remove as much air as possible.
5. Heat-seal the plastic bag.

TEST: To be sure the bag is adequately sealed for shipment, hold the sealed end of the bag in a pan of water and gently press on the bag. No air bubbles should appear. If they do, heat-seal the bag again and retest.

A common "test" that is used to measure progress is a simple picture showing a picture of the final product. Readers can then compare what they see in front of them with the picture. If the two appear the same, everything is fine. If not, they know they have some steps to re-do.

Reviewing to Assure Quality

The best way to find out if your instructions are done is to test them with the reader most likely to use them. If your instructions describe maintenance, ask a maintenance worker to follow your guide while working. If you are writing a set of instructions for assembling children's toys, ask a parent to try the instructions. (See chapter 5, "Quality Reviews," for more information on performance tests.)

In addition to this general test, several aspects of instructions must be specially reviewed.

First, be sure warnings are adequate. Danger comes in two forms. Many procedures involve substantial risk to readers. Even a simple screwdriver can cut a large hole in a hand if held wrong or if it is the incorrect size for the screws. To protect your reader from injury and yourself from lawsuits, you have a responsibility to explain where there is a risk to readers.

The other danger is to the equipment. A surface that is left unsanded, a lock washer ignored, a bolt that is tightened too far, all can result in damage that a reader should be warned of. You cannot prevent people from taking shortcuts, but you should warn them of the possible consequences. Notice how a warning can be built into instructions:

Inner bag

1. Place specimens in a plastic bag.
2. Saturate specimens. Use ten percent neutral formalin.

WARNING: Use of reagents other than formalin may result in damage to the tissue, and may invalidate toxicology tests.

Note that the warning was placed next to the relevant step, and that the text was formatted to heighten its visibility. Other possibilities include red letters, larger or italic print, or recognizable icons. These options generally increase printing costs but should be considered when the magnitude of the danger is great.

A second place for a careful review is illustrations. Instructional material tends to use more illustration than other forms of technical writing, and the illustrations often carry significant weight in communicating clearly the actions that need to be taken. The illustration in Figure 25-4 not only names parts of the assembly, it also conveys warnings to readers. Such an illustration can be valuable in any set of instructions. But because of its special importance, such an illustration should be thoroughly tested with a sample of readers. Can they match the parts in the drawing to the real parts in the assembly? Do they understand what the illustration is trying to tell them? Both performance and comprehension testing of such illustrations should be done.

Figure 25-4. Example of Illustration in Instructions

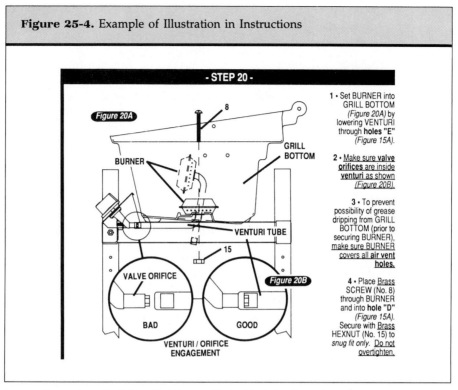

Source: Instructions for "Char-Broil" gas grill. (n.d.). Columbus, GA: Char-Broil. Courtesy of Char-Broil.

Such an illustration can be valuable in any set of instructions. But because of its special importance, such an illustration should be thoroughly tested with a sample of readers. Can they match the parts in the drawing to the real parts in the assembly? Do they understand what the illustration is trying to tell them? Both performance and comprehension testing of such illustrations should be done.

The example in Figure 25-5 shows why instructions must be carefully and thoroughly reviewed and revised. As you read them, imagine yourself in the laboratory actually carrying out the removal of the fish eggs. Where would you put the instructions while you followed them? How comfortable would you be while you worked from step to step? This is not leisure reading. The person following these instructions faces a very difficult situation.

1. " . . . one prepares Quinaldine for anesthetizing the roe fish": Later the passage mentions that the mixture should be one part per thousand, but

Figure 25-5. Example of Instructions before Revision

Removal of Eggs from the Roe Fish

Once the completion of ovulation in the roe fish is verified, eggs should be removed as soon as possible. Efficient operation during this process requires a minimum of three workers. One man should begin capturing bucks, one prepares Quinaldine for anesthetizing the roe fish [1] and the other brings the female to the spawning table. [2]

The roe fish should be held by the head and tail, with the vent covered to prevent egg loss, while Quinaldine is sprayed on the gills.

Quinaldine, at a concentration approximating one part per thousand, has proven a satisfactory anesthetic for use while stripping eggs from roe fish. The Quinaldine solution is sprayed onto the gills with a hypodermic syringe [3] sans needle. Females become sufficiently relaxed within one to two minutes to permit easy removal of the eggs and recover fully within five minutes after artificial respiration is begun in fresh water. Quinaldine was chosen over MS-222 [4] as an anesthetic because of the wider range between lethal and narcotic concentrations (Tatum et al., 1965).

Quinaldine should be handled with extreme care. [5] Gloves should be worn to protect the workers from injury and provide a firm grip on the fish. Attempts to hold fish by the lower jaw invariably produce violent struggling, if the fish is conscious. Therefore, pressure on the lower jaw is a good method for ascertaining when the female is sufficiently anesthetized for stripping. Two men are required to hold a large roe fish, while the third man forces eggs into a spawning pan by applying pressure to the abdominal area. After stripping is complete one man can revive the female in fresh water, while the other two effect fertilization.

Source: Bayless, V. (1987). *Artificial propagation and hybridization of striped bass,* Morone saxtalis. Spartansburg, SC: South Carolina Wildlife Resources Department, pp. 30–32.

nowhere does the description state how much to prepare! Is 5 cc sufficient? 50?

2. ". . . the other brings the female to the spawning table": It may be true that experienced game wardens know what a spawning table looks like, but there is no guarantee the people following this description are experienced. Even experienced people could benefit from a diagram showing the location of materials on the spawning table.

3. "The Quinaldine solution is sprayed onto the gills with a hypodermic syringe": How much is sprayed at a time? What size syringe is to be used? This sentence tells the reader very little.

4. "Quinaldine was chosen over MS-222 . . .": The reasons for the selection of this drug might make a valuable footnote, but the inclusion of this information in the middle of the text just delays the reader's ability to get to the next step of the procedure.

5. "Quinaldine should be handled with extreme care": Nowhere does the description state what risks accompany this drug. It is unclear what it can do to the fish or to the human handlers. Nor is it clear how humans would detect problems from the drug. Should they look for a rash? Dizziness? Also notice that the layout of the page gives no priority to this information. The warning about danger is printed in the same size type as the fact that the fish will recover in five minutes. Anyone just skimming this text (and most people do skim) might totally miss the fact that there is danger. Finally, the danger has been pointed out after the users have already handled the drug. Warnings should be placed early enough for readers to be alerted to possible problems before they happen.

Publishing the Instructions

Because of the way instructions are used, and because they frequently include illustrations, special attention needs to be paid when actually printing instructions.

1. *Assure the quality of reproduction.* Line drawings and other forms of illustration must be reproduced very carefully. Photocopying must be done carefully to ensure that all lines are reproduced fully; more sophisticated reproduction techniques are advisable. Budget to cover higher than normal copying costs, and review copies of instructions carefully to ensure that no illustration was garbled in the printing process.

2. *Make sure they are practical.* Make sure that copies of the instructions are tough enough to be used in the sites where they are needed. The factory floor is often dirty and drafty, and the people using the instructions may have dirt on their hands. Print the instructions on paper that is thick enough to be handled roughly. Most printers have card stock available for such situations.

Also, print the instructions to lie flat. Users will probably need both hands to do their work and will not be able to hold open a set of instructions bound in normal book form. The best alternative is to print shorter instructions on a single large sheet that can be spread out in a work area. Longer instruction manuals should be bound in such a way that they can lie flat.

3. *Pay for two-color printing where necessary.* Color provides a special emphasis. If there are particular places where a reader faces special danger, a warning printed in red will stand out from adjacent text. The second color

will add substantially to printing costs, but it may also prevent injury to readers and save your company legal costs.

Examples of Instructions ─────────────────────

Below are two sets of instructions. The first example (Figure 25-6) is a revised version of the instructions for the removal of fish eggs presented earlier in this chapter. Several improvements have been made. First, the new version breaks the process of extracting eggs into four steps, and both labels and numbers them. There are substeps involving the mixing of the anesthetic and the stripping of the eggs, but it is clear that the process involves just four main steps. This both heightens the awareness of the reader to the importance of these steps and simplifies a complicated process.

Second, there are two new sections, each formatted differently from the rest of the text. One warns of danger and explains it in detail, the other supplies a simple test so that the reader can tell how well things are progressing. Both the warning and the test were present in the previous version but were undifferentiated from the rest of the text and so were easy to miss.

The third change is in the level of detail. Extraneous details comparing two anesthetics are deleted, while crucial details about necessary quantities are included.

The second set of instructions (Figure 25-7) describes the safe application of pesticides. These instructions include a lengthy equipment list, instructions for checking the equipment, a numbered list of safety precautions, and several illustrations showing the safe application of pesticides.

Activities ─────────────────────

1. Reread the original description of fish-egg removal in Figure 25-5, and list the most confusing places. Are all the places clear in the revised version shown in Figure 25-6? If not, rewrite those parts.

With others in your class, discuss the grouping of steps in the revised fish-egg example. In what other ways could the steps have been grouped? Rewrite the instructions using those new groupings.

2. Bring a set of instructions to class—the instructions that came with a consumer product, or instructions from a commercial product. Meet as a group and review all the instructions.

   a. Find the best and worst
      - use of illustrations
      - naming of parts and operations
      - sequencing of steps

- grouping of steps
- formatting to highlight key points

Explain which techniques were used by the best instructions. Discuss the problems with the worst example. How would you fix it?

b. Hold a set of instructions at a distance. Without reading the text, can you tell what places in the instructions contain the most important information? What cues did you use?

c. Find an example of a test step within a set of instructions. Where was the step placed and how was it formatted?

d. Review the instructions for danger warnings. Have they included all that are necessary? How much work (color print, graphics, formatting, and the like) did the authors expend to make the warnings clear to readers?

3. Set up a situation in which three people perform a task. It might be as simple as checking the oil level in a car or putting frosting on a cake. List the things they do similarly and the things they do differently. If you were to create a set of instructions based on what these people did, which of them would be "right"—which would serve as the model for your instructions? Why? Are there places where you could (or must) show more than one way to perform a task?

4. Name three tools that have more than one name. Have the names changed over time, or are they called different names in different regions of the country? How would you include multiple names in an instruction manual?

5. If you were to add a troubleshooting guide to the fish-egg instructions earlier in this chapter, what are some of the troubles you might list? How would you go about getting a more complete list?

6. Bring a poor set of instructions to class. Working in groups of three or four, circle weak areas in the instructions. Then rewrite the instructions using the guidelines of this chapter. Create a review procedure for the revised instructions, and test it to make sure your instructions are better than the original.

Figure 25-6. Example of Instructions

Removal of Eggs from the Roe Fish

Once the completion of ovulation in the roe fish is verified, eggs should be removed as soon as possible. Efficient operation during this process requires a minimum of three workers. One person begins capturing bucks, the second prepares Quinaldine for anesthetizing the roe fish, and the third brings the female to the spawning table.

1. Preparation of Quinaldine:

   DANGER! Quinaldine is a strong anesthetic. It can be absorbed through the skin and can cause dizziness or even unconsciousness if poorly handled. Be sure to wear rubber gloves and use it in a well-ventilated room.

   Roughly 30–50 cc of Quinaldine is required per adult roe fish. It should be mixed at a concentration of one part per thousand.
2. Preparation of the spawning table:
   - A stainless steel table approximately 2 feet by 4 feet is best.
   - You will need a one-gallon bucket to hold the eggs.
   - Place the bucket at one end of the table.
   - Wet the surface of the table to protect the skin of the bass while anesthetizing and massaging.
3. Holding and anesthetizing the fish:
   - The roe fish should be held by the head and tail.
   - Cover the vent (see diagram 1) to prevent egg loss.
   - Spray Quinaldine on the gills with a 50-cc hypodermic syringe from which the needle has been removed.
   - Small amounts (approximately 5 cc at a time) should be sprayed across the entire gill region.

   Females become sufficiently relaxed within one to two minutes to permit easy removal of the eggs, and recover fully within five minutes after artificial respiration is begun in fresh water.

   TESTING: To know when to stop administering the drugs, apply pressure to the lower jaw of the fish. Attempts to hold fish by the lower jaw invariably produce violent struggling if the fish is conscious.

4. Stripping the eggs:

   While two people hold the fish, a third person should force the eggs into a spawning pan by applying pressure to the abdominal area. Pressure should be gentle, yet firm, beginning about the middle of the fish and sliding downward toward the tail.

After stripping is complete one of the workers can revive the female in fresh water, while the other two begin the fertilization process described below.

Figure 25-7. Example of Instructions

SAFETY DURING APPLICATION

Application Equipment. — Selecting application equipment for a specific job will depend on many particulars; i.e., the area to be treated, the area's location and access, pest to be controlled, and equipment available. Application equipment can be either unpowered or powered. Details of equipment for treating land and water weeds, including design specifications, can be found in Bureau publication *Water Operation and Maintenance Bulletin* No. 97. The Bureau's water resource publication, *Herbicide Manual*, discusses equipment requirements, design, and selection.

Manual Applicators. — Generally, unpowered applicators consist of small metal or plastic containers. When pesticide is a spray, it is usually added to the container and diluted with water, and the unit pressurized by hand-pumped air. Some units are carried by a handle with one hand while the spray is applied with a triggered wand with the other; others are backpack models. Probably the most popular unpowered sprayer is the jar-and-siphon device sold in garden centers and hardware stores for home garden and landscape use.

Hand-pressurized applicators are usually used to apply pesticides inside buildings, such as warehouses and storerooms, and other areas where powered units would be too large or cumbersome. For minor infestations, a hand applicator is more practical and efficient.

Power Applicators. — Power applicators are used for rapid application of large pesticide volumes. Using power equipment, a pesticide can be sprayed over many acres in a day's time. With most power sprayers, the concentrate is mixed with a solvent, usually water, and pumped from a storage tank through distribution lines to one or more nozzles that control the size and distribution of the spray droplets.

There are many kinds of power applicators including:

Fogger generators	Low-pressure boom sprayers
Aerosol generators	Ultra-low-volume sprayers
Dusters	High-pressure sprayers
Granule spreaders	Low-volume air sprayers
Air-blast sprayers	

Depending on the individual piece of equipment, they may be mounted on trucks, tractors, trailers, boats, or aircraft. Some smaller units, similar in size to a leaf blower, are designed to be carried by applicators on their back.

Foggers and Aerosol Generators. — The generators break pesticide formulations into fog-like droplets. The fog is created by atomizing nozzles, spinning disks, or heated elements called thermal generators. Their main safety problem is that even light wind can cause long-distance drift and unwanted contamination.

Dusters. — The duster blows fine pesticide particles of dust onto the target. However, like aerosol sprays, dusts are subject to drift.

Figure 25-7. Example of Instructions (Continued)

Granule Applicators. — These devices are designed to apply coarse, dry particles — usually to the soil. Granular applicators offer no special safety hazards.

Air Blast Sprayers. — These units use a high speed, fan-driven air stream to dispense the spray. A series of nozzles inject spray into the air stream, which breaks up and blows the droplets. The sprayers deliver either high or low volumes of spray. They also pose a drift hazard.

Low-pressure Boom Sprayers. — Usually, these sprayers are mounted on tractors, trucks, or trailers as shown on figures 2 and 3. They are designed to deliver 20 to 60 gallons per acre of pesticide at 10 to 40 pounds-per-square-inch pressure (187 to 561 L/ha at 7 to 275 kPa).

Ultra-low-volume Sprayers. — These sprayers apply the chemical concentrate without diluting. Safety disadvantages include increased risk to the applicator and crew from handling and spraying concentrated material.

Figure 2. Vehicle mounted boom spray rig.

Figure 25-7. Example of Instructions (Continued)

Figure 3. Hand-held boom spray rig.

High-pressure Sprayers. — These sprayers deliver various volumes of spray at pressures up to several hundred pounds per square inch. While high-pressure sprayers afford greater reach than low-pressure units, the spray tends to form small droplets and drift, thus creating a potential safety hazard.

Low-volume Air Sprayers. — The low-volume sprayers, often called misters or mist blowers, operate similarly to air-blast sprayers, except they use lower spray volumes. The backpack spray unit shown on figure 4 has a low-pressure metering device that injects spray material into high-speed air which atomizes the liquid. Favorable spraying weather is essential for their use.

Components of Power Application Equipment. — The major features of power application equipment include agitators, nozzles, pumps, and tanks.

1. *Agitators.* — Without agitation, some pesticide formulations will settle in the spray tank, clog lines, and change the concentration of chemical in the spray. In hydraulic agitation, some solution is recirculated through the tank. In mechanical agitation, a paddle or propeller suspended in the spray tank stirs the solution.
2. *Nozzles.* — Nozzles control the rate, size, and uniformity of pesticide droplets during application. Optimum nozzle performance is crucial to safety; they must be cleaned,

Figure 25-7. Example of Instructions (Continued)

Figure 4. Two types of low volume sprayers:
1) backpack spray unit and 2) hand-held mist sprayer.

Figure 25-7. Example of Instructions (Continued)

replaced when worn, and properly adjusted so pesticides are placed where, and only where, intended.

3. *Pumps.*—Many types of pumps are found on application equipment. Generally, there are two safety considerations regarding them. First, the pump must produce the correct operating pressure and volume for the particular application; second, the pump —and especially its nonmetal parts, such as gaskets—must withstand the chemical action of the pesticide without excessive corrosion or leakage.

4. *Tanks.*—Tanks hold the pesticide and diluting liquid. Often, tanks are:

- Constructed with an agitator to keep insoluble formulations such as wettable powders in suspension
- Made of corrosion and rust resistant materials
- Constructed with large openings for easy filling and cleaning
- Fabricated with an accurate sight gauge for determining the liquid level
- Enclosed by a spill-proof cover
- Emptied completely using a bottom drain

Equipment Checkout and Calibration.—To prevent chemical contamination, routinely check application equipment for: leaks, especially in hoses and hose connections; plugged, worn, or dripping nozzles; and defective gauges, pressure regulators, and valves.

Clean and flush sprayers before using them. Often, new equipment contains foreign objects such as metal chips from the manufacturing process. Also, sprayers that have been idle for a time may contain bits of rust or other material. Failure to thoroughly clean such equipment may result in improper operation, increasing chances for pesticide exposure.

Calibrate application equipment so that the proper amount of pesticide is applied. With chemicals, more is not better and less is not necessarily safer. A safety hazard is created when too much chemical is applied, and money is wasted when an insufficient application results in need for a repeat application. Calibration also provides a final operational check on all equipment systems.

Directions for calibrating equipment are usually included in the manufacturers' operating and maintenance instructions. Good discussions of calibration, together with the necessary mathematical calculations, are found in the two Bureau publications *Herbicide Manual* and *Water Operation and Maintenance Bulletin* No. 97.

General Safety Procedures.—Most pesticide applications take several hours or longer to complete, during which time workers are continuously at risk of exposure to pesticide sprays, mists, and vapors. General procedures for minimizing these exposures are:

1. Make sure the application equipment is in good operating condition and has been properly calibrated.
2. Do not apply pesticides that will drift—sprays, mists, or dusts—when the weather is windy or unsettled.

Figure 25-7. Example of Instructions (Continued)

3. Read the safety plan; wear the proper protective clothing. Check the operation of required safety devices.
4. Read safety and *application* information on the product MSDS and label. Make sure safety equipment required by the safety plan agrees with MSDS and label requirements. If it does not, contact the Bureau of Reclamation pest control supervisor or safety officer.
5. Keep first-aid kit, spill kit, and clean water readily available at or near the treatment area.
6. Never work alone; observe partner for symptoms of exposure.
7. Do not contaminate yourself or partner. In a light breeze, apply pesticide from the upwind side of the treatment area.
8. Do not leave equipment unattended. If it has to be abandoned in an emergency, secure it as well as possible.
9. When wearing a respirator, monitor it for pesticide odors. If odors are detected, leave the environment at once; change respirator cartridges or canister or entire respirator before returning.
10. Take rest breaks; do not eat, drink, or smoke around pesticides.

MONITORING WORKERS

Everyone working with pesticides should be checked for signs of fatigue and poisoning. Look for symptoms of poisoning at regular intervals during the day; the sooner symptoms are detected, the less likely a serious incident will result. After a long work period, fatigue can increase the risk of accident. The pressure to accomplish the job should never take precedence over rest when a worker is tired.

Source: U.S. Department of the Interior. Bureau of Reclamation. (1988). *Pesticide applicators safety manual for irrigation systems.* Denver, CO: U.S. Government Printing Office, pp. 39–45.

26

Computer Manuals

Tutorials
Situation
Audience
Guidelines for Tutorials

Reference Guides
Situation
Audience
Guidelines for Reference Guides

On-Line Help
Situation
Audience
Guidelines for On-Line Help

Activities

In the mid-1980s, one of the fastest growing areas of the publishing industry was computer manuals. There were books describing how to use every kind of computer and every major piece of software. None of them should have been necessary. Ever computer comes with an operating manual, and every piece of computer software has a user's guide. However, these manuals are often unreadable. After spending several hours trying to decipher the manual and finding none of it made sense, people went to the local bookstore to get something they could understand.

Sales of computer books are currently down (but still substantial), which indicate that at least some computer companies are getting better at writing. But there is still room to improve.

Consider this common situation: Jennifer buys a word processing package to run on her personal computer. She takes the disk out of the box and puts it in her computer. She takes the manual out of the box and looks for the command to make the program run. The Table of Contents lists chapters on "Installation, Disk Operations, Page Formatting, Spell Checking," and other areas, but nowhere does it say how to start the program. Finally Jennifer just starts clicking her mouse on various program names until the program begins. Score student 1, manual 0.

She types a few paragraphs to test the program, and decides to go back to the beginning of her writing and add a title. She wants to center her title, so she picks up the manual, turns to the Index, and looks for "centering." It isn't there. Being an experienced computer user, she knows centering is a form of formatting, so she looks that up. She finds a page number and turns to it. She finds a section on formatting, and pages through information on left justification, double spacing, underlining, and finally finds a page on centering. Buried in the midst of a paragraph is a reference to a function key that initiates centering. She tries it and it works. Score student 2, manual 0.

As our example shows, computer manuals can often be difficult to read. Locating basic information may seem impossible. Indexes are often incomplete or inaccurate. Important commands can be buried in mounds of text. Key processes may be omitted. In defense of manual writers, computer manuals may represent one of the most difficult forms of technical writing.

Problem number one for manual writers is the novice. Even though digital computers have been with us since the Second World War, and personal computers have been everywhere since 1980, there are still millions of people who are encountering computers for the first time. These people don't know what a disk is for, don't understand how a computer keyboard differs from a typewriter keyboard (and may not be able to use a typewriter keyboard either), think a mouse requires traps, and have no experience with such basic software as word processors, spreadsheets, or databases. To them the machine is a mystery—or an endless agony.

Problem number two is the computer itself. This is not a simple machine.

We are now able to put more computing power on a desktop than nuclear physicists had available to them during the Manhattan Project. And we are adding to the power. Just when it seems the machines will be fairly stable, along comes a new set of features—networks, multimedia, new operating systems—there is always something new to learn.

Problem number three is the expert. Because of the popularity of personal computers, there are millions of people who have used computers for years. Some know how to write their own programs, others use standard software programs. In some schools students start using computers in kindergarten and use them routinely thereafter. For them, computers are hardly a mystery. They understand computers well, and just need information about new computer features or new software. When they pick up a computer manual, they want quick access to pertinent information—not lengthy descriptions of concepts they mastered years ago. Manual writers need to produce writing that recognizes the complexity of computers, yet works for both novices and experts. This is never going to be an easy task.

If this challenge weren't already daunting enough, there is an additional issue that computer manual writers face. Should the "manuals" be on paper, or screen? For a growing number of reasons—convenience, ecology, and economy—manuals are increasingly being placed "on-line." Convenience is served if, as you are using your computer, you need more information, rather than try to find a manual and then the proper page in the manual, the information pops up on your monitor screen. Ecology is served if we cut fewer trees and fill fewer landfills with books. Economy is served if we take the contents of ten inches of books and put it on one floppy disk. Navy repair manuals, for example, are now put on-line. That means that US submarines go out to sea *two tons* lighter.

This movement from paper to on-line text may have its advantages, but it has its problems for writers. Writers have to determine when paper is best, and when screens are required. How should users call up computer-based text? How should such text appear on the screen? Should both paper and on-line manual be supplied with software? If so, should they contain the same material? None of these questions have easy answers. This is a new field, and we all have much to learn.

Nevertheless, computer manuals are instruction manuals, and they are prepared in much the same way as any other instruction manual. What writers must do is recognize the special needs of computer users and present instructions to them in appropriate ways.

Different sorts of documents have emerged to fill different computer instruction needs. The tutorial is typically a learning tool for novices; it emphasizes instruction. The reference guide is more of an informational document for the expert. On-line help screens provide both general information and explicit instructions for anyone who needs them.

Tutorials —————————————————————————————

A common device for teaching the novice user is the tutorial. A computer tutorial leads a new user step by step through a program or a specific process within the program. Longer tutorials are typically broken up into a series of lessons.

The information in a tutorial lesson is presented sequentially; the user is given explicit and detailed instructions to follow, down to which keys to press. The assumption is that the reader will follow the tutorial lesson completely and exactly, beginning at the beginning and following dutifully through to the end rather than jumping from point to point.

Several aspects of the instructions in Figure 26-1 are typical of tutorials. First, each action is presented on a separate line and set off from other text. The format of the text tries to make it clear when a reader is to perform an action. An explanation of each action follows, but it is indented and separated, and put in a less important position on the page.

Second, each action is numbered. It is clear what the sequence will be. Readers follow step by step, just as if they were following a recipe or an auto repair manual.

Third, vocabulary is carefully chosen. Some of the commands tell users

---

Figure 26-1. Page from a Computer Tutorial

1. Press CTRL-HOME.

   This moves the cursor to the top of your document. Now you will be able to insert your title.

2. Type "Here is the title of my report."

   The words you are typing will appear in the upper left hand corner of the screen.

3. Press CTRL-C.

   Your title will automatically be centered on the screen.

4. Press ENTER three times.

   Three blank lines will be inserted after your title. Use this method to add blank lines anytime you wish. The screen should look like Figure 1-9.

to "type," others to "press." Since the most likely reader of a tutorial is a computer novice, the tutorial assumes that the reader is not aware of the importance of certain keys on the computer keyboard. ENTER is a crucial command key telling the machine to begin an operation. Novices don't often know that. When instructed to "type Enter," many would just type the letters "E-N-T-E-R" and then wonder why nothing happened as the book said it would. They don't know they should be looking for a special ENTER key. The example above avoids this problem by instructing readers to "type" when they are to type individual letters, and "press" when they are to look for a special key. Other manuals further distinguish between these two operations by putting boxes or the drawings of keys around special command keys. The fact that so much effort is put into distinguishing between regular letters and special keys gives you some sense of how important it is to make such distinctions for novices.

Situation

A tutorial may be presented as a printed manual, or it could be presented as an on-line document. In either case, a tutorial is helpful when computer users are new to a process. The tutorial gives a complete description of a process, says how to format text, and helps the user learn how to perform the process.

Because the tutorial teaches a process, it may be fairly lengthy and require substantial time to complete. For this reason, it is fairly demanding of users. It assumes they are willing to take the time to learn a process completely, and it can only be used well if this is the case.

Audience

Tutorials represent a style of learning that is often associated with novices, but that is not their exclusive domain. The attraction of tutorials for readers is that the writer is in control of the learning process. The reader assumes that the writer, as an expert, knows what information is central, will present everything at its proper time, and will be complete. All the reader has to do to learn a new computer process is follow along.

While such a learning style is common among novices, other users want information that is most useful to their immediate needs. They are more willing to take responsibility for their learning and find tutorials much too confining.

Should a writer respond to learning style or to level of expertise? In practice, writers often do a little of both. Assuming that even a self-directed learner can benefit from tutorials on occasion when encountering new material, a writer looks for ways to make tutorials more useful (and palatable).

Let's return to the tutorial lesson in Figure 26-1. The strength of this tutorial is that it is clear about what to do and in what order. If the reader follows instructions, there won't be any problems.

The weakness of the tutorial is that it is very rigid, giving the user no options and few explanations. To appeal to self-directed learners, the tutorial should do everything possible to make the learner feel in control of the learning situation. One way to achieve this is to add a clear introduction and flexible responses.

In the introduction, readers should be told what this section will do and how it will do it. This helps readers understand why they will later be asked to perform certain operations. Here is one way the example tutorial could be changed to achieve this:

Centering

Centering titles or other headers essentially takes four steps:
1 – move to the desired position
2 – type the title or heading
3 – press the centering command
4 – insert a blank line or two to set off the title

This introduction provides an overview of the process. Now the reader has a general sense of how centering works, and understands why each step in the following tutorial will be necessary.

The next step is to adjust the commands to allow some flexibility where it won't interfere or confuse. These changes are underlined to make them easier for you to see.

1. Press CTRL-HOME.
   This moves the cursor to the top of your document. Now you will be able to insert your title.
2. Type <u>a word or phrase to serve as a title</u>. "Here is the title of my report."
   The words you are typing will appear in the upper left hand corner of the screen.
3. Press CTRL-C.
   Your title will automatically be centered on the screen.
4. Press ENTER <u>two or</u> three times.
   <u>Each time you press ENTER it inserts a blank line and helps set off your title from the body of your text</u>. Three blank lines will be inserted after your title. Use this method to add blank lines anytime you wish. The screen should look like Figure 1-9.

These changes are minor, but they give the reader more control over the lesson. There is more "ownership" since the text on the screen is based on what the reader prefers. In essence, the changes show more respect for the reader.

Guidelines for Tutorials

From our past discussion, five of the major guidelines for tutorial writing should already be clear.

1. *Provide an overview of each tutorial lesson.* This way readers know what will be covered and what they can expect to learn.

2. *Be very careful in the wording.* "Type" and "Press" are examples of words selected carefully for novices.

3. *Break each activity into simple steps* and explain clearly what readers are to do at each step.

4. *Use page layout techniques to separate actions from explanations.* Skip lines, use indentation, try bold print and underlining as a way to highlight actions.

5. *Provide sample screen layouts* as often as possible. Novice computer users will be reassured if they can produce on the computer what they see in the book.

There are two additional guidelines. The first makes tutorials more useful to self-directed learners; remember that these people want to determine for themselves what they will learn and in what order. The last guideline reflects the fact that most tutorials are, in fact, used by novices.

6. *Keep tutorial lessons short and focused on a single activity.* In our example, the tutorial lesson exclusively taught users how to center titles. Users interested in this ability could focus on this skill alone. Users not interested in centering could skip over the lesson. The author could have created a large lesson on all formatting skills, putting centering in the same lesson as spacing, indentation, underlining, and so on. But that assumes that all readers want to learn about all formatting commands at one time. This is seldom the case. Novice readers take time to assimilate information, whereas experts will go directly to the one new skill they lack. Tutorials will be much more accessible if broken into reasonable units.

7. *Test your tutorial by having a novice use it.* As careful as you may be, there will be places where someone new to computers will misunderstand. The jargon is new, the keyboard is new, the activities are new. The only way you can be sure you are being clear is to watch a novice try to navigate through your prose.

The following example explains a portion of Windows Write by Microsoft. Notice its combination of explanation, illustrations, and instructions.

Figure 26-2. Computer Tutorial Lesson

228 *Editing a Document*

Finding Text

Using the Find command on the Search menu, you can look for a character, word, or group of characters or words in a Write document.

When you choose the Find command, Write starts a search at the insertion point or at the end of any text that is currently selected. When it reaches the end of the document, Write automatically continues searching from the beginning of the document to its original starting point.

▶ **To find text:**

1. Position the insertion point where you want to start your search.
2. Choose Find from the Search menu.
   The Find dialog box appears.

3. In the Find What box, type the text you want to find.
4. Select the Whole Word check box if you want to find only separate occurrences of the search text.
   Otherwise Write finds, for example, *main* in *remainder*.
5. Select the Match Upper/Lowercase check box if you want to match capitalization exactly.
   Otherwise Write finds, for example, both *Main* and *main*.
6. Choose Find Next (or press ENTER) to start the search.
   Write searches for the text and selects the first occurrence. If you want to edit the text, point to it and click the mouse button to switch from the dialog box to the document. (If you are using the keyboard, press ALT + F6 to switch from the dialog box to the document and back again.)
   If there are no occurrences of the text in the document, Write displays a message telling you that it could not find the search text.
7. Choose Find Next (or press ENTER) again to find the next occurrence of the search text.

Reference Guides

Reference guides are a common type of instruction manual, but they are especially useful for computer users. Because computer systems are so complex, large numbers of commands or procedures need to be explained, even though they may be rarely used.

Figure 26-2. Computer Tutorial Lesson (Continued)

When Write finds the last occurrence of the search text, it displays a dialog box telling you that the search is complete.

8. Choose Close from the Control menu (or press ESC) to close the Find dialog box.

To search for text that includes wildcard characters, spaces, tabs, paragraph marks, and page breaks, type the following characters (where ^ means the caret symbol) in the Find What box:

Type	To represent
?	Any character in a file (called a *wildcard* character). For example, if the search text is *hea?*, Write finds every word with *hea*, such as *headstrong*, *figurehead*, *heal*, *heap*, or *heat*.
^w	A space anywhere in your document—between words, between paragraphs, and so on.
^t	A tab character.
^p	A paragraph mark.
^d	A manual page break. For more information, see "Breaking Pages Manually" later in this chapter.

Changing Text

The Change command on the Search menu allows you to find text and replace it with new text. You can find and change all occurrences of the search text automatically, or you can decide whether to change each occurrence as Write finds it.

When you choose the Change command, Write searches forward from the insertion point. When it reaches the end of the document, Write automatically continues searching from the beginning of the document to the insertion point. If you select a section of your document before choosing Change, however, the command searches only within that selection.

▶ **To find and change text:**

1. Position the insertion point where you want to start your search.
    If you want to make changes to only part of your document, select that part.

2. Choose Change from the Search menu.
    The Change dialog box appears.

Source: *Microsoft® Windows™ user's guide.* (1990; pp. 228–229). For the Microsoft Windows graphical environment version 3.0. [Computer Program]. Portions © 1985–1992 Microsoft Corporation. Reprinted with permission from Microsoft Corporation.

Take, for example, the disk operating system (DOS) of personal computers. Typical operating system functions include file naming, copying, renaming, deleting, and so on. Users rarely perform all these functions at the same time. They typically want to perform one function in DOS and then do something else in a different program.

Reference guides make this easy. Because they are organized with each function in its own section, the user need only turn to the first page of the description, get the necessary information about the function, do what the reference guide says, and then put the guide away.

Figure 26-3 shows an entry in a reference guide for DOS. Notice how the page layout makes subsections of the entry easy to find. Notice also the brevity of the entry. There is a short statement of purpose, one line giving the standard form for the command, a single example, and a short explanation.

Notice what is *not* explained. Near the end of the entry, "ASCII" and "binary files" are mentioned. Neither are defined. There are two reasons for that. One is the expectation that most people using a reference guide are relatively expert around computers. The second reason is the need to keep entries short. To serve as an effective guide, sections need to be easy to look up, and need to be focused on a specific activity. A description of the COPY command should concern itself only with the copy command, and not with the meaning of other terms like ASCII or binary. Users who don't understand either of those terms are better served by using other reference entries to find explanations.

Figure 26-3. Entry in a Computer Reference Guide

COPY

PURPOSE	Copy allows you to copy a file from one disk to another or to copy a file onto the same disk.
FORM	copy[pathname][pathname][/v][/a][/b]
EXAMPLE	copy c:\writing\letter a:let1
COMMENT	The first listed file is found on the given directory and copied to the second location. In the example, file "letter" located on drive C, directory "writing," is copied to drive A. It is given whatever name is listed in the second pathname, "let1" in the example.
OPTIONS	/V Verify. If this option is used, DOS will check for recording errors as it copies the file. /A ASCII. Causes the file so designated to be treated as an ASCII (text) file. /B Binary. Causes the file so designated to be treated as a binary file.

Situation

Reference manuals are used in very specific situations. Typically the user is in the middle of some activity and needs a specific question answered about a procedure. The user stops using the computer for a moment and looks up the necessary information in a manual. With the manual open to the necessary page, the user types in the appropriate command, then puts away the manual and continues with the work at hand. In short, reference manuals are not read for pleasure. They are opened only when needed, and kept open only until a problem has been resolved.

Audience

This manner of using reference guides underlines two important facts about guides and how they are used. First, reference guides are for experts. Novices are virtually excluded from reference guides, since in order to use them they have to know the command they want. A DOS reference guide with the entries "COPY, DIRECTORY, ERASE, FORMAT" is virtually useless to someone who doesn't know what "copy" means in the computer world. Some novices will use reference guides anyway since that is their learning style, and writers can make reference guides more helpful to those users by supplying simple examples, but the principal readers will be experts.

Second, in contrast to tutorials, reference guides do not primarily teach. Reference guides supply information, and can sometimes teach by example, but in general they supply detailed information that few people want to (or can) remember.

Guidelines for Reference Guides

Visualize a computer user working away on a program. The user stops, reaches over to a bookshelf, pulls out a reference book, thumbs through to an explanation of a command, tries out what the book says, and then puts the reference book away. That vision underlies all the guidelines below.

1. *Keep entries short.* Entries should describe a single command only, and should typically be one page in length—seldom more than three or four pages. If you find you are writing longer entries, check to make sure you are not really describing more than one concept or command. If you are, break those additional descriptions out into separate entries.

2. *Put entries in a logical order.* Think of a reference guide as a dictionary. Readers should be able to find information without using a table of contents or an index. For that reason, commands are often placed in a reference manual alphabetically. A user can then find a command as simply as using a telephone directory. But there are exceptions.

On occasion you will have to group entries and then alphabetize within groups. An example is integrated software that contains major subprograms such as a word processor, spreadsheet, and database. A reference guide for this program would group all word processor commands and alphabetize them, group and alphabetize all spreadsheet commands, and group and alphabetize all database commands.

Another approach is to order commands the same way as they appear on the program's on-line menus. For instance, a "File" menu might contain the following commands:

FILE
    Load
    Save
    Delete
    Print

A reference manual might do well to have a chapter on file commands, with information on load, save, delete, and print, presented in that order.

3. *Place each entry on its own page.* Many published manuals put several commands on a single page to save paper, but this adds to the work of readers. Entries should be as easy to find as possible. Placing each on its own page facilitates that. A reader need simply look at the top of each page to locate a command.

4. *Use a standard format.* Readers seldom want to read an entire entry for a command. They usually want just the example, or just the options. If you format your entries so that each has the same contents and the same order, readers will be able to jump to the information they want. You can help this process by labeling sections and by using reverse indenting and bold print to set off key words. In the example below, the command includes the complete command name, a description, and examples. If this approach is taken with one command, it should be taken with all commands.

AVE(range)

Purpose   The AVE function automatically returns the average of all numbers in a range of cells. Items in a range should be numbers.

Example   AVE(rlc4:rlc9)
          AVE will return the average of all cells in row 1 from column 4 through column 9.

5. *Be sure each entry is complete.* In the DOS example given earlier in this chapter, these sections were used for the entry:

Purpose
Form

Example
Comment
Options

Each section supplied information in a slightly different form and helped insure that a reader got all that was necessary to perform the command.

- *Purpose* explains what the command does so the reader can determine if this is the appropriate command.
- *Form* shows how the command is to be entered.
- *Example* helps clarify the form to readers. It is especially useful to novice readers. Although most readers of reference guides are experts, novices too will use them from time to time.
- *Comment* gives you a chance to explain the details of the command or to highlight special areas. It might include such statements as, "Be sure to put a space after the comma."
- *Options* describes special capabilities of the command. Because options are used less frequently, they have to be explained carefully, and often have their own examples.

6. *Make sure your guide is complete.* Reference guides aren't needed for common information, they are needed for the unusual — for the commands readers might need once in a year. Every command has to be in the guide. Be sure you have included every function a program can perform, every feature that a user might want to try.

7. *Include an index and table of contents.* These should not be necessary very often if you have organized your guide well, but they will still be needed for commands or key terms that appear in multiple entries. They also give a reader another access path to information.

8. *Bind your manual so it can be laid flat on a table.* It is hard to type in a command with one hand while trying to hold open a book with the other. Your binding should reflect your understanding of how the book will be used.

On-Line Help ━━━━━━━━━━━━━━━━━━━━━━━━━

If someone is working on a computer and has a question, why make him or her get up and find the right manual and then thumb through the book for the right entry? Why not just put the information right on the computer screen?

As it turns out, this is not as simple as it sounds, since it means integrating a series of text screens with software and figuring out ahead of time which information will be needed where. But the approach is so valuable that more companies are taking this approach and providing on-line help.

Situation

While placing information right on the computer screen of the user has some advantages in ease of use, there are also some problems. The biggest is room on the screen. You can put more words on a piece of paper than you can on a computer screen. For that reason, on-line materials tend to be used for items that can be explained briefly. That means that rather than being used for lengthy tutorials, on-line materials generally serve as reference materials with a few succinct instructions.

Rather than being referred to as reference manuals, however, typical programs refer to on-line material as "help" screens. A user stops in the midst of some activity, and presses a special key that brings help in the form of information put on the screen.

Audience

It would seem that both novices and experts could use this kind of help, but it turns out that because of limitations in the amount of information that can be presented on the screen, true novices are often disappointed in help screens. The kind of help they need requires more than what can conveniently be placed on a computer screen. For that reason, the principal users of help screens tend to be expert users who are unfamiliar with certain aspects of a particular program; they use help screens as a substitute for a reference manual.

Such help is very handy, but the fact that the information is presented where and when it is places a number of restrictions on how writers should create such screens. Those restrictions are described below.

Guidelines for On-Line Help

There are several main concerns you should have when designing on-line manuals to incorporate into a software package.

1. *Use a common access to help.* It is fairly typical for computers with function keys that F1 is the help key. Most programs use it, so users are in the habit of pressing that key to get help. If users are already confused enough to need help, you don't want your help to add to the confusion. So, whenever possible, follow the conventions and use the same key for help as other programs, and certainly always use the same key within your program.

2. *Make your help "context sensitive."* When a computer user refers to a book for help, the first problem is finding the right page. A computer program can find the right "page" automatically if it is programmed carefully. For instance, if the program keeps track of the last key that has been struck, it should be able to find the "help" most pertinent at the time. By tracking the last command used, the program would know, for example, that the user is printing. If the user asks for help at this point, he or she probably wants to know about printing rather than about disk formatting. So the program automatically brings up the help screens on printing. In the unlikely event that the user wants to know about something other than printing, the screen allows for that possibility and lets the user "quit" or "move" to some other help screen.

Notice that this facility requires careful programming. If you choose to include such a capability, it will have to be designed into the program from the start.

3. *Facilitate navigation through help screens.* Here is where on-line manuals come up short against books. With a book, it is hard to get lost, since you can turn back a page or forward. You have a sense for where you are—in the middle, toward the end, and so on. On-line information is just one screen after another, and they can all look the same. If readers are to have a sense of where they are and where they want to go, you will have to help them. Look at the example in Figure 26-4.

Do you see how much of the screen is devoted to helping users get located? They are told which page they are on (page 1 of 3), and they are given a menu to take them forward or back a page. They are also given a chance to get out of help, or to transfer to the main help menu where they can get information on other topics. All those options are "navigation," and they are crucial if readers are to use your on-line materials with ease.

4. *Size your help screen to less than the full screen.* Users seem to be more comfortable when new materials are presented as "layers" over part of their

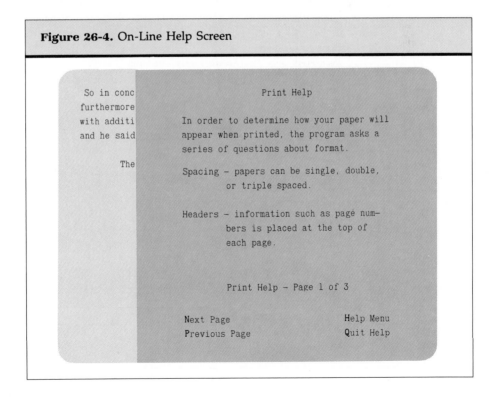

Figure 26-4. On-Line Help Screen

old material. In Figure 26-4, some of the words from the word processing document are still visible along the left edge, so the user won't worry that the material has been lost or destroyed. The help screen is clearly just a temporary addition and the user can be fairly confident that when it goes away the document will be just where it was seen last.

All this reassurance comes at a price. The computer screen is already smaller than a piece of paper, and now we are using part of it as a security blanket, leaving even less room for messages. This means you will have to be very succinct in your prose.

5. *Don't clutter the screen.* You always want to keep the screen uncluttered, but on-line helps provide even more challenge than normal. The help window takes up only part of the screen, so you have little room; other materials are probably on the screen "under" and to one side of the helps; and a user needing help may already be frustrated and confused. To make your information as readable as possible, leave plenty of room around the edges, and use headings, colors, and underlining as a way of highlighting key terms.

The screen in Figure 26-5 presents its information completely but succinctly.

Figure 26-5. On-Line Help Screen

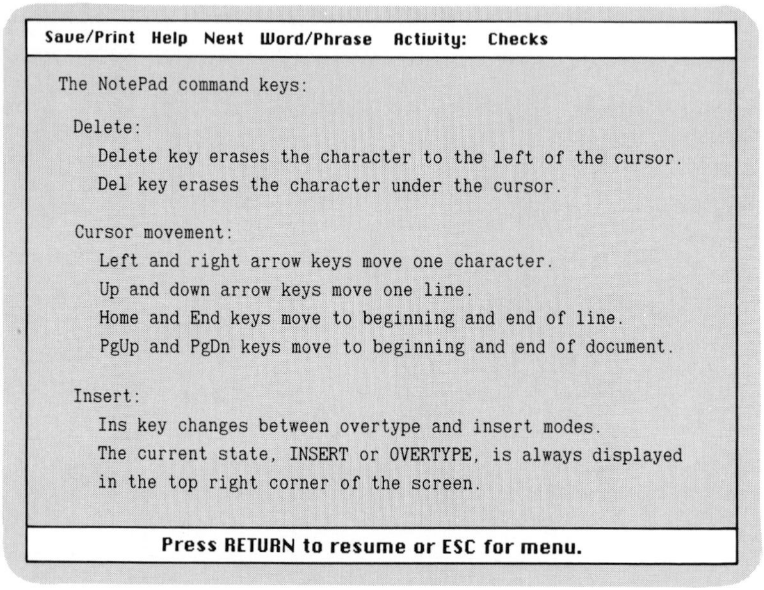

```
 Save/Print  Help  Next  Word/Phrase   Activity:   Checks

   The NotePad command keys:

    Delete:
        Delete key erases the character to the left of the cursor.
        Del key erases the character under the cursor.

    Cursor movement:
        Left and right arrow keys move one character.
        Up and down arrow keys move one line.
        Home and End keys move to beginning and end of line.
        PgUp and PgDn keys move to beginning and end of document.

    Insert:
        Ins key changes between overtype and insert modes.
        The current state, INSERT or OVERTYPE, is always displayed
        in the top right corner of the screen.

             Press RETURN to resume or ESC for menu.
```

Activities

1. Remembering that tutorials are generally used by novices, look back at the example tutorial in this chapter (Fig. 26-2). What words or phrases might be difficult for a novice to understand? Are there any that could lead to problems if misinterpreted?

2. As a group, pick a computer tutorial for any piece of software.
   a. Before opening it, make a list of five to ten things you would expect to find in its index. Now look and see how many are there.
   b. Make a list of five things a computer novice would have to know to operate the program. Now review the first few pages of the manual to see how many are there.
   c. In the first two pages of the manual, circle all words that only an experienced computer user would understand.
   d. Create a cloze text for sections of the tutorial. Have a novice fill in the blanks. Rewrite the sections where the novice received a score of less than 60%. Create another cloze test for the new section and test it with another novice. Keep rewriting the manual until readers consistently score 60% or better.

3. Assume your campus has been using WordPerfect as its official word processing software for the last two years. An upgrade comes out from WordPerfect. Will most users want a tutorial describing the changes, or a reference manual? Which users will want which form of instruction? If you were ordering the manuals, what portion would be tutorials, what portion reference manuals?

4. Your campus budget is cut, so you drop WordPerfect and go to a new program called "Free Words," which is downloaded from an electronic bulletin board and costs nothing. In training people to use this new program, do you rely on tutorials or reference manuals? Which people get which?

5. Review a reference manual.
   a. Take a four- or five-page section and see how many of the first six guidelines for reference manuals are followed. What is your reaction to the manual? Could you think of ways of improving it?
   b. Ask an experienced computer user to perform an activity covered in the reference manual. Before the user begins, list which parts of the manual you think will cause problems. Now watch the user. How good were you at predicting trouble spots?

6. Use the "help" screens of at least three different programs. List the navigational aids each of them uses. Which seems to be the easiest to use? Suggest improvements for the others.

7. In a group of three or four, pick a piece of software that is commonly used on campus. Write an introductory tutorial that would help students

learn to use the software. The tutorial should be twenty to thirty pages long and should be desktop published using good page design. After your group has written the first draft of the tutorial, thoroughly test the manual with at least three students. Revise the manual based on the results of that testing. Publish your tutorial by duplicating a small number and giving them to students who will benefit most from the document.

27

Electronic Documents

Features of Electronic Documents
- Speed
- Larger Audiences
- Media Integration
- Flexibility
- Authorship Ambiguity
- Increased Access
- Technical Limitations

Electronic Mail
- Situation
- Audience
- Guidelines for Electronic Mail

Hypertext
- Situation
- Audience
- Guidelines for Hypertext

Activities

All of us still associate writing with paper. Maybe we put words on paper with a pen, or maybe we put words on paper through a computer and printer, but the end result of writing is words on paper. But is the paper really necessary?

Consider company memos. Imagine you want to send your supervisor a short description of some equipment you need. You sit down at your word processor and write up a memo. Now you have two choices. You can print that memo out on a printer, put it in an interoffice mail pouch and wait two days for it to arrive, or you can use your company's electronic mail system to send the memo to your supervisor instantly. If you use e-mail, you not only save time but the ecological cost of the piece of paper.

Consider reports. Writers can use a good desktop publishing program to merge text plus appropriate graphics and print out a nice looking report, or they can use a good hypertext program to prepare a report that contains not only text and graphics, but animation and even video clips. If they prepare a report using hypertext, they not only save paper but have the power to show processes in motion. (The CD-ROM versions of several popular encyclopedias now include animation. You might want to try one at your university library.)

In both these situations and many others, writers are choosing to abandon paper. For a growing number of writing tasks, electronic documents are a superior alternative.

Features of Electronic Documents

Electronic document technology is still new. Nevertheless, even in this rapidly growing field, some features common to these new forms of communication have emerged. Each of these features has important implications for us as writers and readers.

Speed

Electronic words can move faster than paper words. The first moves at the speed of light, the second moves at the speed of the Postal Service. Facsimile machines convert paper words into electronic words and improve the speed of transmission, but FAXing still requires that you take the time to put words on paper.

While getting memos and reports to readers quickly has some obvious advantages, there is also a serious disadvantage. Writers who draft their work on paper have time between each draft to reconsider their work. Even as words came up on a printer, a writer could reread and reconsider. As the report was being duplicated there was more time to reconsider. Even after a report had been mailed, a writer could make changes by telephoning. With electronic communication, a reader can have the document in seconds. If what was written was incomplete, ill-considered, or ill-mannered, a reader may have read the words and already be reacting to them before the writer even realizes there is a mistake. This is especially common with electronic mail, where the practice of firing off ill-mannered memos is so common it even has a name — "flaming." Writers must be careful to keep the speed of their thought processes equal to the speed of electronic communication.

Larger Audiences

Another surprise for people is how far electronic words can wander. Mail networks are set up so that one communication can be broadcast simultaneously to dozens or thousands of readers. This eases the burden of reaching large numbers of readers but creates problems for writers, as each reader can instantly pass these words along. This often becomes an unpleasant shock for people when they discover that a memo intended for one audience has been transported to very different readers. Clearly, paper can also be forwarded without the author's knowledge, but electronic transmission seems to be making such forwarding more common. As a writer, you will have to give more thought to those invisible audiences down countless miles of wire.

Media Integration

Computers are equal-opportunity storehouses. In large measure, they are equally adept at handling written words, images, sound, or even full-motion video. This area of technology is evolving under the name "multimedia," and already has developed to the point that software is available to present dramatic combinations of words, music, and images. This enables some creative opportunities for communicators. Information can be presented as text, sound, spoken language, images, and motion. Writers now have many more tools for communication available — tools that can be used in amazing combinations. This opportunity for writers also poses new problems: learning how to use all these new tools, and learning how to use them in effective combinations.

Flexibility

Another feature of electronic documents is that they are read in a computer. The disadvantage is that readers need a computer to read the documents; the advantage is that they have the power of their computer as their read. This power is already being used to empower readers to move through text in any order that is meaningful to them. The most common example is hypertext. Hypertext documents are typically nonlinear and nonsequential. Although electronic documents all have the potential to supply this flexibility to readers, it is the author that must build the flexibility into the document. The author must set up access paths and supply supplemental materials. This is not an easy task for writers, but it is a marvelous opportunity for readers.

Authorship Ambiguity

Electronic documents create interesting questions about authorship. Reports arrive at readers' desks in a form that can be changed very easily by them. They can access your document and change it—delete a paragraph, insert a reference—and retransmit the document as their own work. Who is the real author? What if the same document is modified by one reader after another? Are they all co-authors? How much change is required before a person becomes the "legitimate" author of a new communication? There are currently no rules about any of these questions. All we know with certainty is that electronic documents can be changed more easily by more people than ever before. We are less and less certain about what it means to be an author.

Increased Access

To get a report that's on paper, it either has to be sent to me, or I have to walk to it. Electronic documents can come to me anywhere in the world with adequate phone service. Librarians are already seeing the effect of this new technology. They are deciding which journals should be put on their shelves in printed form and which should be accessible electronically from some source, possibly across the country. As a writer you may find that even if your job takes you to a small town with a limited library, you have the power to access electronically the best libraries in the world. There are thousands of articles that can be sent to you electronically in microseconds. Electronic documents are accessible documents.

Technical Limitations

Unfortunately, there are still some technical restrictions to the creation and communication of electronic documents. The varying capabilities and requirements of hardware and software complicate all of this. An electronic document that an author creates with stunning graphics may not even run on the computer of someone with an older model computer or a cheaper monitor. Computers vary dramatically in processor speed, memory size, operating system, disk storage, disk format, screen resolution, and capacity to access both networks and ancillary resources such as CD-ROMs or optical disks. As a result, electronic document preparation requires writers not only to think about the readers' needs but also about their office equipment.

A good way to think about these features of electronic documents and their impact on writers and readers is to examine closely two popular examples: electronic mail and hypertext.

Electronic Mail

Conceptually, electronic mail systems are simple. They require networked computers and a central facility to store and route messages. A writer sits down at a computer terminal, calls up a mail program, types in an address, and enters a message. To send a message, the writer enters a command on the computer, and the message goes to some network hub and is routed to the computer controlling the mail. There the message is stored and the address is interpreted. The computer now sends a message to the receiver saying a message has arrived. When the person receiving the message asks for mail, the message is moved to the computer and displayed on the screen.

As Figure 27-1 indicates, e-mail programs clearly mark the kind of information needed for messages. This is important since this form of mail is becoming more and more a business standard, even with people who otherwise dislike computers.

Figure 27-1. Piece of Electronic Mail

```
Date:  Tue Jan  7 09:29:50 1993
Subject: Meeting
To: Jennifer Johnson
Content-Length: 189

Our task force meeting will be discussing your
quality assurance project. Could you give an
overview of your progress? Something on the order of
15 minutes should do it.

Margaret
```

How important is e-mail as a communication medium? "IBM alone connects 355,000 terminals around the world through a system called VNET, which in 1987 handled an estimated 5 trillion characters of data. By itself, a single part of that system—called PROFS—saved IBM the purchase of 7.5 million envelopes, and IBM estimates that without PROFS it would need nearly 40,000 additional employees to perform the same work" (Toffler, 1990).

It may seem that IBM would be a special case, but in fact many other corporations are receiving similar benefits from the use of e-mail. Given the opportunity to save on the employment expenses of mail room employees, and the opportunity to save the ecological disruption from millions of envelopes and letters, we can assume e-mail will become a permanent part of business procedures.

Situation

Electronic mail is a system of linked computers. Typically those links are within a single place of employment. Sometimes people are linked through several mail services to resources outside the company. In any case, e-mail links are increasingly common. This is because e-mail becomes more valuable as more people are connected to it. The situation is equivalent to phone service in this regard. How good would it be to own a phone if only 5% of your friends had phones too? There wouldn't be many people you could call, so your phone would have little value. Electronic mail is growing for the same reason. Every time one more person is linked to e-mail, the value of all the other mail links increases. For this reason, it is safe to assume that more and more companies will provide this resource to more and more employees. Electronic mail is becoming ubiquitous.

Audience

Electronic mail systems are set up so that you can identify a single reader or a group of readers. Addressing readers can be fairly simple if you are writing to someone at your location, but it can be more complicated if you are writing to someone across the country. Here are typical addresses:

Local:	bsmith
Distant:	bsmith@majoru.engdept.bitnet
Discussion Group:	QA__List@company.ulink.edu

For more distant connections, the address needs to include the transfer nodes that are being used by the system, plus some information about which network is being used (such as "bitnet" for Bitnet or "edu" for Internet).

How do you know someone's address? Companies often publish directories of e-mail addresses, and increasingly professionals put their e-mail address on their business card.

But you also have to remember that there are hidden audiences. Text that has been transmitted electronically can easily be forwarded electronically.

Guidelines for Electronic Mail

Given the prevalence of this communication medium, people are already considering how best to use it. What tips are there for writers using e-mail? Experienced e-mail users often make the following suggestions.

1. *Keep messages short.* Electronic documents aren't normally read on paper, they are read on computer screens. This means that people describe message lengths in the number of screens used, rather than the number of pages. And there are relatively few lines per screen: twenty-four lines per screen, versus forty to fifty lines per page. As a result, while a 100-line message still seems a relatively short message on paper (two pages), it may seem very long on a monitor (five screens). The general response of e-mail users is to keep messages short. Messages that require more than two screens may be better put on paper.

2. *Make references clear.* Electronic mail arrives in large volume and in random order. By the time you answer an electronic message, the sender may have already sent many other messages and have received many as well. It is best to begin with a line or two explaining which message you are responding to.

3. *Take time to cool off.* Any message you send may arrive across the building or across the country in a microsecond. Beware of "flaming," the sending of unfriendly messages that might be regretted later. It is often valuable to write a message but not send it immediately. Go on to other work and return to your message later. Give it a second or third reading before finally sending it. Once you press the SEND key, you will not be able to recall your message.

4. *Remember, a message can end up anywhere.* Any reader can forward your message anywhere in the world. This may be to your advantage. If what you said was brilliant, your message may help you establish a reputation. On the other hand, if what you said was unkind, your words may damage you greatly. As you write, ask yourself, "What if my supervisor read this? What if our competitors got hold of it?"

5. *Electronic privacy doesn't exist.* Your words on paper have some legal protection; your mail can only be opened under special circumstances. Electronic words are much harder to protect. For one thing, they are usually copied into temporary storage by each computer that transmits your message. This makes it easier to retransmit your message if there is a breakdown down the line. But this also means that anyone who has access to that networked computer can read all the messages it holds. Companies and computing specialists are working to improve the level of protection for e-mail, but there is only limited protection currently. As you write, imagine your words appearing in large letters on a screen in the main office. Real privacy for electronic communication is years away.

The following example e-mail message shows how several of the previous guidelines can be employed.

Figure 27-2. Piece of Electronic Mail

```
Date: Tue Jan  7 09:42:02 1993
Subject: Meeting
To: Margaret Ohare
Content-Length: 484

I would be happy to discuss our quality assurance
project at your task force meeting. Here are the
topics I believe I could cover in 15 minutes:

Background of the problem-Customer complaints-
Competitive responses

Technology Survey-Vendor responses-Our reviews-
Site visits

Procedures Reviewed

Let me know if you'd like something different.

Jennifer Johnson
```

Hypertext

In the 1950s Theodor Nelson proposed a new kind of text. This text would let readers move through it via a variety of paths. Rather than just reading from page to page, for instance, a reader could move "into" a text, by identifying a word or phrase and having more information on that word or phrase made visible. And texts could be linked so that a reader might move from one to another without ever being aware of leaving one and going to another.

This concept of linked and dynamic texts remained theory until the proliferation of computers. Computers have the power and the memory to present large texts and jump from one text to another. Software to control this process is necessary, but in the last few years a number of vendors have developed easy-to-use, inexpensive programs for the creation of hypertexts.

The sample screens in Figure 27-3 show what such a text might look like.

The first drawing shows the original text as it would appear on a computer screen. It includes some imagery and uses multiple type sizes for screen display, but otherwise looks indistinguishable from any desktop-published work.

In the next drawing, a reader has moved a cursor to a phrase. The phrase becomes highlighted, indicating that additional information is available if the reader wishes. The reader presses a button on the mouse (or other computer control device) and the screen changes.

The third drawing shows the screen that appears. It contains additional information about the phrase. The reader now has a choice: to learn even more about the phrase (accomplished by highlighting the MORE button) or to return to the original screen (accomplished by highlighting the RETURN button).

By pressing MORE, the reader moves into a new screen (shown in the last drawing) that contains additional information plus a new illustration. Notice that it was the reader's choice that made this screen appear. The reader determines which information to view and in what order. The writer supplies the opportunities, knowing that most additional screens will never be seen by readers.

Situation

Hypertext is the first kind of electronic document that can never be put on paper. Electronic mail is meant to be seen on screen, but it can be printed on paper if a reader wishes. Hypertext cannot. Each initial screen may correspond to a page, but the additional screens of supplemental information make no sense on paper. How would a writer order them? How would a

Figure 27-3. Hypertext Screens

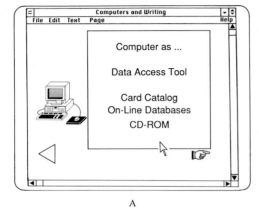

A

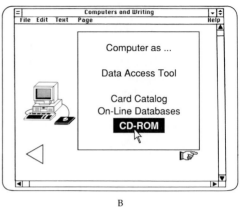

B

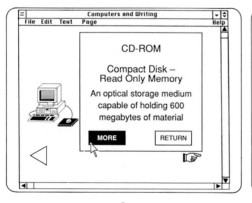

C

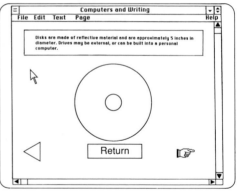

D

reader turn to them? Sequencing and access, while easy on a computer screen, are impossible on paper.

A second important feature of hypertext is that movement through the text is associational rather than linear. This means that rather than move from page 1 to page 2, a reader might read the first line of screen 1, find an interesting phrase, and read the screen associated with that phrase, followed by another associated screen, and so on. Theoretically, there is no limit to the number of screens that might be associated with that first phrase. A reader might never return to the first screen. Meanwhile, another reader might pay no attention to the first screen, move to the second, and start a series of screens associated with some key word there. Each reader is "reading" information and they both started out with the same text, but they are having very different experiences. There is no predetermined sequence for them to follow, so there is no predetermined result that they will achieve.

Briefly, hypertext creates a situation in which text can only be presented via computer and in which the "document" read varies substantially from one reader to another. Where could we possibly use such a strange beast?

A common early use has been training materials. Companies often have substantial experience in training new employees with computer-based education (CBE) materials, so they are used to putting new employees on computers. The problem with many training materials is that employees arrive with widely varying backgrounds. Some may be new to a technical field, while others have many years of experience. Given this range of backgrounds, consider the problem faced by the writer of the text in Figure 27-4.

This text is full of technical terms. Readers with a good technical background would not need any definitions, while those with little background would be confused by words such as "circuit," "load," and "pressure."

The writer could create a glossary for such terms in the hope that the confused reader would look them up, but that includes its own problems. How much definition is required? Some readers might just need to be reminded of a definition they have forgotten, while others are new to the concept. How will readers know which terms are included in the glossary? Creating one text that meets the needs of all readers will require a great effort and may ultimately be impossible.

In hypertext the writer would leave it up to the reader to determine which terms were given added definition, and how much definition was enough. The writer would set up the text to "mark" words or phrases that might need elaboration. For each word or phrase that is marked, there would be an additional screen. This screen might contain a written definition, a picture, or in some cases a video clip. There could be levels of definition, so that the first level was fairly simple for those who needed little help, but additional levels would be available for those who wanted more. These additional helps would sit unseen unless a reader selected one of the marked words. In our example above, the text would appear just as we have it, but if

---

Figure 27-4. Diagram of Hypertext Training Manual

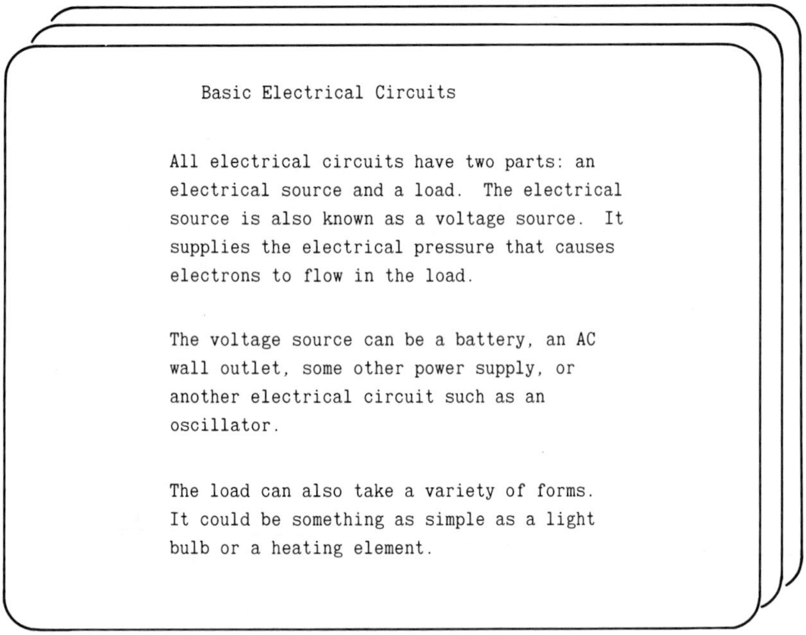

```
                    Basic Electrical Circuits


          All electrical circuits have two parts: an
          electrical source and a load.  The electrical
          source is also known as a voltage source.  It
          supplies the electrical pressure that causes
          electrons to flow in the load.


          The voltage source can be a battery, an AC
          wall outlet, some other power supply, or
          another electrical circuit such as an
          oscillator.


          The load can also take a variety of forms.
          It could be something as simple as a light
          bulb or a heating element.
```

---

a reader selected a word such as "circuit," the first level would disappear temporarily, and the reader might see the screen in Figure 27-5.

Readers needing even more help could choose to see a picture of a circuit. If this brief definition was enough, they could return to the main level and continue reading.

Such flexibility isn't possible with paper. There is only one level, the page. With a computer there can be as many levels as the author creates. Each is accessed quickly and easily as the reader wishes. In a sense, each reader "creates" the text while reading, since no two readers will see the same text. Some will see only the most superficial level, while others will see much additional information. It is up to the reader whether the document is a one- or two-page synopsis or a lengthy technical description.

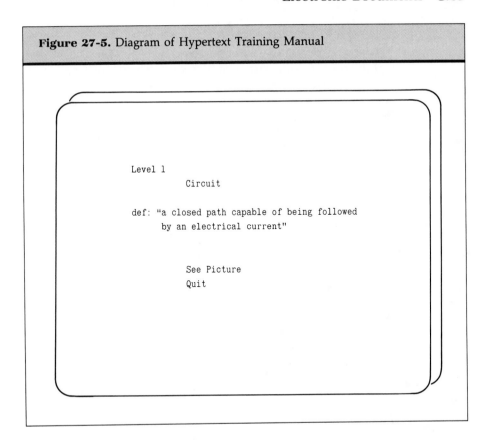

Figure 27-5. Diagram of Hypertext Training Manual

```
Level 1
        Circuit

def: "a closed path capable of being followed
        by an electrical current"



        See Picture
        Quit
```

Audience

Hypertext is ideal for presenting information to readers with a wide range of backgrounds and interests. New employees are one common audience. Readers of some technical reports are another, since readers tend to have a wide range of technical backgrounds. Readers of business documents are another, since most readers only want the main ideas of a report, but some may want to see the details underlying a particular portion of a report.

When considering which audience is appropriate for hypertext, though, remember the one restriction on the use of hypertext: hypertext documents can only be created and read on a computer. Therefore they can only be used by people with ready access to a computer and with at least a basic background in computer operations.

Guidelines for Hypertext

Hypertext documents are powerful and flexible, but they create unique problems for both writers and readers. These problems are reflected in the following guidelines.

1. *Study your readers.* As a writer, you will be creating layers of detail either to clarify points for readers or to substantiate your main ideas. Words and phrases you choose to substantiate depend on your understanding of your readers. If you choose not to create additional layers for a word that turns out to be confusing, you are leaving your readers without help they may come to expect. If, on the other hand, you create lengthy aids for words or phrases that no one ever selects, you have wasted your time. Hypertext forces you to study your readers as you never have before.

2. *Keep screens short.* The idea behind hypertext is to let readers move through a text at their own speed and at their own level of need. It is better to have short screens with additional layers available on request than to have one long screen.

3. *Present information in new forms on each layer.* Each layer should clarify and elaborate on the preceding layer. This should be done by adding new information, but should also be done by presenting the information in new forms. Move from text to graphics to animation to video if necessary to clarify a key term. Each time you use a new presentation form you increase the odds that you will hit the principal learning mode of your reader.

4. *Make navigation clear to your reader.* If on-line help can cause some readers to get lost, hypertext can be even more confusing. There are many levels, many access paths to those levels, and many forms of information. You will have to help your readers by labeling their current position, and by clearly marking keys for moving "up" a level or "down," or back to the main text. The example in Figure 27-6 includes this "navigation" information.

Activities

1. The speed of transmission of electronic documents is an advantage, but it also leads to some problems.
   a. "Flaming" is the electronic mailing of unfriendly messages. If there are e-mail users in the class, have they ever encountered flaming? What were the circumstances? What did they do in response?
   b. A more benign problem occurs when documents are "mailed" before a writer has fully reviewed the document. As a group, create a review strategy for electronic documents. What steps or procedures would you recommend?
2. How private are documents sent by e-mail? What is the process by

Figure 27-6. Hypertext Screen

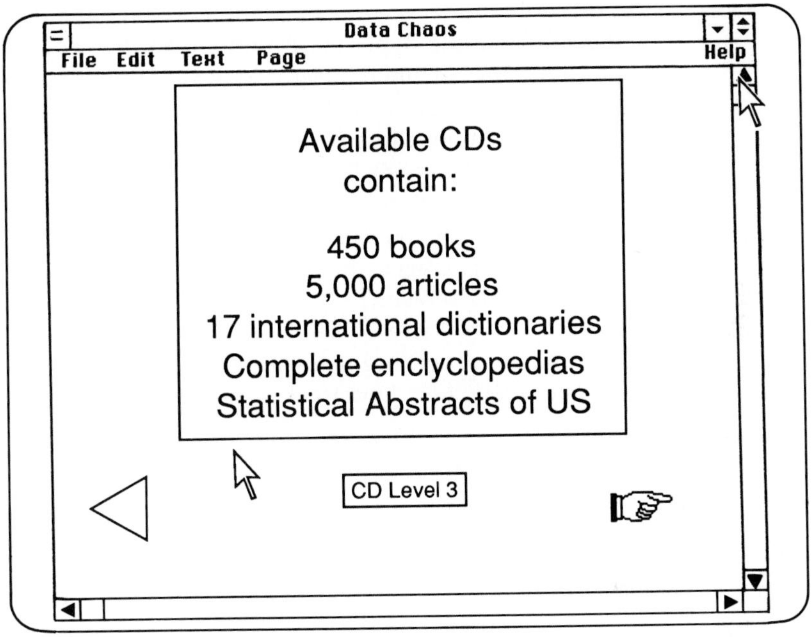
Data Chaos
File
Edit
Text
Page
Help
Available CDs
contain:
450 books
5,000 articles
17 international dictionaries
Complete enclyclopedias
Statistical Abstracts of US
CD Level 3

which such mail can be read by others? Interview someone on campus or at a local company that uses e-mail, and ask what procedures they use to protect the privacy of such correspondence. As a class, create a list of recommendations to safeguard privacy.

3. As a group, pick one of the subjects below:

a. A complaint

b. A question about a grade

c. A request for a course syllabus

d. A job reference

Two students should write an e-mail message about the subject, and two students could write a message to be sent by paper mail. Discuss the differences between the messages.

4. If e-mail is available on your campus, set up a mailing list for the class and have each person send at least three messages.

5. Make a list of the hypertext software available on your campus. Which program is used the most? Why?

6. Outline the process that would be used to convert one of the chapters of this book to hypertext. Photocopy the chapter, identify the portions of the chapter that would be the main text, and circle words that would be used as links to detailed sections. Identify the level at which you would place each paragraph.

7. Create a simple hypertext document for training. Pick a concept that could normally be explained in eight to ten computer screens (such as how to format a computer disk, how to access the college library, or how to get a parking permit). Write those screens and then test them with two readers, one totally unfamiliar with the subject and one a relative expert. Have the novice circle words that are confusing. You might also create a comprehension test to see which concepts are unclear after reading the text. Have the expert point out passages that are "obvious." With both sets of results, rewrite the training screens in hypertext so that the expert can read only the information he or she needs, while the novice has access to additional help. Develop a review procedure to ensure that you have made the best possible choices, and revise your hypertext as necessary.

VI

OTHER TECHNICAL COMMUNICATION TASKS

28

Oral Presentations

Situation
  Work Group Reports
  Customer Presentations
  Professional Meetings

Preparing for an Oral Presentation
  Adapting to a Location
  Creating an Organization
  Creating Note Cards or Prompts
  Choosing Information to Include in Visual Aids
  Selecting the Kind of Visual Aid to Use
  Practicing the Presentation

Giving an Oral Presentation
  Oral Presentations Are Flexible
  Oral Presentations Are Interactive
  Oral Presentations Present You
  People Don't Listen Well

Following Up the Presentation

Activities

This country has a particular affection for oral presentations. A century ago people would come from miles and stand hours in the sun to hear formal discourses. In the most famous of such presentations — the Lincoln-Douglas debates — Douglas spoke for three hours, the audience took a break for a meal, and then Lincoln spoke for three more hours. And thousands came to listen (Postman, 1985).

This tradition of oral presentations has continued to be a major part of our culture. For all the computers and phone lines that link us, we still feel the need to stand personally before a group of people and talk. Listeners want to hear the words, but they also want to see the face, watch the expressions, determine for themselves the honesty and sincerity of the presenter.

So all of us give talks. For some people, this is an invitation to terror. They think back to grade school and standing in front of the class to give book reports. Or they recall high school assemblies and presentations to a huge crowd of bored teenagers. But it need not be so terrible. In fact, for many people, presentations can be an opportunity.

For one thing, adults make much more pleasant audiences. Adults pay attention (or at least pretend), they usually show respect for a speaker's ideas, and they generally listen in the first place because they think the speaker has something useful to say. With few exceptions, you will find adults a much more forgiving and supportive audience than those you remember from childhood. This is good, because you may find yourself giving talks quite often.

Situation

Although most people think of oral presentations as formal occasions on which a presenter stands at a lectern before a large group and uses a microphone, in fact oral presentations fall into a wide range — from the very informal to the most formal, from a small audience of peers to a large audience of strangers, from a three-minute update to a lengthy discourse lasting hours. Described below are several common situations that illustrate the range of oral presentations.

Work Group Reports

From the chapters on technical writing tasks, it should be clear that much of the writing you will do will be addressed to co-workers. So will many of your oral presentations. Reports to your work group are a good way to keep others up to date on the progress of a project and problems, and gives them an opportunity to ask questions and voice opinions.

Work group presentations tend to be relatively informal, to involve just a small group of people, and to be relatively short. The purpose is to gather all the principal participants in a project in one room and quickly inform them of the present situation, and possibly ask them to recommend some action. In such situations the presenter usually speaks for five to twenty minutes, with the expectation that a much longer period of questions and discussion will follow.

One of the things that make such talks relatively easy is that there are usually unwritten rules for such presentations. The corporate culture quickly defines whether presentations are to be short or elaborate, whether they involve visual aids (and if so, which ones), and whether questions are to be encouraged. Although these rules may not be written down, they can easily be observed. After attending your first presentation or two you will understand the conventions of the company.

Corporate styles are not permanent, however, so watch for changes. There may be periods in which peers play games of "one-upsmanship," especially with visual aids. One person brings overhead transparencies to a presentation, then the next week someone has full-color transparencies, and the week after that someone tries computer presentation graphics. You may not want to join in a useless competition, but do watch for changing expectations and new formats.

Another factor that eases work group presentations is that most of the people in the room will be familiar with the project or at least will understand any technical terms you use. They will be one of the most educated audiences you will ever face.

On the other hand, much of the impression your co-workers and supervisors will form of you will be based on these talks. This is your chance to show through the content of your presentations how much you know about the business. And it is through the style of your talk that they will decide whether they would be willing to work for you in the future or have you on their team for subsequent projects. Both the content and the style of your presentations, then, will largely determine your future in the company.

Customer Presentations

There will be many occasions on which reports will need to be made to customers. Initial feasibility studies, operating instructions, proposals, and progress reports are all common. Take progress reports as an example. Customers will have a general sense of progress—if you are erecting a building they can see walls go up, or if you are working on pollution abatement they can see test wells going in—but much of what you are doing, and why you are doing it, may be invisible to your customers. Even if they have read the written progress reports, an oral presentation gives them a chance to ask questions.

You may want to schedule regular presentations to customers. It is good policy to set up a schedule for such meetings at the beginning of a project so that people know when they can get answers to their questions. If presentations become too irregular, it can become easy to get in the habit of calling them only when there is a problem, and the atmosphere at such meetings quickly becomes unpleasant.

Customer presentations are often fairly formal. In part this is to show respect for the customer you are serving. You will want to dress appropriately, prepare thoroughly, and use attractive visual aids.

Unfortunately, no project ever goes perfectly. Every contract contains flaws, no schedule is perfect. Sooner or later you will have to give a progress report that reports no progress, higher costs, or unexpected changes. Making that presentation will be difficult. How well the presentation goes will depend in large part on your credibility. If you have given regular reports, have been honest and specific in the past, and have not promised what you couldn't deliver, you can minimize the hard feelings you will encounter. Each oral report you give builds on the last.

Professional Meetings

It is common for professionals to have meetings to report on recent developments. The purpose of such meetings is to report on research, give people updates on important projects, and describe strategies that seem to be working. These meetings can be local, state, or national. They may last for an afternoon or two to three days, and may involve both major keynote speakers and a number of "breakout" or concurrent talks given to smaller groups.

You need not be a nationally known expert to speak at a professional meeting; in fact, some sessions are designed to give exposure to younger people just establishing a professional reputation. So you might be closer to giving such presentations than you imagine.

What is it like to give such a talk? At first glance it might seem difficult because everyone in the audience is a professional in the field. Some people are bothered by the possibility of making an error in front of a room full of experts. But in some ways fellow professionals are the ideal audience. You can assume that they are interested in your subject—it is what they do for a living—and they have a good background in the subject, so you can use technical terms without fear of confusing them. In addition, since this is a professional presentation, your listeners will be looking for information, not entertainment, so you can concentrate on what you have to say and less on how you say it.

As a result of the forum and the audience, giving a professional talk can be quite simple. Speakers prepare extensive notes in advance, sometimes use simple visual aids, and speak in a direct manner. If highly detailed information is to be presented, some speakers write their entire presentation in advance and read it to the audience. Reading a paper is common for technical presentations, as it is one way to be sure that detailed information has been covered accurately. In this situation it is content that matters, not style.

There are many other occasions for oral presentations, but work group meetings, customer presentations, and professional meetings each represent a common presentation situation.

How do you prepare for such presentations? There are a number of techniques that seem to work for most people.

Preparing for an Oral Presentation ⎯⎯⎯⎯⎯⎯⎯⎯⎯

Adapting to a Location

Initially you prepare for an oral report as you would prepare for a written report: by considering who your audience is, what it knows, what you want it to know when you are done, and so forth. But some preparations are unique.

For instance, if you plan to use visual aids such as slides, will the projection equipment be available when you need it? Is the room in which you will be speaking adequately equipped with outlets? If you need to dim the lights, can windows also be covered? Is there any place to set up a screen? Clearly, you will want to find out where the meeting will take place and what kind of equipment is reasonable for that facility.

Rooms vary in size. Will you need a microphone? Will people in the back of the room be able to see diagrams you may have? Will they be able to see any visual aids through all the people that may be seated in front of them? Some rooms lend themselves nicely to splashy presentations, others are poorly designed for this purpose. If possible, visit the room before your talk and know what you are getting into. Expert speakers come prepared to make a presentation in any of several ways, depending on the physical circumstances of the room.

Creating an Organization

The way you organize and deliver an oral presentation depends on the situation. Each of the situations described earlier in this chapter calls for a different level of formality, different use of presentation tools, different levels of specificity, and even different lengths of presentation. There is no single best speech that fits all situations.

There are, however, some common responses you can make to the speaking situation. Those responses are based on the needs of people who are listening (as opposed to reading) for information.

USE REPETITION

Because reception of the spoken word is different from reception of the written word, speakers generally use more repetition than writers. Readers can reread a paragraph if they are confused. Listeners don't have that option. Repetition is often built into the organization used by speakers. They often choose an outline something like this:

> Introduction listing key points
> Description and illustration of each main point
> Conclusion repeating key points

This organization is sometimes summarized as "tell them what you are going to say, say it, and then tell them what you told them." This organization recognizes that listening is not easy, that an audience can get confused, and that a speaker should do everything possible to keep things clear.

STATE YOUR PURPOSE

What are you trying to accomplish, what point do you want to make? If you state your goals clearly and early in your talk, your listeners will better understand where the rest of your talk fits. It is not just a steady stream of information, but talk directed at a goal they recognize.

CREATE ORGANIZATIONAL GRAPHICS

Organizational graphics list the main points to be covered. The speaker puts them up at the beginning of the talk and then begins describing each. It is a good way to help listeners build expectations of what will follow. Occasionally a speaker will return to the organizational graphics at the close as part of his or her summary.

MOTIVATE ATTENTION

Asking listeners to focus exclusively on what you are saying even for ten to twenty minutes is not easy, but it is not impossible, either. The listeners must have had some interest in your talk or they wouldn't have come in the first place.

Some people recommend that you start every speech with a joke, to focus the audience's attention. This is probably bad advice. For one thing, some people don't tell jokes well. For these people, starting a speech with a joke only makes them and their audience uncomfortable. And it may not be appropriate. In a technical presentation on ground-water pollution, where's the humor? Or if your point is to explain why a project is over budget and behind schedule, a joke may be interpreted as showing your lack of concern.

A more valuable way to focus attention may be to match your opening to your purpose. If you want people to relax and enjoy your talk, start with a joke. But if you wish the audience to take some action based on your presentation, it may be more useful to begin with a particularly impressive fact or statistic. For instance, "For twenty-nine of the last thirty weeks this project has been on schedule. Unfortunately, we are now in week 30." Or, "Nitrate concentrations as low as ten parts per billion have been shown to affect the health of children under two. Our first test well found levels three times that."

Another way to hold an audience is to use a narrative style rather than an expository style. This means that you present your talk more as a story than an essay. Talk about particular people and real events as illustrations, rather than relying on numbers. Elected officials seem to understand this when they give speeches including portions of letters they have received. A politician might say, "I recently received a letter from Mrs. Edna Farberg asking me how she could make her social security payments last, what with inflation increasing the price of hamburger to $.89 a pound." She could have said, "We need to respond to our current inflation rate—4.9% is far too high," but that is much more dry.

It takes work to put all these features into your talk, but if you listen carefully to the presentations of professional presenters, you will hear these organizational techniques used again and again. They work for other speakers and will work for you.

Creating Note Cards or Prompts

Once you know what you are going to say and have organized your basic ideas, you need to decide if you will write out the report and read it, speak from notes, or give the entire talk unaided.

Although it may seem stuffy and dull, there is nothing wrong with reading your report. John Ackers, the president of International Business Machines, reads his presentations word for word from prepared reports. It is a common practice, but it is not as simple as it might seem. For example, you should look up at your listeners from time to time. When you look back at the text, you will have to find your place again, and that can lead to uncomfortable delays. The same situation occurs if you leave your text to put up a new visual aid. Then there is the problem of reduced light. If the room is darkened to allow projection of graphics, you may be unable to read your report at all.

This doesn't mean that you should not write out your report, but it does mean you should come prepared. Be sure to practice and time your report; it will usually end up being longer than you expected. You will want to check on lighting in the room. You may want to print out your report in large type. And you may want to separate paragraphs by several lines and use those break points to establish eye contact.

Some people prefer to speak using small note cards, both to include key information they do not want to forget and to hold organizational keys for themselves. Such cards have several advantages. They are small, so they don't cover your face while you speak, and they can be reordered quickly if you decide at the last minute to alter your presentation. But they can be a problem if you lose your place in them, and you will need enough light to read.

A technique that many people find helpful is to use organization graphics (such as transparencies or slides) as their own cue cards. As the graphics inform the audience of the key points the speaker will make, they also remind the speaker of those points. In a sense, the graphics function as a shared outline.

If you try this technique, you must put some more care into your presentation graphics to be sure that key points are highlighted and your organization is complete. Here is an example of the text a speaker might choose to use for such a graphic:

Four key problems with current procedures:
  Slow turnaround
  Inadequate tracking
  Poor quality assurance
  Insufficient customer service

Notice that the key points are reduced to a word or short phrase. The graphics are not meant to provide a complete explanation but to function as an organizational cue. Using this approach assumes that you know your topic well enough that a key word or phrase will remind you of everything else you wanted to say on that subject. For instance, when you see the graphic, you will have to remember what you meant by "slow turnaround."

You can test yourself by putting the graphics away for a day or two and then seeing if you remember what each point was about. With enough practice you can use this technique for general presentations. If your presentation contains substantial detail that must be presented without error, note cards or a full text might be a better approach.

Choosing Information to Include in Visual Aids

Oral presentations are not ideal for learning. Listeners must assimilate what is said, remember it, and make judgments about it, all as the speaker continues. This problem is well known, and a number of solutions have been tried. One of the most popular is to use visual aids as a way of helping readers see the point you are making and where this point fits into the general scheme of your presentation. Several ways to use graphics beneficially in this context are described below.

EMPHASIZE MAIN POINTS

A visual aid strongly reinforces a point made orally. If you want to be sure that people remember your main points, create a graphic for them. Here's an example:

> We are losing market share.

Saying it is one thing. Flashing it on a screen in letters one foot tall is entirely different.

DISPLAY IMPORTANT DETAILS

If you are trying to communicate specifics of a project, it may be difficult for your listeners to retain all the detailed information you want them to know. Putting details on a screen in front of them will help. Consider this example:

> Nitrate pollution has reached 10 ppm in Portage County.

Even though you said what the concentration levels were, a graphic showing the specific number and location will help your listeners remember.

PRESENT PRIMARY VISUAL INFORMATION

Some ideas translate well into words. Others don't. If you want to report on a machine's maintenance problems, you can talk about the place where breakdowns are occurring, or you can bring up a picture and point to the location. Which will be quicker and easier?

PRESENT THE ORGANIZATION OF YOUR PRESENTATION

If you have three key points to make, the easiest way to help listeners know what you are going to say and follow your organization is to lay out the points on a screen. Here is an example:

> Four steps in formatting a disk:
>
> 1. Insert the disk into the disk drive.
> 2. Type FORMAT A:.
> 3. Enter a name for the disk when asked.
> 4. Remove the disk and label it.

You can now talk about each of the actions listed on the graphic, and listeners will have a clearer sense of the sequence.

PROVIDE BREAKS OR TRANSITIONS IN LONGER TALKS

If you plan to give a talk that lasts more than fifteen minutes, you will probably have several sections to your presentation. In addition to your initial outline telling listeners of these main sections, you may want to create a separate graphic to show at the start of each new topic. Such a graphic might include this text:

> Problem 4: Unexpected vibrations under peak loads.

With such a graphic you can remind listeners of where you are in the report ("Problem 4") and tell them what you will now be reporting on.

SUMMARIZE YOUR REPORT

A final slide showing the list of problems you have covered or the solutions you are recommending will help listeners organize the information you have presented. Such a graphic also helps during a question-and-answer period after your talk.

Selecting the Kind of Visual Aid to Use

Once you have decided what information should be presented visually, you should next decide which of the many presentation media to use. Each has advantages and disadvantages.

SLIDES

Slides are an older medium but are still popular, for a number of reasons. With slides you can project an image to varying size: the image can be large enough for a room of 400 to see or small enough so that a group of three or four can study it. Also, because slides have been widely used for a long time, projection equipment is plentiful, as are screens. The technology is stable and reliable; there are few last-minute disasters because some part of the system isn't working. It is also hard to beat the price of making slides — usually about twenty-five cents each.

Convenience and cost do matter, but high resolution is another advantage. Slide images can hold a great amount of detailed information. In addition, they add a realism that can be very persuasive. It is one thing to say that poor maintenance is leading to collapsing bridges. Much more dramatic is a picture of girders completely rusted through.

As a presentation medium, slides have the advantage of variable pace: you can pause on one and talk as long as you need, and then quickly move through others that are less important. You can back up for another look at slides shown earlier. Since slides are full-color, they are attractive visually, and they are easy to see.

The disadvantage to slides is that you must dim the lights so they can be seen. This can be an invitation to the audience to sleep. They can also draw attention away from you: listeners may be so engaged by the slides that they ignore you. One last point: with lights dimmed, it is hard for people to take notes.

There are several ways to combat these problems.

1. Select slides carefully. Make sure they are clear and that each is essential to your presentation.
2. Use a pointer or other device to clarify what part of each slide is important. Highlight main points.
3. Explain carefully what each slide shows. Direct your listener's attention.
4. Keep it short. Even the most well-intentioned audience will respond to a darkened room.

VIDEOTAPE

Videotape cameras and tape players aren't quite as common as slide projectors, but they are widely available. There are a number of reasons why videotapes are becoming popular for presentations. First is the power of the technology: videotapes show a process in motion and allow the audience to hear it. This adds a level of clarity that helps in many presentations. There is also the advantage of immediacy. If you are trying to explain a new development, you can tape it one minute and show it to a group the next; there is no need to develop film or create transparencies. Videotapes can also be used to bring extra talent to a meeting. An expert who cannot attend a meeting can tape a short presentation, which you can then show as part of your talk. This technique is routinely used to lower the costs of presentations.

But there are also problems with videotape presentations that are often overlooked. The first is size of display. Although large-screen projectors are available, they tend to be expensive and clumsy, leaving people to use regular TV monitors. This is fine for small groups but poses visibility problems for large groups. Videotape also has its own pace. Once you start the tape, it is very disruptive to stop it and answer questions, or go back, so speakers tend to let the tape go, even when it is clear that listeners would like to know more about a particular point.

Another disadvantage is cost. Videotape can be very badly made. A common complaint is that speakers on tape are "talking heads," a face and a voice that talks without interruption or animation. It takes expertise to make good videotapes. This means professionals, and professional prices.

A third disadvantage is that videotaped material takes over a presentation. Once you start the tape unit, you are extraneous until the tape stops.

Some speakers use this time as a chance to rest. When the tape stops you resume your position at the front of the room and continue your talk. Because you are supplanted as a speaker while the tape is running, you should be careful about using very long videotape clips.

To get the most out of videotape, follow these guidelines:

1. Select tape carefully. Few tapes are clear in a large room or hold an audience's interest for long.
2. Introduce all tapes carefully so that listeners know what they are supposed to be seeing.
3. Keep segments short. Even four or five minutes can seem long.
4. If you want to use more than a few minutes of tape, stop the tape every few minutes and comment on it, or ask for questions about it. Keep listeners actively engaged.

TRANSPARENCIES

Overhead transparencies are clear plastic sheets displayed by an overhead projector. They are popular for a number of good reasons. They are relatively inexpensive to produce and are fairly portable. Overhead projectors are a simple technology, so that parts can be replaced quickly if there is a problem. They are also powerful enough that it is often possible to use them without dimming the lights (and inviting your listeners to sleep). Furthermore, the technology for creating transparencies is simple; any photocopying machine can do the job.

For a presenter, transparencies have a number of advantages. They refocus attention on the presenter. It is the presenter who supplies the description, points to areas on the transparency, and determines when it is time to move on. The speaker retains control of the presentation. It is also easy for a speaker to modify a talk quickly if things aren't going well. The speaker can just skip transparencies that suddenly seem unimportant.

Transparencies are not, however, the perfect presentation tool. They lack the animation and sound of a videotape. In general, transparencies involve either words or simple pictures.

A serious potential problem is the quality of the transparencies themselves. Because any piece of printed material can be converted to a transparency, many people take simple typed pages and make them transparencies. Standard typewriter type is too small to project visibly. Others put too much text on a screen. The result is hard to read.

Here are several guidelines to follow when creating transparencies:

1. Use large type, at least 24 point or its equivalent.
2. Leave wide margins. Do not crowd the text.
3. Use only the top seven inches of the page. The overhead projector will only display that much at a time.

4. Use illustrations carefully. Fine detail will be lost in the projection process.

There are many simple desktop publishing programs that will generate large letters and even add simple graphics so that transparencies can be easy to read and attractive. Alternatively, many photocopiers will allow you to enlarge text (and graphic images).

PRESENTATION GRAPHICS

A powerful new presentation tool is the computer. It can be used in one of two ways. In the first case, its desktop publishing capability is used to print lively combinations of text and graphics, which then are converted to transparencies or photographed for slides.

In the second case, the computer is used to create, store, and then generate visual information during a presentation. The presenter uses computer software to make screens that contain various sizes and kinds of type as well as simple graphics. The inclusion of graphics is becoming more common with the availability of "clip art" — large collections of images that can be copied from computer disks. The resulting screens can be quite sophisticated and can include color, simple animation, and highlighting.

Some of the newer programs of this kind can also run other programs "in the background," so that during the course of a presentation a speaker can move quickly into a spreadsheet program to do some recalculations, or use a sophisticated modeling and simulation program. The speaker need not interrupt a presentation to load other software. The programs can be loaded prior to the presentation, and without an auditorium full of people watching.

Computer presentation programs can be very effective. The speaker has control over the presentation, stepping through one screen after another, and often has control over such things as sequencing and highlighting. In response to a question, for example, the presenter can bring up one series of screens, then another. This doesn't mean that computer presentation programs are perfect. For presentations to large groups, they require projection systems that are expensive and clumsy, and lights must be dimmed. In addition, creating these presentations can be very time-consuming. You would not use this tool to put something together at the last minute.

To get the most out of this tool, you should observe several guidelines:

1. Don't put too much on a screen at a time. Information should be limited to what would be included on a transparency.
2. Avoid too much flash. Simple animation, which may appear attractive at first, can become distracting.
3. Give yourself several ways to move from screen to screen. In the course of answering questions you may need to use a screen in a way you hadn't planned.

4. Know your audience's information limit. A long series of screens might be very pretty, but your listeners probably stopped retaining the information after the first half-dozen screens.

POSTERS AND FLIP CHARTS

Despite all the drama of the latest in electronic display media, some professional speakers swear by the most simple media—posters and flip charts. Either can be created by computer plotters, by professional artists, or by the speakers themselves.

Why use simple posters or flip charts in the electronic age? Because you never have to worry about a bulb burning out in a projector, or a computer crashing, or being assigned to a room with no electrical outlets. With posters and flip charts, nothing can go wrong.

On the other hand, there are real limits to the effectiveness of these tools. Neither is much use with a large audience, as they would be too hard to see. There is also the problem of transporting posters and flip charts without them getting folded or crushed. And there is not much drama with these tools. If you want to dazzle your listeners, this is not the approach to take.

CHALKBOARDS

Many meeting rooms, especially those used for routine work group meetings, have chalkboards or white boards. The point is to provide a means for participants to create lists or sketches quickly. Chalkboards work well for informal, working meetings. One speaker can write down initial ideas while others add to points. Words or phrases can be erased or rearranged as the discussion moves along. The chalkboard is a very flexible medium and, once installed, is essentially free.

The disadvantage of chalkboards becomes apparent in a more formal setting. They require speakers to have fairly good handwriting, and few people do. They also require time, as the speaker stops a talk to write. And they may not be visible to many people in a larger audience. As a result, chalkboards should be reserved for informal presentations involving a good deal of discussion.

HANDOUTS

One simple way to help people remember what you have said is to give them a handout, either with an outline of your remarks or with the complete text of your talk. It eliminates the need for listeners to take notes while you talk and reduces the possibility of your being misquoted.

The disadvantage is that listeners may stop listening to you entirely and spend the time reading your handout. You have essentially passed out competition for their attention. For that reason, it is more common for speakers to provide an outline of their remarks so that the listeners will have the key points in front of them for reference but still will want to listen to you to get elaboration on the ideas presented in the handout.

The other disadvantage of handouts is that they limit your talk. Once you have a handout prepared, you will not be able to add large sections to your talk, or to delete comments you later feel are unnecessary. You lose flexibility.

Practicing the Presentation

Once you have organized your talk, selected what materials should be presented as visual aids, and selected which visual medium to use, it is time to practice. What are you hoping to achieve through practice?

TIME

Most speakers are surprised at how long it actually takes to give a presentation. They start out thinking that ten minutes is an eternity and they could never fill it, and then end up talking for half an hour. Give your talk several times, complete with visual aids, and time yourself. You will normally find you will need to delete material.

BUGS

The more visual aids you use, the more likely you will run into mechanical problems: the overhead projector doesn't focus well, your computer doesn't link to the projection system, or there is no place to hang your flip chart. Better to find the problems before you have a room full of people staring at you as you try to make balky equipment work. For this reason, try to practice in the surroundings you will have for your real talk and using the same equipment you will use during the talk.

CONFUSION

A co-worker who listens to your practice presentation can spot places where you are not clear. Have a questionnaire ready for the end of your practice session. Ask your trial listeners to name the major points you made, or describe an important point you tried to illustrate. This will let you know how clearly you came across.

CONFIDENCE

Every time you give your talk without a computer crashing or a projector bulb burning out, you will feel more comfortable about your presentation.

When you know the talk takes exactly the alotted time, you will feel more confident of your success. When you feel more comfortable you will look more comfortable, and more professional.

Giving an Oral Presentation

As you give the presentation, remember the four principal differences between an oral report and a written report. First, an oral report is flexible: you can change it as you go. Second, oral reports are interactive: listeners can stop you and ask questions. Third, you are there in person: it is you they see, not just your words. And fourth, listeners are faced with significant burdens as they listen: they are not doing something that humans do well. Each of these points affects how you give your presentation.

Oral Presentations Are Flexible

An advantage of speaking over writing is that you can make changes in a speech during the speech itself. You don't have to wait until you're finished to find out whether changes need to be made. You can see it in the eyes of your listeners.

What cues should you watch for? Some are obvious. If no one laughs at your first joke, make it your last joke. If eyes start wandering, people are bored or confused. If their eyes snap shut, you may have talked them to death. (Horatio Alger, a famous writer of boy's novels, actually had an old friend fall asleep and die while Alger read him a chapter from his "best" novel.)

But also remember that you are giving a technical presentation. With the exception of the speech to community groups, you will normally be describing technical information. Listeners will normally respond by taking notes. Watch their pens. If they are scribbling madly away, you are going too fast. If they stop taking notes, either you are being unclear or you are covering information they already know.

Besides watching your audience, be prepared for what might happen. For instance, you may want to be ready with several examples in case the first one seems inadequate. Rather than making one chart showing cost estimates, you may want to have several at more detailed levels. You may never use the ones that are more detailed, but at least you have them if the audience seems to want more. In essence, you prepare to change your talk if necessary.

Of course, none of this is easy. If you have typed your entire talk, you may be reluctant to vary from your text. Some visual aids are hard to change on the fly. But by preparing for possible change and being sensitive to your listeners, you can at least respond in some ways to your audience.

Oral Presentations Are Interactive

Depending on how formal your talk is, you may be interrupted frequently with questions, or you may only answer questions at the end of the talk. In either case, you should expect some give-and-take with your listeners.

To a large measure you are in control of this interaction. If you feel comfortable doing so, you may tell listeners they are free to ask questions as they wish. Or you may ask them to hold questions until later. Whichever you prefer, you should state at the beginning of your talk what the "rules" will be.

Questions during your talk can be an advantage for you since they give you a quick way to find out if listeners understand what you are saying, and let you clarify points immediately. On the other hand, some questions might be tangential to your topic, or take so long to answer that you risk not covering the material you wished to cover. If you do not want to answer questions until the end of your talk, a good way to handle interruptions is to remind people you will be happy to answer all questions at the end, and to explain that you have much to cover and that some questions will be answered by information later in the report. But be flexible. If it appears a large number of listeners want an answer *now*, it may be best to stop and handle the questions. Better a few minutes' delay now than to have your listeners confused during the remainder of your talk.

Oral Presentations Present You

Oral presentations give listeners a chance to hear not only what you say, but how you say it. If you appear uncomfortable, listeners will be less certain they can believe you. If you appear hesitant, they will be less certain you fully understand your topic. In short, how you present yourself will determine how well listeners will accept your ideas.

Presentation skills, like any other skills, get better with practice. If you are new to giving talks, you will have to work much harder than those who have years of experience. To give yourself the best chance of looking good, consider three areas of your presentation: your appearance, your composure, and your style.

Appearance matters. At one time or another we have all been distracted by a speaker's hair or odd clothes. You can minimize problems by wearing fairly conservative clothes, and ones you are comfortable in. You will have enough on your mind without having to worry about whether your suit fits well or is a good color. Your attire should be conservative so that it does not distract attention from what you are saying.

To improve your composure, there are two simple tricks you can try. First, arrive early for your talk. Allow yourself twice the amount of time it should take to get set up. You will be nervous enough without having to rush around in the seconds before you are supposed to start. Second, take the time to talk to at least one person in the audience prior to your talk. That way you will never be speaking to total strangers — there is someone there you know. This kind of casual conversation will relax you greatly.

To improve your style, move toward your audience. Most people who are nervous create a wall between themselves and their listeners. They seem to hide behind a lectern or computer. They keep distance between themselves and their audience. As a result, they come to see their listeners more and more as a distant (and potentially hostile) group. Then they wonder what to do with their hands, how to put their feet, how to make eye contact, and where to hold their note cards. If you move into your audience, you know where to put your feet (one in front of the other) and where to put your hands (swinging at your sides, as they always do). You cut down the physical distance from your listeners, which also cuts down the psychological distance. Move toward your audience, and good things will follow naturally.

People Don't Listen Well

We have already described the problems people face as they listen and the help you can give them through visual aids and a careful organization of your report. Now consider three other helps you can give your listeners as you speak.

The first is voice control. People can teach themselves to speak clearly and loudly enough so that an audience can hear and understand them. But the way you speak can do a great deal more to help your listeners. If there are important terms that your listeners must remember, you can emphasize them. In the sentence, "Paper machine three has been halted for maintenance four times this week," you want to be emphatic about the word "three," so listeners don't confuse the problem with another machine, and "four," so they understand how serious the matter is.

You can also use pauses to signal changes in your talk. If you have four main points to make, a slight halt between each one of them will help signal listeners that you have finished one point and are about to start on a new one. The halt in this case serves much the same function that a heading and several blank lines would serve in a printed report — a cue to change. Such cueing can be especially effective if you combine it with a graphic to highlight the point you have just made, or the place you are within the context of the speech.

You can also use your voice to help with short attention spans. Whatever volume level you use, listeners will grow accustomed to it quickly. To get their attention, change. Novices sometimes change by getting very loud for a few sentences. A much more effective strategy is to speak more quietly for a sentence or two. People will bend forward for a whisper. They will look up at the sound of silence. Popular speakers use this technique all the time.

Following Up the Presentation

After you finish your presentation you have another opportunity to learn how well you did and then revise your talk for future use (or at least prepare better for future presentations). Unfortunately, few speakers take advantage of this opportunity. Whether from modesty or from fear of disappointment, most speakers wait for the applause to die down, pack up, and go home. They pass up the opportunity to get some free advice.

There are a number of short evaluation techniques they could be using. For instance, many organizations create a short, four- to five-question evaluation form that they ask listeners to fill out after a presentation. Questions are worded so that they are easy and quick to answer (the hope is that more people will take the time to complete them), and include such questions as

> Was the presentation content valuable to your work?
> Was the content clearly presented?
> Did the presentation style hold your attention?
> Would you like to know more about this subject?

Often such questions are followed by a space for the listener to comment. To make the process even simpler, there may be answers that the listener only has to circle such as

Very valuable   Somewhat useful   Neutral   Not useful

It is very simple for you to create such questionnaires yourself and collect them at the end. They may not always say what you would like to hear, but they are a valuable tool in helping you learn an important skill.

Activities

1. Remembering that a corporate culture affects the ways presentations are made, can you tell something about the corporate culture of your university? Do presentation styles change from department to department? Are visual aids more typical in some departments than others? What distinguishes the university culture from the culture of a business you have worked at?

2. Assume that you have to give a talk about a groundwater pollution project or a new screen design you are creating for a computer program. Discuss the advantages and disadvantages of giving the talk in each of the three major presentation situations described in this chapter: workshop presentation, customer presentation, and professional meeting.

3. Look around your classroom. If you were to give a talk in this room, which of the visual aids described in this chapter would work well? Which wouldn't? List the special problems the room would create for a speaker. Is there a better room for presentations on campus? List features you would like to see in the classroom of the future.

4. Try to read the first page of this chapter out loud as if it were a speech. Look up occasionally to make eye contact with your audience. Could you hold your place? As a group rewrite the page in a way to make it easier to follow during a speech.

5. Discuss the presentation tools you would use in each of these situations. Give reasons for your choices.

   a. A lecture on paper making at the local Rotary Club.
   b. A progress report to the local building commission.
   c. An explanation to the County Board about why the new county jail will be $500,000 over budget and 6 months late.
   d. A report to a homeowners' group about oil your pipeline company spilled in their backyards.
   e. A report to your management group about a new computer system that is almost complete.
   f. A presentation to a logging group about a spotted owl whose wooded habitat your society wants to protect.
   g. A talk to a taxpayers' group opposed to building a new sewage treatment plant.

6. As you listen to a lecture in one of your classes, evaluate the performance of your professor as a speaker and of your classmates as an audience.

   a. List the devices the professor uses to help the class understand and remember. Make another list of the techniques that could have been used. Compare your list with those of some of your classmates and discuss what major improvements in presentation style could be made. As a group, write a letter to the professor recommending these techniques.
   b. Watch your classmates during a lecture. What cues are they giving the professor about their listening and understanding? Which cues does the professor seem to notice? How does the professor respond? Which cues does the professor seem not to notice? As a group, discuss ways you and your classmates could better signal your needs.

7. Take any of the reports you wrote earlier in the semester and convert it into a ten-minute oral report.

    a. Begin by determining how much information can be presented in that amount of time.

    b. Select the portions of the report you wish to use.

    c. Revise your introduction and conclusion to meet the needs of the listeners.

    d. Determine the kinds of visual aids that would be available and effective in your classroom.

    e. Create the visual aids.

    f. Create an evaluation questionnaire to give your listeners.

    g. Give the report.

    h. Collect the evaluation forms and summarize changes you would make if you were to give the report again.

29

Correspondence

Memorandums
   Types of Memorandums
   Elements of a Memorandum

Letters
   Types of Letters
   Elements of a Letter

The Writing Process
   Assess the Writing Situation
   Determine the Solution
   Collect Information
   Organize the Information
   Evaluate the Communication
   Distribute the Communication

Activities

Although documents such as reports, proposals, and manuals are a form of correspondence in that they communicate ideas from one person to another, correspondence usually refers to memorandums and letters. However, there is a difference between correspondence in general and technical correspondence. Technical correspondence refers to memos and letters that address technical matters, such as reporting a solution to a problem, describing services available, interpreting laboratory analysis, providing instructions, or requesting information.

Do not underestimate the role and importance of correspondence in the workplace. In addition to functioning as a method of communicating among people, agencies, and organizations, correspondence also functions as a method of documenting transactions. Correspondence requires the same level and degree of care and precision as a research report, a proposal, a feasibility study. Correspondence is an integral part of the technical writer's world. One study of a group of technical writers concluded that a little more than 50% of technical writing consisted of memos and letters (Casari & Povlacs, 1988).

Memorandums

The memorandum is the most frequently used form of written communication within an organization. A memo is a short document that can range from a one-sentence notice of a meeting to a four- or five-page notice of a new procedure. If a five-page memo initiating a new procedure sounds suspiciously like a report, that's because the audiences and purposes of memos and reports are often similar. Often the information of a report is conveyed in the format of a memo. Thus, the document really is a report (because it does what reports do), but it is conveyed in a memo. The differences are that reports are often longer, more detailed, and more formal than memos. However, as in all writing tasks, the level of formality depends more on the audience and purpose than on the type of document. Memos are (normally) sent to someone *within* an organization, either conventionally (by paper) or electronically (through electronic mail).

Types of Memorandums

Memos fill a range of functions. They supply information, help readers make a decision, and instruct. The most common types of memos are informational memos (those that describe a new procedure to be implemented), cover memos (those that accompany a report, study, or proposal within a company), reminders (those that remind people of meetings or visits), inquiries (those that request information), and clarifications (those that explain rules and policies).

Elements of a Memorandum

The standard memo has either four or five parts, although the order of the parts varies widely.

DATE

This is the date the memo is sent, not the date it was written.

TO

Be clear here. Memos are written and used by people in their professional roles, which are therefore essential components of the To/From information. Some memos are "blanket memos," which might go out to all field researchers merely with the identification "field researcher." However, if a specific person is to receive your memo, be clear. Although it might be obvious to you to whom you are sending a memo, it might not be obvious to whoever is sorting the mail, especially if two people in the same company have the same or a similar name.

FROM

Again, be clear. It is surprising how many people send memos from "Jim" or "Nancy." Unless it is absolutely clear who is sending the message, type or write in your full name and title. Include the title or position within the organization both to enhance delivery and to provide a historical record. Although roles often continue, their occupants may change.

SUBJECT

Although this heading used to be standard, many writers prefer to omit it. The advantage of a subject line is that it helps the reader understand the content of the memo by alerting him or her to the purpose of the memo, and it allows for quick sorting and retrieval of memos. If you choose to include a subject line, be precise:

Too broad:

Subject: Pipe Connections

More precise:

Subject: Withdraw/insert pipe connections to CRD

BODY

Depending on the nature of the information, typically there are two ways to present information in a memo. One way is to present your information in a straight narrative form. If you do, keep paragraphs short, generally no longer than seven lines, and strive for several short paragraphs rather than a single long one. They will be easier to read and digest.

If you have several items, consider using an enumerated form, listing the items in a series of points. If you use an enumerated form, use a clear and consistent order, and clear captioning techniques to make the order apparent (see chapter 9, "Document Design").

"CC" LIST

This list takes its name from the old practice of sending people carbon copies of the original document. Of course, photocopiers and word processors have made this actual technique a thing of the past, but the term lives on as a way to indicate who, other than the addressee, should get a copy of the memo. These are generally people who should be aware of a situation or project but who aren't directly involved.

Compare the two memos in Figures 29-1 and 29-2, written from the supervisory project manager to a project engineer at a nuclear power plant. The first memo is presented in a narrative form; the second memo is presented in an enumerated form.

Regardless of which form you use, the main point should be stated early in the body of the memo. Even if you need to give some background information on the situation or the problem, the main point should come as early as possible. If your memo is more than one page long, consider using subheadings to separate parts of the memo; subheadings will make the task of reading the memo easier for the reader. (To review strategies for document design, see chapter 9.)

Letters

Just as memos are generally sent to people within an organization, letters are normally sent to someone *outside* an organization, either conventionally (by paper) or electronically (through electronic mail). And, just as memos range from very short to four or five pages, letters range from one page to four or five pages, although, depending on the audience and the purpose, a letter can run to ten or even twenty pages.

Figure 29-1. Memorandum in Narrative Form

Date: January 14, 1988

To: Allen Young, Project Engineer

From: Fred Buehler, Project Manager

Subject: NRC Bulletin 88-04 ''Potential Safety-Related Pump Loss''

The minimum bypass flow rates given in the current Limerick design documentation for the HPC1, Core Spray and RCIC system are valid and can be used to develop the final response to NRC Bulletin 88-04. Consolidated Electric's position is based on the fact that the minimum flow line is used only briefly during the few seconds necessary to start the pump and open the valve to the full flow line.
In addition, for the LPC1 and core spray pump, the minimum flow line would be used for a postulated small break LOCA or degraded isolation event until vessel depressurization occurs. This is estimated to be less than 30 minutes. Therefore, because of the very limited time of operation on minimum flow, only insignificant pump wear or performance degradation could occur.

Two more reasons are that Consolidated Electric believes that required surveillance testing assures proper pump operation, and there has been no reported instance of ECCS pump performance degradation due to pump damage caused by minimum flow operation.

cc: Marilyn Santos

Types of Letters

Generally the four main types of technical letters are cover letters (also called letters of transmittal), which often accompany reports, studies, and proposals; inquiry letters, which request information; responses to inquiry let-

Figure 29-2. Memorandum in Enumerated Form

Date: January 14, 1988

To: Allen Young, Project Engineer

From: Fred Buehler, Project Manager

Subject: NRC Bulletin 88-04 ''Potential Safety-Related Pump Loss''

The minimum bypass flow rates given in the current Limerick design documentation for the HPCl, Core Spray and RCIC system are valid and can be used to develop the final response to NRC Bulletin 88-04.

Consolidated Electric's position is based on the following:

1. The minimum flow line is used only briefly during the few seconds necessary to start the pump and open the valve to the full flow line.
2. For the LPCI and core spray pump, the minimum flow line would be used for a postulated small break LOCA or degraded isolation event until vessel depressurization occurs. This is estimated to be less than 30 minutes. Therefore, because of the very limited time of operation on minimum flow, only insignificant pump wear or performance degradation could occur.
3. Required surveillance testing assures proper pump operation.
4. There has been no reported instance of ECCS pump performance degradation due to pump damage caused by minimum flow operation.

cc: Marilyn Santos

ters, which answer both written and oral requests for information; and instruction letters, which provide directions.

Figure 29-3 shows an example of a cover letter that accompanied a 100-page report from two paper products technicians to the general manager of a paper making plant. The report includes instructions for designing a lightweight coated paper machine and for analyzing the quality of paper

---

Figure 29-3. Cover Letter

June 27, 19--

Mr. Gerry Ring
General Manager
Paper Products Division
Consolidated Industries
265 University Drive
Ithaca, NY 14850

Dear Mr. Ring:

Enclosed is our report on designing a lightweight
coated paper machine and analyzing the quality of paper
required to produce lightweight coated paper.

The report is divided into the following sections:

    1. Fiber Properties and Base Stock Preparation
    2. Headbox Design
    3. Forming Table and Related Equipment and Design
    4. Drying, Coating, and Supercalendering
    5. Paper Properties

Please contact us if you have any questions.

Sincerely,

Frances Hardeston
Technical Support Associate

Enclosure

---

required to produce lightweight coated paper. Using a standard block letter format, with all the letter's elements flush against the left-hand margin, the writers of the cover letter first introduce the subject of the report, then describe the contents of the report, and, finally, encourage the manager to contact the writers if he has any questions.

Figure 29-4 shows a letter written in response to an inquiry letter. In this

Figure 29-4. Response to a Letter of Inquiry

May 3, 1989

Mr. Frank Jessic
Consolidated Electric Company
2301 Main Street
Philadelphia, PA 19101

Dear Mr. Jessic:

The following is a complete reply to your letter no.
PRC-3645, dated February 19, 1989, our corresponding
Document Control Number 16534.

The above PECo letter requested PacMor to evaluate the
desirability of using the level 8 trip signals of HPC1
and/or RCIC to separate completely the reactor feedwater
controls from the reactor feedpump turbine high level
trip and also to provide a complete cost estimate for
performing the change.

We examined the electrical circuits related to the above
change. Although either the HPCI or RCIC trip signals
would have been technically acceptable, we believe the
HPCI rack is too crowded to incorporate the change without
complications. Therefore, we selected the RCIC level 8
trip signals.

The estimated direct engineering manhours associated
with this change are as follows:

PacMor	265 manhours
General Dynamic	285 manhours

The construction estimate for the cost of the
modification is $37,000. This includes
roughly 630 manhours and the cost of materials.

The estimates and evaluation above, and the
record of changes that have to be made are
noted in deferred PCR 20915.

Very truly yours,

John Golub
Project Engineer

instance the writer used a modified block format, which is similar to the full block above except that the date line and the closing signature are placed on the right-hand side of the page. Because memos and letters are an integral and important part of almost all organizations, many large corporations that receive and send hundreds of letters a day use control numbers to keep track of the heavy volume. In this letter the writer uses organizational control numbers to refer to specific documents, including the original letter of inquiry from Mr. Jessic.

Elements of a Letter

The conventional technical letter has six elements.

HEADING

The heading includes both the writer's full address and the date the letter is sent.

INSIDE ADDRESS

The inside address includes the name, title, and address of the person getting the letter.

SALUTATION

Though it is still traditional to include a salutation, such as Dear Mr. Cohen or Dear Ms. Wolf, some professional organizations, in order to avoid the dilemma of titles and the inadvertent use of sexist language, suggest that the salutation be omitted and that the body begin directly after the inside address.

If you choose to do this, use a subject or reference line instead of the salutation.

BODY

The body is the heart of the letter and should be clear and concise. As in the memo, keep paragraphs short, generally no longer than seven lines. Strive for several short paragraphs rather than a single long one. Consider itemizing a series of points, rather than using a narrative form. The purpose of the letter should be stated in the first paragraph. Most workplace letters are presented in a standard block letter format, as in the letter above directed to Mr. Gerry Ring.

COMPLIMENTARY CLOSE

Choose a close that best conveys your relationship with the reader. Standard closings range from formal to informal.

Sincerely (The most standard)
Very truly yours
Respectfully
Sincerely yours
Yours
Yours truly
Cordially
Best wishes

SIGNATURE

Some writers, for whatever reason, prefer not to sign their letters. Because most technical letters are public documents, and because you are liable for what you say, in order to prevent the possibility of error, and as a courtesy to the reader, *always sign your letters*. Also, the signature should be followed by your name and position, typed.

Figure 29-5 shows a format for letters of inquiry that was prepared by a career services office for people interested in getting a job with a specific employer. The letter is included here both to illustrate the simplified letter format and to describe a strategy for inquiring about a position (to review letters of application, see chapter 30, "Job Applications"). The simplified letter format, endorsed by the Administrative Management Society, is similar to the full block, except that it omits the salutation and the complimentary close. In place of the salutation, the simplified letter format includes a headnote line in all capital letters.

The Writing Process

Although you are writing a memo or a letter, and although the writing is relatively short and likely addressed to one person, do not be lulled into a false sense of security. Memos and letters are every bit as much a reflection of you and your organization as are reports and proposals. Although you might compress the writing process, if you want your document to be effective, follow the stages of the process (described in chapter 2) as they apply to memo and letter writing.

Figure 29-5. Simplified Format for Letter of Inquiry

Your Present Address
City, State and ZIP Code
Today's Date

Person's Name (if possible)
Person's Title
Organization's Name
Street Address
City, State and ZIP Code

HEADNOTE IN ALL CAPITAL LETTERS (e.g., YOUR OPENING
FOR . . .)

Tell why you are writing; name the position, field, or
general area in which you are interested. If applicable,
tell how you learned of the opening or organization.

Show why you are qualified for the job. Identify specific
achievements you have made which would be of particular
interest to your prospective employer, slanting your
remarks to their point of view. If you had related
experience or specialized training, be sure to point it
out, but do not reiterate your entire resume. Refer the
reader to the enclosed resume or application form.

Explain why you want to work for this employer:
—Specify your reasons for wanting to do this type of
  work, and
—Identify what you can do for this employer.

Close by requesting an interview and indicating the
action you will take to make arrangements for that
interview. If, instead of wanting an interview, you want
further information about openings, enclose a
self-addressed, stamped envelope. In your closing,
elicit a specific action from your reader.

[Signature]

YOUR NAME TYPED (ALL CAPITAL LETTERS)

Enclosure (if you are enclosing a resume or application
form)

Assess the Writing Situation

TIME CONSTRAINTS

Do you have a deadline? Does the recipient of your memo or letter have a deadline?

ROLE

What is your organizational role in the document you are preparing? Are you reporting the results of a meeting? Are you trying to persuade your supervisor to institute additional security measures? Are you requesting information about a product?

AUDIENCE

Who, exactly, is the primary audience for whom you are writing? If you have "copied" others, who exactly is the secondary audience for whom you are writing? How much does the reader know? How much does the reader want to know? Do you need to give any background information? Do you need to define any terms? How much technical data should you give, if any? What can you do to make your memo or letter interesting to the reader?

PURPOSE

Are you writing to inform, to help a reader make a decision, or to help a reader learn how to do something? Or is your memo or letter intended to fulfill some combination of these three purposes? Exactly what do you want your memo or letter to achieve?

Determine the Solution

Once you have a clear notion of the situation, your role, and the intended audience and purpose, consider what constraints you might be under, and set a schedule.

KNOW YOUR CONSTRAINTS

Memos and letters often are not subject to the same constraints as reports and proposals; you will not often have to visit a site or conduct lengthy interviews or prepare a questionnaire before writing a memo or letter. However, do not be misled. Though the amount of time you will have available to complete a writing project is often not as critical with memos and letters as with reports and proposals, that does not mean that time is not a factor. It is surprising how often technical writers put off writing a memo or letter until the last minute, only to find out that they need a vital piece of information from a colleague or from a library.

SET A SCHEDULE

Even if you are writing a memo to a colleague informing her of a meeting, set a schedule for yourself. Your colleague will not appreciate getting the memo less than an hour before the meeting. If you are writing a memo to a department head attempting to persuade him to initiate a new procedure, or a letter to an executive in a large corporation requesting authorization to conduct a series of interviews, setting a schedule might make the difference between a well-done document gotten out in time and a hastily prepared document sent out at the last minute.

Collect Information

Unless you have all the information at your fingertips, the time to gather information is before you sit down to write. If you are the source of information for the correspondence you are preparing, do some brainstorming to find out just what you do know. In writing memos and letters, you will often be the primary source of information. But do not overlook colleagues and the library as additional sources of information. For example, if you are preparing a letter to a foreign pharmaceutical company requesting information on a vaccine, you might check a reference book to make sure the terms and symbols you are using are the same in both languages.

Organize the Information

In some ways, the organization of a memo or a letter is more critical than the organization of a report or a proposal. Because memos and letters are often brief, a weak organization will be more evident.

SELECT DETAILS

Regardless of the type of document or the writing task, details will be essential to your writing. The detail can be as simple as including the time and room number in a meeting notice. Though it is often difficult to determine whether you included all relevant details, one quick but effective method for testing memos and letters is the reporter's formula: will your reader be able to answer whichever of the following questions are applicable to the subject of the memo or letter you are writing?

Who?
What?

Where?
When?
Why?
How?

DETERMINE THE PROPER FORMAT

Most organizations have evolved their own formats for memos and letters. If your company has preestablished formats for memos and letters, use them. If they do not, you will have to choose those elements that you consider most effective. The formats of letters and memos and the formats of technical writing in general have a similar function. As with all technical writing, the purpose of these conventionalized formats is to highlight the most important information and present it in the most useful form possible.

Evaluate the Communication

Even the simplest memo or letter is liable to error, misunderstanding, or ambiguity. While it is unlikely that you can ask the recipient of the memo or letter to review it in advance, you might ask a colleague to look it over: Are there any errors? Is the point clear? Does the message make sense? And, because self-review is always part of quality review, be sure to read your memo or letter and proof it yourself before you sign it.

Distribute the Communication

As with other steps, this one is briefer and less complicated than for other technical documents, but still essential. Memos and letters often have long distribution lists that are included as part of the format. It is unlikely that you will have any budget concerns unless you are distributing a memo to a very large number of people. Even in that case, duplication is relatively inexpensive—but the costs of *sending* the letters or memos may be considerable. Be sure to make a copy for each person who is to receive one, and then send them using a reliable and economical method.

Activities

1. The following information appears in a manual for bank employees. Because a large number of errors resulted from workers not knowing this information, you were asked to rewrite the information as a memo reminding employees of the procedures described in the manual.

Data Set and Line Problem Information

This information will clarify any questions regarding the data sets and their status. Any deviation from the normal condition of the status lights may be an indication of a possible telephone line or data set problem. The condition of the status lights during normal transmission is as follows:

ON	Lighted	CS	Blinking
MR	Lighted	CO	Lighted
RS	Blinking	TM	Off

If you suspect a data set or telephone problem and you find it necessary to put in a trouble call to First Data Processing, you must be able to give the First Data representative the following information:
  —The condition of the status lights
  —The FD number (this is on a silver piece of tape usually located on the top of the data set—e.g., FD 3495-16).
The representative may ask you to do an LT test. This should only be done at First Bank's direction. When asked to do so, depress and hold the LT button for approximately 15 seconds. (All status lights should light.) If the TM light goes out or flashes while the LT button is depressed, the data set has detected an internal error. Report this to First Bank. Don't forget to release the LT button at the end of the test.

2. Rewrite the memo as a letter responding to an inquiry from the manager of a bank for which you supply services. In a brief oral report to the class, describe the decisions you had to make and what changes those decisions led to.

3. With two or three other students, write a letter to an appropriate person (the manager of the bookstore, the president, the department chair) requesting a solution to a problem on your campus (too few popular paperbacks in the bookstore, the lack of students on university committees, the need for departmental scholarships); then, exchange letters with another group of students and write a letter responding to that group's letter.

4. Find a letter that you believe is poorly written. Rewrite the letter, and prepare a memo to the author describing the changes you made and why you made them. If you have difficulty finding letters, look in the "Letters to the Editor" section of a newspaper, magazine, or journal.

5. Because members of your writing team cannot find a convenient time for all members to meet, the group has decided to exchange ideas through memos. You have been asked to devise a standard memo form for use by the group. As you plan the form, consider both the kinds of writing the group might be doing and the various steps in the writing process that will require an exchange of ideas.

30

Job Applications

Securing an Interview

The Résumé
  Types of Résumés
  Elements of a Résumé
  Guidelines for Résumés

The Letter of Application
  Elements of a Letter of Application
  Guidelines for Letters of Application

The Interview
  How You Look
  How You Communicate
  Follow-Up

Getting the Right Job

Activities

IBM Corporation processes approximately 1 million applications a year; each year several hundred people are hired. TWA regularly receives approximately 750,000 applications each year and hires about 300 people a year. These figures suggest that you must work to give yourself an edge in an increasingly competitive market.

The first thing to do is to keep your purpose in mind. Remember that your immediate purpose is to secure an interview. Your long-range purpose is to get the right job. Once you know what job and employers you are interested in, you are ready to apply effectively. Being effective means doing whatever you can to give yourself an edge in the job application process.

Securing an Interview

It is possible to circulate your unsolicited résumé to employers for potential positions. Sending a résumé and a letter stating, "I'm looking for a job in marketing; please consider me for the next one that becomes available," does sometimes pay off, since personnel workers do sometimes look through their files to fill a position. At the very least, once a position opens up, you have two advantages: they have already heard of you, and they have your résumé in hand. This last point can be of great importance when openings must be filled quickly. All you need do is call the personnel office, tell them you are interested in the newly posted job, and let them know they already have your résumé. You should always send a letter (and another résumé), too, of course.

The job search is actually a two-pronged attack: blanket the market with your résumé, but also apply promptly for any relevant job openings.

The more traditional way to conduct a job search is to watch known sources of vacancy information, such as want ads in newspapers, job listings in professional journals, postings in company employment offices, and notices in your own university career service office.

These postings should be checked as often as possible — very often, the early bird really does get the worm. If Consolidated Industries, Inc., sends out a weekly bulletin listing job openings that arrives at the university career service office on Wednesday, you should try to stop by the office every Wednesday, or, if you can, go by the personnel office of Consolidated Industries on Monday or Tuesday, when the listing probably first becomes available.

You should also consider other sources of information about new positions: job fairs, employment agencies, job clearinghouses, contacts within the company or the industry, or any other people who might have some sort of inside information. Even something as general as "I hear that Consolidated is thinking of hiring a couple of new assistant technicians next summer" can sometimes be enough of an edge.

This kind of informal contact is often referred to as "networking," an important feature of the job search. Networking is a good way not only to hear about job openings, but also to learn more about potential employers, potential positions, and even potential fields. Networking doesn't have to be grueling: often talking to family members, professors, other students, career-office workers, and perhaps the personnel-office employees with whom you have already had contact is enough. The important thing is never to let a potentially important contact shrivel up.

When you do find ads that appear promising, read them carefully. Employers place ads in newspapers and journals to generate a pool of potential candidates for specific positions. Understand exactly what is being asked for. If the ad is for an "entry-level" position in telemarketing, you will spend most of your time on the telephone trying to sell someone something. If the ad is for a managerial position, you will most likely be expected to have had managerial experience already.

Once a position opens up, respond to it promptly. Often the only applications considered for a desirable job are those that come in two to three days after the ad was placed. Either the company has already found someone by then or it has so many applications it doesn't need any more.

If you are going to respond to a want ad, do it properly. Using all the information you have accumulated so far, you will need to develop a résumé and cover letters designed for the employers and positions appropriate to your skills and interests.

The Résumé

If you are looking for a job, you should have a résumé ready to go at all times. The *first* thing to do in responding to a want ad is to prepare a résumé. You will need a résumé in order to write the cover letter. The résumé is a concisely worded, brief introduction of your qualifications. A résumé allows you to present yourself in the best light and differs from application forms and questionnaires in that the applicant is in charge of the form. It contains information about you that a prospective employer would likely find useful in determining whether you should be interviewed. It is the combination of an effective cover letter and an effective résumé that will get you an interview.

There are no hard and fast rules about what makes a good résumé. However, there are some general guidelines. While you would like to think that employers will spend hours analyzing your résumé, most spend less than 20 seconds scanning a résumé to determine if the applicant is worth further consideration. To pass this initial screening, your résumé must be immediately effective. Therefore, be certain your résumé is organized and attractive, and identifies the following:

Who you are.

What you would like to do.

What you have accomplished.

What skills and knowledge you can offer to the employer.

Types of Résumés

There are two basic types of résumés: *chronological* and *functional*. Regardless of which type you elect to use, preparing your résumé will take considerable time and effort. The résumé is your sales tool and should be prepared as a persuasive as well as informative document. Therefore, select the style that will best present your qualities and be visually appealing to the reader.

THE CHRONOLOGICAL RÉSUMÉ

The chronological résumé is the easiest to prepare and thus is quite commonly used. It presents information in descending chronological order (most recent events listed first) under each heading. The emphasis is on those schools, organizations, and employers you were involved with, what you did, when you did it, and what you accomplished. This form is usually best for recent graduates or those new to a field. Figure 30-1 shows a chronological résumé.

The chronological résumé usually includes information in this order:

1. Name
2. Address and telephone number
3. Objective (optional)
4. Education
5. Job experience
6. Skills, interests, activities

THE FUNCTIONAL RÉSUMÉ

The functional résumé de-emphasizes chronological listings and focuses on competencies or abilities that can be applied to a number of situations. Skills are organized into categories that tell employers what you will be able to do for them. This form may work best for applicants with substantial experience. Figure 30-2 shows a functional résumé.

The functional résumé usually includes information in this order:

1. Name
2. Address and telephone number
3. Objective (optional)

Figure 30-1. Chronological Résumé

Jane A. Smith

Address until June 1, 19__   Permanent address
411 James Ave.   25 Temple St.
Boston, MA 01830   Ames, MA 01950
(508) 462-3489   (508) 467-9532

Professional Objective

Entry-level sales or marketing position with long-range interest in management.

Education

Bachelor of Science degree, May 19__
Boston University, Boston, MA
Major: Business Administration, emphasis in Marketing

Experience

Hostess   June 19__ to present
Mayflower Hotel, Boston, MA. Interacted with restaurant customers. Greeted and seated restaurant
patrons.

Research Assistant   Academic year 19__
Student Marketing Association, Boston University. Gathered data necessary for development of entry
for national student marketing competition.

Salesperson   Summers 19__ and 19__
J. C. Penney Company, Ames, MA. Handled sales and customer relations in women's and junior
apparel departments.

Student Assistant   Academic year 19__
Mugar Library, Boston University, Boston, MA. Filled book orders and stacked returned books.

Camp Counselor   Summer 19__
Camp NoTac, Derry, NH. Coordinated activities for and supervised groups of 12 adolescent girls
during live-in summer camp.

Activities and Interests

Treasurer of Student Marketing Association, member of the American Advertising Federation Student Chapter, member of Residence Hall Council, participated in intramural sports. Community
activities: served as a United Way volunteer. Interests include team sports, music, travel.

Figure 30-2. Functional Résumé

Jane A. Smith

Address until June 19__: 411 James Ave.   Permanent address: 25 Temple St.
                                       Boston, MA 01830   Ames, MA 01950
                                       (508) 462-3489   (508) 467-9532

Career Objective
Entry-level sales or marketing position with long-range interest in management.

Education
Bachelor of Science degree, May 19__, Boston University, Boston, MA
Major: Business Administration, emphasis in Marketing

Skills and Qualifications
Promotion and Sales
- Provided information and assistance to retail customers.
- Designed and created in-store displays of women's wear.
- Publicized nonprofit organization and solicited contributions from area businesses.

Leadership
- Managed $10,000 organizational budget of Student Marketing Association.
- Represented interests of 45 college students in student senate.
- Supervised groups of 12 adolescent girls during live-in summer camp.

Communication
- Interacted with hotel restaurant customers; greeted and seated patrons.
- Surveyed area businesses to gather data for marketing strategy.
- Composed letters to inform 48 campers of scheduling and activities for summer camp.

Employment

Hostess, Mayflower Hotel, Boston, MA   (June 19__ to present)

Research Assistant, Student Marketing Association, Boston University   (Academic year 19__)

Salesperson, J. C. Penney Company, Ames, MA   (Summers 19__ and 19__)

Student Assistant, Mugar Library, Boston University, Boston, MA   (Academic year 19__)

Camp Counselor, Camp NoTac, Derry, NH   (Summer 19__)

Activities and Interests
   Treasurer of Student Marketing Association, member of the American Advertising Federal Student Chapter, member of Residence Hall Council, participated in intramural sports. Community activities: Served as a United Way volunteer. Interests include team sports, music, travel.

4. Education
5. Skills and qualifications
6. Job experience (not already covered)
7. Interests, activities

Elements of a Résumé

NAME

Use your full, formal name. Do not use a nickname.

ADDRESS AND TELEPHONE NUMBER

If you are still a student when you apply for a job, two addresses may be supplied: a temporary one on campus (with date until which it is valid) and a more permanent one (either your parents' house or your new lodgings).

EMPLOYMENT OBJECTIVE

Include information about both your immediate and your long-term objectives. If you have diverse interests, either tailor-make individual résumés or leave this statement off your résumé entirely and address it in your cover letter. It's generally best not to tailor your objective too blatantly to the job you're applying for.

EDUCATION

Identify your high school and college degrees, the dates they were conferred (or on which you expect it to be conferred), and the institutions from which they were received, as well as your academic major and minor. For those with advanced degrees, list the most advanced degree held.

Grade point data may be included if you feel it would be beneficial. Also mention honors, awards, scholarships, and organizational memberships if you feel this information would be helpful.

Identify and explain upper-level course work, specialized study, independent study, internships, or practicum experiences if they are recent and relate to the position you are seeking.

EXPERIENCE

Identify your paid work history by including summer employment, part-time positions held during college, and any full-time positions. It is typical to include information back to the time you began college, but such information may be condensed if your work history is quite diverse or extensive. Include the name of the company or organization, your position, dates employed, responsibilities, and accomplishments. This is the part of the résumé with the most variation.

Describe related experiences, special projects, or volunteer work. This provides a place to record significant experiences that relate to your objective or intended use of the résumé. If they are not described elsewhere on the résumé, these experiences might include internships, practicums, academic projects, or part or full-time positions. These experiences may be paid or nonpaid assignments.

SKILLS AND QUALIFICATIONS

Summarize the abilities, knowledge, and attributes which are relevant to your objective, such as computer programs or foreign languages you know. This information may be presented in a narrative paragraph or list form as its own section, or may be addressed through the descriptions found in other categories of the résumé. In a functional résumé this becomes the focus of the résumé; in chronological résumés it may be a brief section at the end. Identification and development of skills information is one of the most difficult aspects of résumé writing but one of the most important.

ACTIVITIES

Most employers consider the total person and would welcome information on your activities outside the classroom. Identify your involvement in various community or school organizations, highlighting any activities that are closely related to your career goals or the needs of the employer.

INTERESTS

You may want to identify some of your personal interests if they are pertinent to your career goals or the needs of the employer. On the other hand, do not include this information if it could be taken in a negative light or if it takes up space needed for more relevant information.

REFERENCES

You may include the line "References furnished upon request," although it's usually assumed anyway. Generally, lists of references are not included with application letters. If you're applying to 15 companies, the people on your list might be called 15 times. It is likely they would rather you waited until you got an interview and then gave your prospective employer the list.

However, many employers prefer immediate access to reference information and appreciate a listing of the people who may be contacted for further information (including their titles, addresses, and telephone numbers). Three to five previous employers, academic personnel who know your professional strengths, or people who know you well and who have achieved prominence in their field may be of interest to potential employers. *Always secure permission from individuals before using them as references.*

Additional categories may describe awards, distinctions, honors, professional affiliations, publications, military service, travel experience, or other areas relevant to the objective or intended use of the résumé. Choose categories or headings that will highlight your background and qualifications for the position you seek.

Guidelines for Résumés

1. *The résumé should be typed, laser printed, or typeset.* If you need to add an update to the résumé and can't get the whole thing retyped, add it to the cover letter instead of writing it in by hand on the résumé. Always use a high-quality bond paper.

2. *Do not make the résumé fancy.* Although it is standard practice to underline, boldface, and capitalize important headings, use a standard typeface in black ink and a conservative color of paper (see below). Do not enclose the résumé in a plastic binder.

3. *Do not include too much information.* This is not your life story. The fact that you delivered papers as a teenager, while admirable, doesn't help a personnel manager decide whether you are a potential candidate for a technical writing position.

4. *Do not include too little information.* Just listing employers and dates is not enough. The résumé should highlight accomplishments and responsibilities, especially those that relate most clearly to the position for which you are applying. Normally, the most recent accomplishments receive the most attention.

5. *Be honest about your accomplishments.* Because employers are now examining the background of potential employees, your résumé should be able to withstand close scrutiny. If you give the impression that you have a degree in business, and you don't, your chances with that company will be severely diminished. Be clear. For example, if you earned 64 credits in business-related courses, say so: Cornell University, Ithaca, NY, 64 credits toward a degree in business administration.

6. *Don't give your reasons for leaving previous jobs.* Such detail gives a negative tone to what should be a positive piece of writing. Save explanations for the interview.

7. *Avoid wordiness and strive for conciseness.* Avoid the pronouns "I," "me," and "my." Use phrases instead of lengthy sentences, and start such phrases with "strong" or "action" verbs in the past tense:

achieved	maintained
administered	operated
built	organized
coached	planned
conducted	prepared
coordinated	represented
designed	selected
developed	showed
directed	supervised
evaluated	tested
guided	trained
improved	tutored
inspected	worked
installed	wrote

Wherever possible, use verbs rather than nouns. Use the active voice ("took inventory," not "inventory was taken").

8. *Your résumé should probably be one page in length.*

9. *The layout of your résumé should make it easy to skim.* Because most employers will spend only seconds scanning your résumé, an attractive résumé will provide a positive first impression. Use generous spacing and clearly identified components to aid readability. Highlight through underlining, capitalization, and, if you are using word processing software or typesetting, boldface, italics, and even different typefaces. For a fuller discussion of ways to make your résumé visually attractive, see chapter 9, "Document Design."

10. *There must be no mechanical errors in the résumé.* Make it error-free or don't bother to develop a résumé at all. In an increasingly competitive job market, employers can afford to be very sensitive to details. If there are even one or two errors in the résumé, the reader will question everything in the entire résumé. It is not uncommon for a candidate to be dropped from consideration because of errors in spelling, grammar, punctuation, or typing. If necessary, use a ruler to proofread your final draft word by word and line by line to check for errors. To double-check your proofreading, read the résumé backward, from the end to the beginning. Get a friend to help. Remember, an error could cost you an interview, and without an interview your chances of getting a job are almost zero.

11. *Use good paper.* Select a quality paper of at least 20-pound weight to avoid having your résumé appear limp or flimsy in the employer's hands.

Papers with rag bond content of 25% or more are generally good choices. Knowing your audience is your best guide to the selection of paper and color. In general, employers are cautious and conservative in their views and prefer résumés in white, off-white, ivory, light tan, or light gray colors. When in doubt, stick with white or off-white.

12. *If possible, have your résumé printed.* Although it is not essential, having your résumé printed will save you considerable time. Select a duplication process that will deliver dark, crisp, clean copies to make the best possible first impression. Offset printing and high-quality photocopying are commonly used to reproduce résumés. Multiple copies of résumés may also be produced from word processors linked to letter-quality and laser printers. (Dot matrix printers may be used to develop your rough draft copy, but *not* for producing résumés to be sent to employers.)

The Letter of Application

The letter of application, also called a cover letter, might be the most important letter you ever write. The letter of application is your first chance to make an impression, and, as conventional wisdom has it, "You never get a second chance to make a first impression." The immediate purpose of a letter of application is to tell the reader how to read your résumé.

Elements of a Letter of Application

Letters of application generally have three parts. In the first part, usually one paragraph, tell why you are writing; name the position, field, or general area in which you are interested. If applicable, tell how you learned of the opening or organization.

The second part, often two or three paragraphs, summarizes your strongest educational, professional, or technical qualifications for the position. Show why you are qualified for the job. Regardless of your experience level, if you have ever done anything well or out of the ordinary, stress it here. Identify specific achievements that would be of particular interest to your prospective employer. If you have had related experience or specialized training, be sure to point it out, but do not reiterate your entire résumé. Refer the reader to the enclosed résumé or application form. Explain why you want to work for this employer: specify your reasons for wanting to do this type of work, and identify what you can do for this employer. And show that you know something about this company or its business.

Close by requesting an interview and indicating the action you will take to make arrangements for that interview. Also, thank the addressee for her or his consideration.

Remember that what you are trying to convey is not a sense of your needs but rather a sense of the company's needs. This letter, especially, must be reader-based, not writer-based. Figure 30-3 shows an example of a letter of application.

Guidelines for Letters of Application

Whereas a well-prepared letter of application and résumé might not guarantee you an interview, a poorly written letter of application will almost certainly ensure that you will not be called for an interview. Here are some general guidelines that will help you prepare an attractive, effective letter of application.

1. *You do need a letter of application.* The résumé by itself is not self-explanatory. A busy personnel manager will not take the time to try to figure out why a résumé is being sent. Especially in large companies, a number of positions may be open at the same time. The letter of application should be addressed to a specific person — not "Dear Personnel Officer." You can usually get this information via a phone call. If you are not sure whether a woman is Mrs. or Miss, use Ms.

2. *Avoid the generic letter of application.* This is where the name, address, and position are filled in in clearly identified blank spaces. You should tailor the letter to fit each position for which you apply.

3. *The letter of application must be typed or printed.* If you don't have access to a typewriter, find someone who does. There's no excuse for an application being handwritten. Any employer would seriously question the professional qualifications of any applicant sending a handwritten letter.

4. *There must be no mechanical errors in the letter.* Even small mistakes in spelling, grammar, or punctuation, or getting the name of the company wrong, will reflect badly on you. If you are writing a letter to the Bates School of Hotel Management and the school receives a letter addressed to the Baits School of Hotel Management, you will create the impression that you are not accurate (which is true), that you do not take the time to check details, and that you do not care about the reader. A letter with mistakes in it will almost assuredly hurt your credibility with personnel managers and people on hiring committees. They are looking for people with certain skills, and accuracy is one of those skills.

Use a good-quality bond stationery. Never use erasable bond, paper with your last employer's letterhead, or personal stationery with cute graphics.

5. *The letter of application should be brief.* Generally it should be no more than one page. It should tell what job you are applying for, where you found out about the job, and why you feel you are qualified for the job.

Figure 30-3. Example of a Letter of Application

4236 Commonwealth Ave.
Boston, MA 01830
March 15, 19--

Ms. Elyssa Mosbacher
Personnel Manager
Jonathan Company
42 Bay State Road
Boston, MA 01830

Dear Ms. Mosbacher:

Please consider me for the marketing position with
Jonathan Company that is advertised in the Boston Globe.

I expect to graduate from Boston University in June with a
Bachelor of Science degree in business administration,
with an emphasis in marketing. As you will see from my
résumé, I have practical experience in marketing. I was
active in the Student Marketing Association, both as a
research assistant and as treasurer, and I have worked as
a salesperson with J.C. Penney Company.

I believe my academic experience, my practical training,
and my interest in marketing will benefit Jonathan
Company. Also, I have always enjoyed shopping and
browsing in Jonathan Company, and would look forward to
working in what appears to be such a pleasant and friendly
atmosphere.

I would like to discuss my qualifications in an interview.
Please write me at the above address or call me at home
(508) 462-3489. I look forward to hearing from you.

Sincerely,

Jane A. Smith

Enclosure: Résumé

The Interview ————————————————

You have achieved your immediate goal. Now it's time to move on to the next goal—getting the job. Interviews range from 15 minutes to all-day affairs. They can be with personnel workers, potential supervisors, potential peers, or groups of people. They may include some sort of tour. They usually include a brief period during which the interviewer talks about the job and company, followed by a series of questions from the interviewer (the meat of the interview), followed by a period for the interviewee to ask any questions.

One valuable source of help with interview practice is career services offices. They often provide tape recorders and videotapes for practicing interviewing.

Although practicing for an interview can help, the real challenge is at the interview. You can reduce the stress and tension that inevitably accompany interviews by adhering to some simple rules of interview etiquette: smile, greet the interviewer by name, shake hands firmly, allow the interviewer to signal the end of the interview, and thank the interviewer for his or her time. Also, bring extra copies of your résumé, lists of references, and any letters of reference that you already have. Ultimately, how you look and how you communicate during an interview can often determine whether you get that job.

How You Look

Most corporations of any size value a neat and professional approach to work and apparel. According to Phyllis Macklin, an employment specialist, it isn't enough to *be* competent. You also have to *look* competent. For instance, according to Macklin, men make the best impression in a navy blue suit, single-breasted, with a white shirt with a spread collar and long sleeves, laundered and ironed. The tip of the tie should come to the top or center of the belt buckle. The back of the tie should go through the label so that the ends stay together. A belt should show no signs of wear or weight loss or gain. Shoes should be well-shined and recently heeled. Choose lightweight navy or black over-the-calf socks. Wear no more than one ring on each hand, and no other jewelry—no tie tack, no bracelet, no necklace. Nails should be cut short and filed and clean. Hair should be trimmed above the ears.

According to Macklin, women have a little more flexibility, but only a little. They should wear navy, gray, or dark maroon conservative suits, with a long-sleeved, high-necked blouse in a complimentary solid color—white or cream or a pastel, "not red or fuchsia." The skirt should cover the knee when you are seated during the interview. Choose medium-heel leather pumps (no open toes or backs), and always wear stockings or pantyhose, even in the summer. Wear simple earrings, not dangling, a pin or a necklace, a watch, no more than one ring per hand, no more than one bracelet. Makeup should be simple and soft. Hair should be short and professional—"not teased, not bouffant, not bleached." Nails should be manicured, with light or clear polish. If you wear a scarf or a necklace, make it short enough to focus attention on your face.

Macklin tells men and women not to wear any scent to an interview—no perfume, no cologne, no aftershave, no strong-smelling deodorant—because what smells lovely to you might smell awful to the interviewer. And don't smoke or chew gum, even if the interviewer does. Don't even smoke on the way to the interview, because the odor can be objectionable.

How You Communicate

Although how you look is important, the number one key to a successful interview is how you communicate, both verbally and nonverbally.

Listen carefully so that you will know exactly what is being asked. Although most interviewers consider educational background and attainments, extracurricular activities, and communication skills, in general the interview is designed to answer three questions:

1. Can the candidate do the job?
2. Will the candidate do the job?
3. Is the applicant compatible with the company?

Answer just as carefully as you listen in order to ensure that you answer the right question. Your answer should not be either too long or too short.

It is also important for you to use questions and answers to your own advantage. Although a direct answer should never be avoided, there is always room for embellishment. Many employment counselors recommend that applicants decide ahead of time on two or three important facts or features about themselves that they want to highlight and then mention them sometime during the interview: "Yes, sir, my first experience with inventory control was at Consolidated Enterprises, and that's also where I first had a chance to work with the Zowie database program."

Employment counselors also recommend that interviewees should ask questions, both to show that they have thought seriously about the job and to get a good idea of what is expected of them. For example, you might ask:

1. What skills or talents are most essential to being effective in the job?
2. What are the toughest problems I might face?

3. What is most rewarding about the work itself, apart from salary or fringe benefits?
4. Will there be overtime? How much?
5. Will I have to travel? How much travel?
6. Are there any opportunities for continuing education?
7. What is the evaluation procedure?
8. What are the opportunities for advancement?

Be aware that nonverbal communication is important in the hiring process. You want to appear competent, confident, and personable. Experts on body language and public speaking agree that there are several techniques to help. For example, because posture is important in the job interview, avoid slouching (sit on the edge of the chair), maintain eye contact, and appear confident (but not cocky or aggressive).

Follow-Up

Follow-up is extremely important. Every interview should be followed by a letter thanking the interviewer again for her or his time and restating your interest in the job. Figure 30-4 shows a typical follow-up letter.

Remember that interviews are learning experiences: your poise, confidence, and knowledge increase with each interview, whether or not they result in being hired.

Getting the Right Job

You have done everything, and you have gotten an interview. In addition to making a good impression, you should do one more thing: be sure that this job is the right one for you. The only thing worse than not getting the job you want is to get a job you didn't want. Although you can never be absolutely certain that you and the job will work out, you can do a few things to minimize the risk of getting the wrong job. You can remember to get a good idea of what is expected of you by asking the kinds of questions listed above during the interview. Even if you do everything right, there are no guarantees that you will get the job you want or that the job you do get will be the right job for you. But, by preparing yourself carefully, your chances for finding the right job will improve dramatically.

Figure 30-4. Example of a Follow-Up Letter

```
4236 Commonwealth Ave.
Boston, MA 01830
April 15, 19--


Ms. Elyssa Mosbacher
Personnel Manager
Jonathan Company
42 Bay State Road
Boston, MA 01830

Dear Ms. Mosbacher:

Thank you for the interview last Wednesday. Everyone was
so pleasant and helpful that I am now more than ever
interested in working for Jonathan Company.

I look forward to hearing from you.

Sincerely,



Jane A. Smith
```

Activities

1. Prepare a chronological résumé for a job in your area.

2. Prepare a letter of application and a résumé for a job in your area.

3. With a group of students who have similar career goals, find an ad in a newspaper or a trade journal for a position you might apply for. Try to determine exactly what the employer is looking for. Is the ad specific? Are there any terms in the ad that need clarification?

4. You are scheduled for an interview for the job described in the ad above. Prepare a list of the information you will need before going to the interview.

5. In a group of three or four, conduct an employment interview for each member of the group. After the interview, prepare written evaluations of how each interviewee looked and communicated.

References

Chapter 1

Gas barbecue grills. (1991, July). *Consumer Reports, 56,* 491–497.

Lewis, R. S. (1988). *Challenger: The final voyage.* New York: Columbia University Press.

Naisbitt, J. (1982). *Megatrends: Ten new directions transforming our lives.* New York: Warner.

NASA's response to the committee's investigation of the Challenger accident. (1987, February 26). Hearing before the Committee on Science, Space, and Technology. U.S. House of Representatives.

Presidential Commission. (1986). *Report to the President on the space shuttle Challenger accident.* Washington, DC: U.S. Government Printing Office.

Chapter 2

Casari, L. E., & Povlacs, J. T. (1988). Practices in technical writing in agriculture and engineering industries, firms and agencies. *Journal of Technical Writing and Communication, 18*(2), 143–159.

Chapter 3

Rosenbaum, S. (1991, spring/summer). Writing for the computer world: A conversation with Stephanie Rosenbaum. *Issues in Writing, 3*(2), 113–129.

Chapter 4

Prentice-Hall author's guide (5th ed.). (1978). Englewood Cliffs, NJ: Prentice Hall.

Queijo, J. (1985, February-March). Medical tests for hire. *Bostonia, 59*(2), 17–23.

Standards for manual preparation. (1984). Stevens Point, WI: First Financial Bank.

Chapter 5

Allen, P., & Watson, D. (1976). *Findings of research in miscue analysis: Classroom implications.* Urbana, IL: National Council of Teachers of English.

Estes, T., & Vaughn, J. (1978). *Reading and learning in the content of classroom: Diagnostics and instructional strategies.* Boston: Allyn and Bacon.

Instructions for "Char-Broil" gas grill. Columbus, GA: Char-Broil.

Rivers, W. E., & Carr, D. R. (1991, January). The NCR-USC document validation laboratory: A special collaboration between industry and academia. *Journal of Business and Technical Communication, 5*(1), 88–103.

Chapter 6

Casari, L. E., & Povlacs, J. T. (1988). Practices in technical writing in agriculture and engineering industries, firms and agencies. *Journal of Technical Writing and Communication, 18*(2), 143–159.

Forman, J. (1991). Novices work on group reports: Problems in group writing and in computer-supported group writing. *Journal of Business and Technical Communication, 5,* 48–75.

Likert, R. (1984). The nature of highly effective groups. In D. Kolb, I. Rubvin, & J. McIntyre (Eds.) *Organizational psychology: Readings on human behavior in organizations.* Englewood Cliffs, NJ: Prentice Hall.

Chapter 7

University of Wisconsin. (1977). [Catalog]. Stevens Point, WI: University of Wisconsin.

Chapter 8

Felker, B., Pickering, F., Charrow, V., Holland, V., & Redish, J. (1981). *Guidelines for document designers.* Washington, DC: American Institutes for Research.

Lutz, W. (1989). *Doublespeak.* New York: Harper and Row.

Tufte, E. R. (1983). *The visual display of quantitative information.* Cheshire, CT: Graphics Press.

Wresch, W., Pattow, D., & Gifford, J. (1988). *Writing for the 21st century.* New York: McGraw-Hill.

Chapter 9

Kostelnick, C. (1990, fall). The rhetoric of text design in professional communication. *The Technical Writing Teacher, 17,* 189–202.

White, J. (1988). *Graphic design for the electronic age.* New York: Xerox Press.

Chapter 10

American Medical Association. Feeling Fine Programs. (1988). *What you should know about high blood cholesterol.*

Asimow, M., & Bosticco, I. L. M. (1958, September). Cut corners with conveyors. *Technical Aids for Small Manufacturers, 63.* Washington, DC: Small Business Administration.

Burton, L. V. (1958, August). Protective packaging problems. *Technical Aids for Small Manufacturers, 62.* Washington, DC: Small Business Administration.

Pavesic, E. J. (1958, August). Surface hardening practices. *Technical Aids for Small Manufacturers, 62.* Washington, DC: Small Business Administration.

Request for proposal: Telecommunications system. (1984). Stevens Point, WI: University of Wisconsin.

Smith, R. E. (1969, March). Using adhesives in small plants. *Technical Aids, 92.* Washington, DC: U.S. Small Business Administration.

Summar, P. (1984, February). Bat guano brokers strike pay dirt. *New Mexico Magazine.*

U.S. Department of Health and Human Services. National Institute of Health. (1985, October 2). Request for proposal no. NHLBI-HV-86-01. *Fabrication of cardiovascular devices.* Washington, DC: U.S. Government Printing Office.

U.S. Environmental Protection Agency. (1987, September). *Radon Reduction Methods* (2nd ed.). OPA-87-010.

Chapter 11

Kendrick, M. (1988). The Thames barrier. *Landscape and Urban Planning, 16,* 57–68.

Two reservoir types in system. (1990). *Hard Worker Newsletter.* Wausau, WI: Wisconsin Valley Improvement Co.

Chapter 12

Bersekas, D., & Gallager, R. (1987). *Data networks.* Englewood Cliffs, NJ: Prentice Hall.

Mainstone, K. A., Extrustion in web processing. In D. Satas (Ed.) *Web processing and converting technology and equipment.* New York: Van Nostrand Reinhold.

Massaro, D. (1975). *Experimental psychology and information processing.* Chicago: Rand McNally.

Nelson, M. E. & Mech, L. D. (1981). Deer social organization and wolf predation in Northeastern Minnesota. *Wildlife Monographs, 77.* Washington, DC: The Wildlife Society.

Schaffer, E. R. (1984). Die cutting. In D. Satas (Ed.) *Web processing and converting technology and equipment.* New York: Van Nostrand Reinhold.

Titus, J. R., & Van Druff, L. W. (1981). Response of the common loon to the recreational pressure in the Boundary Water Canoe Area, Northeastern Minnesota. *Wildlife Monographs, 79.* Washington, DC: The Wildlife Society.

Chapter 13

Ausubel, D. (1960). The use of advance organizers in the learning and retention of meaningful verbal material. *Journal of Educational Psychology, 51,* 267–272.

Feasibility of a Portage County waste-to-energy project. (1988). Stevens Point, WI: Portage County.

Forestry Abstracts. (1991). Farnham Royal, England: Commonwealth Agricultural Bureaux.

Naisbitt, J. (1982). *Megatrends: Ten new directions transforming our lives.* New York: Warner Books.

Presidential Commission. (1986). *Report to the President on the space shuttle Challenger accident* (Vol. 1). Washington, DC: U.S. Government Printing Office.

Chapter 14

Wurman, R. S. (1989). *Information anxiety: What to do when information doesn't tell you what you need to know.* New York: Bantam.

Chapter 15

Rogers, C. (1951, October 11). *Communication: Its blocking and facilitation.* Centennial Conference on Communications.

Sternberg, R. J. (1988). *The psychologists's companion: A guide to scientific writing for students and researchers.* Cambridge, England: Cambridge University Press.

Zinsser, W. (1989). *Writing to learn.* New York: Harper and Row.

Chapter 16

Becket, A. L., Pike, K. L., & Young, R. E. (1970). *Rhetoric: Discovery and change.* New York: Harcourt, Brace.

Burke, K. (1945). *A grammar of motives.* New York: Prentice Hall.

D'Angelo, F. J. (1975). *A conceptual theory of rhetoric.* Cambridge, MA: Winthrop.

Chapter 20

U.S. Department of Health and Human Services. National Institute of Health. (1985). *Guide for the care and use of laboratory animals.* Washington, DC: U.S. Government Printing Office.

Chapter 21

Fagan, W. T., Jensen, J. M., & Cooper, C. R. *Measures for research and education in the English language arts, 2.* Urbana, IL: National Council of Teachers of English.

Grace, J. (1990). Cuticular water loss unlikely to explain treeline in Scotland. *OECO-LOGIA, 84*(1), 64–68.

Loban, W. (1976). *Language development: Kindergarten through grade twelve.* Urbana, IL: National Council of Teachers of English.

McAllister, C., & South, R. (1988, December). The effect of word processing on the quality of basic writer's revisions. *Research in the Teaching of English, 22,* 417–427.

Presidential Commission. (1986). *Report to the President on the space shuttle Challenger accident* (Vol. 2, appendix L). Washington, DC: U.S. Government Printing Office.

Chapter 22

Request for proposal: Telecommunications system. (1984). Stevens Point, WI: The University of Wisconsin.

U.S. Department of Health and Human Services. National Institute of Health. (1985, October 2). Request for proposal no. NHLBI-HV-86-01. *Fabrication of cardiovascular devices.* Washington, DC: U.S. Government Printing Office.

Chapter 24

Committee on Foreign Relations. (1978, February). *Gas generator research and development bi-gas process.* 77th Monthly Progress Report. Washington, DC: U.S. Senate.

U.S. Department of Energy. (1976, December). *Industrial applications of fluidized-bed combustion.* Quarterly Progress Report. Washington, DC: U.S. Government Printing Office.

U.S. Department of Energy. (1978, February). *Industrial applications of fluidized-bed combustion.* Quarterly Progress Report. Washington, DC: U.S. Government Printing Office.

Wisconsin Department of Natural Resources. (1984). *Bear population indices (survey and inventory).* Wildlife Research Projects. Annual report.

Chapter 25

Armed Forces Institute of Pathology. (1976). *Instruction booklet for submission of pathologic material.* Washington, DC.

Gas barbecue grills. (1991, July). *Consumer Reports, 56,* 491–497.

Quasar VHS owner's manual: VH6200/VH6300/VH6400 (p. 23). Elk Grove Village, IL: Quasar Company.

Chapter 27

Toffler, A. (1990). *Powershift: Knowledge, wealth, and violence at the edge of the 21st century.* New York: Bantam Books.

Chapter 28

Postman, N. (1985). *Amusing ourselves to death: Public discourse in the age of show business.* New York: Viking.

Index

Abstracts. *See also* Informative abstracts
  content of, 258
  descriptive, 249, 253, 257
  examples of, 247–48, 259–61
  indexes of, 248
  in progress reports, 457
  in research reports, 398, 399
  types of, 249–50, 258
Academic libraries, 305
Accuracy
  computer shortcuts to aid, 122
  in process descriptions, 231–33
  quality reviews for, 81–82, 90
Achtert, W. S., 363
Active voice, 58–59
Agency, 300–301
Agent, 300–301
Alternatives, comparison of
  in feasibility studies, 437–38, 444–45
  in proposals, 422–23
American Chemical Society style manual, 362
American Institute of Physics style manual, 362
American Mathematical Society style manual, 362
American Medical Association style manual, 362
American national standard for bibliographic references, 362
American Psychological Association (APA) publication manual, 362–64, 366–68
Analysis. *See also* Information analysis; Technical analysis
  of group writing tasks, 104, 111
APA publication manual, 362–64, 366–68
Appeal to authority, 286
Appendixes
  in feasibility studies, 438, 446
  graphics in, 164

  in progress reports, 459
  in proposals, 420–21
  in research reports, 398, 404, 407
Applied Science and Technology Index, 309
Argumentation. *See* Technical argumentation
Aristotle, 298
Assignment, writing, 20
Audiences. *See also* User testing
  adapting document to, 36–37, 47–52, 222
  checklist for assessing, 52
  for computer manuals, 499–500, 505, 508
  for correspondence, 563
  for electronic documents, 515, 518–19, 525
  for feasibility studies, 436
  information processing by, 174–75
  for instructions, 471–72
  involving in document, 57–59
  for literature reviews, 378
  needs of, 44–47, 77
  for oral presentations, 532–35, 536–37, 546–49
  for progress reports, 454
  for proposals, 414–15
  types of, 19–20, 42–44

Background information
  in feasibility studies, 437, 443, 444
  in instructions, 475
  in progress reports, 457
  in technical arguments, 284
Bar graphs, 136, 141–44, 148, 149–50
  in research reports, 401, 402
  spreadsheet-based, 160, 161, 330–32
Begging the question, 285
Bibliographic Retrieval Service, 311

Binding of documents, 32
  computer reference guides, 507
  instruction manuals, 485
Biological and Agricultural Index, 309, 310
Blocking text, 179–81
Boilerplate, 126
Book citations, 366–67, 368–70
Boolean strategy, 310
Boxes, 179, 187–88
Brainstorming, 303–4
Browse Search, 310
BRS-After-Dark (Bibliographic Retrieval Service), 311
Budget, 21, 32, 79. *See also* Economic feasibility
  as proposal section, 419–20
Bullets, 186
Burke, Kenneth, 300
Burke's pentad, 300–301, 324–25
Business Periodicals Index, 309

Card catalog, 306
Careers and technical communication, 11–14
Casari, Laura, 20
Case studies
  group writing, 111–13
  writing process, 22–23, 25–26, 27–29, 31, 33
Catalogs
  card, 306
  on-line, 116, 306–8
CBE style manual (Council of Biology Editors), 363, 365, 368–71
CC list, 555
CD-ROM (compact disk—read-only memory), 116–17
  full-motion display in, 237
  Wilsondisc, 308, 310
Chalkboards, 544
Charts. *See* Diagrams; Graphs
Chicago manual of style, The, 363
Chronological organization, 218, 339–40
Chronological résumés, 570, 571
Circumference, 300–301
Citation of sources, 361–73. *See also* Reference lists
  books, 366–67, 368–70
  experts, 286
  for graphics, 158, 163
  guidelines for, 372–73
  in-text, 362–65, 371, 372
  for literature reviews, 379, 380, 384–85
  number system, 371–72
  periodicals, 367–68, 370–71
  plagiarism, 122, 361
  for scavenged documents, 123
Clarification memos, 553
Classifications
  examples of, 273–74, 276

role of, 265–66, 272
  writing of, 267–69
Cliches, 62
Clip art, 543
Closing of letter, 561
Cloze tests, 92–93
Collaborative writing. *See* Group writing
Colleague-based reviews, 30, 90–92
  of oral presentations, 545
  of proposals, 427
Collecting information. *See* Information collecting
Color, use of, 189
  for instructions, 485–86
  for résumés, 575, 577
Columns
  in tables, 158–59
  as text format, 178–79
Communications
  breakdown of, 7–11
  careers and, 11–14
Company standards, 80–81
Company style guides, 56
Comparative Biochemistry and Physiology, 345
Completeness
  of information, 323–25
  of literature reviews, 380
  pattern to aid, 121
Completion tests, 95–96
Complimentary close, 561
Comprehension questions, 94–95
Comprehension tests, 92–95
Compuserve, 311
Computer-Aided Design (CAD) programs, 161
Computer manuals, 495–512
  audiences for, 499–500, 505, 508
  on-line help, 499, 507–10
  problems of, 496–97
  reference guides, 502–7
  situation, 499, 505, 508
  tutorials, 498–502
Computer tools, 116–20. *See also* On-line information; Software
  laptop computers, 319
  scanners, 160–61, 162
Computer writing techniques, 120–28. *See also* Desktop publishing; Word processing
  boilerplate, 126
  invisible writing, 304
  management of, 110, 112
  networks, 108
  scavenging, 122–23
  shortcuts allowed by, 27, 120–22, 128
  structured document processors, 124–25
  style tests, 85–87
  templates, 123–24, 349–50
Conclusions
  alternative, 336

development of, 325–35
in feasibility studies, 438
in progress reports, 458
reasonableness of, 335–36
in research reports, 403, 405–6
in technical arguments, 287, 293
Condensing text, 135–36
Conflict management, 110–11, 112
Content-driven organization, 25
Context, 20–21
Controlled experiments, 320
Correspondence, 552–66. *See also* Letters
audience for, 563
memorandums, 553–55
writing process for, 561–65
Cost. *See* Budget
Council of Biology Editors (CBE). *See* CBE
style manual
Cover letters, 556–58
for job applications, 577–78
for questionnaires, 317
Cover memos, 553
Credibility, 287
Customer needs survey, 318
Customer presentations, 533–34
Cutaway drawings, 151, 153, 220–21

Databases, on-line, 311–12
Deadlines. *See* Time constraints
Decision-oriented writing, 38–39, 40
audience for, 45
feasibility studies as, 436
organization of, 341
proposals as, 414–15
research reports as, 395–97
Decision tables, 138–39
Definitions. *See* Technical definition
Description. *See* Technical description
Descriptive abstracts, 249, 253, 257
Design of document. *See* Document design
Desktop publishing
document design using, 119–20,
192–93, 194–96
of presentation graphics, 543
type size choice in, 181
Details, presentation of, 25
in instructions, 486
in oral presentations, 539
in process description, 234, 236, 239, 240
in proposals, 423
Diagrams, 153–56
in descriptions, 219, 221–22
examples of, 137, 156, 157
labels for, 477
word processor production of, 158
Dialog's *Knowledge Index*, 311
Dissertation Abstracts International (DAI),
311
Document design, 27, 172–97
graphic balance in, 191–92

principles of, 173–75
tools for, 192–96
visual aspects of, 175–92
Documenting sources. *See* Citation of
sources
Document management, 108, 112
Document types, 20, 47–48
Doublespeak (Lutz), 166
Drawings, 150–54
Computer-Aided Design (CAD)
programs for, 161
in descriptions, 219, 220–21
in research reports, 407

Economic feasibility, 325, 433, 434–35,
441–42
Educational Resources Information Center
(ERIC), 309, 311
Effectiveness of document, 90
Either/or fallacy, 286
Electronic documents, 513–28. *See also*
Electronic mail; Hypertext
features of, 514–17
hypertext, 514, 521–26
privacy of, 520
Electronic mail, 514, 517–20
group writing use, 106, 110, 112
Emotional appeal, 281, 282
Endnotes, 361
Engineers Joint Council style of references,
363
Equipment descriptions, 233, 235–37
Ethical appeal, 281, 282
Ethics of graphs, 166–70
Evaluation. *See also* Quality reviews
of correspondence, 565
of information sources, 335
of oral presentations, 549
as proposal section, 419
surveys, 96–98
of writing situation, 18–23
Evidence, 286
Executives
acceptance of document by, 78
as audience, 19, 43, 48
Executive summaries, 250–51, 256, 257–58
Expanded definitions, 207–9
Experiments, controlled, 320
Experts
as audience, 19, 42, 48, 51–52
used as authority, 286
Exploded drawings, 152, 154

Facts, 25, 286
False analogy, 285–86
Feasibility studies, 432–52
arguments in, 281
completeness of, 323–24
conclusions in, 325
example of, 446–51

Feasibility studies *Continued*
  form of, 436–38
  information collecting for, 439–42
  oral presentation of, 533
  organization of, 341, 344, 442–45
  publishing, 446
  quality review of, 446
  situation, 433, 435–36
  types of, 433–35
Field observations. *See* Observations
Field questions, 299
File conversion programs, 110
Flip charts, 544
Fonts, 182–83, 386
Footnotes, 211, 361
Formal definitions, 205
Format, 21. *See also* Document design
  computer manuals, 498, 506
  letters, 560–62, 565
  management of, 108–9, 112
  memos, 554–57, 565
  progress reports, 455–59
  proposals, 415–21, 423–26
  quality review for, 82–83, 84–85, 90
  research reports, 397–404
  résumés, 573–75
  templates, 123–24, 348–50
Format-driven organization, 25, 345–48
Free writing, 304
Full-motion display, 237
Functional organization, 218
Functional résumés, 570–73

Gantt chart, 107
Generalizations, 284
General Science Index, 309, 310
General-to-specific organization of
    document, 342, 343
Gibaldi, J., 363
Glossary, 251
  definitions in, 211–14
  of equipment descriptions, 236–37
Glosses. *See* Marginal glosses
Government Printing Office Index, 309, 310
Grammar quality review, 82–83, 84
Grant proposals, 413
Grantsmanship Center, 413
Graphics, 27, 133–71. *See also* Diagrams;
    Graphs; Illustrations; Tables
  balancing in document design, 191–92
  clarity of presentation in, 165–66
  in classifications, 269
  to condense text, 135–36
  definitions illustrated by, 208–9
  in descriptions, 219–22
  in instructions, 479
  labels in, 163
  manual production of, 158
  for oral presentations, 536, 538–40,
    543–44

  in partitions, 271
  patterns shown by, 136
  placement in document, 163–64
  in process description, 237–39
  production methods, 156–62
  relationships shown by, 137
  in research reports, 401–2, 407
  source, 157–58
  spreadsheet, 160, 330–35
  text integrated with, 162–63
  titles in, 163
  when to use, 134–37
  word processor production of, 118,
    158–59
Graphs, 136, 140–50
  bar, 141–44
  ethics of, 166–70
  guidelines for, 148–50
  labeling of, 163
  line, 144–46
  pie, 147–48
  spreadsheet-based, 160, 161, 330–35
Greeking, 179–80, 186
Group writing, 21, 101–14
  creation of group for, 111
  management tasks for, 102–11
  of proposals, 415, 421
  reasons for, 102
  software for, 108, 110, 112, 120
  structured document processors for,
    124–25
Gutters, 179

H. W. Wilson Company, 308
Halftones, 158
*Handbook for authors of papers in the
    journals of the American Chemical
    Society*, 362
Handbooks. *See* Style guides
Handouts, 544–45
Hasty generalizations, 284
Headings, 69–70, 88, 183–85. *See also*
    Labels
  icons in, 189
  letter, 560
  lines separating, 187
  running head, 175–76, 187, 189
Heuristic questioning techniques, 298,
    324–25
Highlights, 190–91
Horizontal bar graphs, 144
Hypertext, 514, 521–26

Icons, 176, 189–90
Illinois Researcher Information System (IRIS),
    413
Illustrations, 150–53
  for computer tutorials, 501
  for descriptions, 219–21

for instructions, 474, 475, 478, 482–83, 485, 489–91
publishing, 485
in research reports, 407
Indexes
abstract, 248
for computer reference guides, 507
on-line, 116, 308–11, 379–80
periodical, 306
Index to Legal Periodicals, 309
Informal definitions, 204–5
Information. *See also* On-line information
accuracy of, 47, 231–33
analysis of. *See* Information analysis
collecting. *See* Information collecting
as evidence, 286
organization of. *See* Organization of document
sources of. *See* Information sources
user processing of, 174–75
Informational descriptions, 217
Information analysis, 322–37. *See also* Technical analysis
checking completeness, 323–25
checking conclusions, 335–36
developing conclusions, 325–35
Information Anxiety: What to Do when Information Doesn't Tell You What You Need to Know (Wurman), 266
Information collecting, 297–321. *See also* Information sources
for correspondence, 564
determining information needs, 23–24, 298–302
for feasibility studies, 439–42
for instructions, 477–79
for literature reviews, 379–80
for progress reports, 459
for proposals, 421–22
for research reports, 404–5
Information sources, 21–22, 24, 302–20. *See also* On-line information
brainstorming, 303–4
citation of. *See* Citation of sources
controlled experiments, 320
evaluation of, 335, 380–81
free writing, 304
incorporating source material in document, 355–60, 373
inquiry letters, 314–15
interviews, 313–14
invisible writing, 304
libraries, 24, 305–12
observations, 231–32, 319–20, 336
on-site visits, 319
primary, 24, 302–3
questionnaires, 315–19
secondary, 24, 232–33, 302–3
self as source, 24, 303–4

Informative abstracts, 250, 253–56, 257, 258, 399
Informative writing, 38, 39. *See also* Informative abstracts
audience for, 44–45
feasibility studies, 436
letters, 556–57
memos, 553
organization of, 341
process descriptions, 230
Inquiry letters, 314–15, 556, 558–60, 561, 562
Inquiry memos, 553
Inside address, 560
Instructional descriptions, 217
Instructions, 39–41, 470–94. *See also* Computer manuals
audience for, 45
chronological organization of, 339–40
examples of, 486–93
form of, 473–77
illustrations in, 474, 475, 478, 479, 482–83, 485, 489–91
information collecting for, 477–79
letters of, 557
oral presentation of, 533
organization of, 67–68, 479–81
performance tests for, 95–96
process descriptions as, 230–31, 240
publishing, 485–86
quality review of, 482–85
situation, 471–73
technical descriptions as, 217
tests in, 481, 486
Interviews, 313–14
for feasibility studies, 439
for instructive writing, 477
for jobs, 580–82
verification of, 446
In-text citations, 362–65, 371, 372
Introduction
in progress reports, 456
in proposals, 417–18
in technical descriptions, 217, 224–26
Invisible writing, 304

Jargon, 60–61, 203
Job applications, 567–84
application letters, 577–80
evaluating opportunities, 582–83
interview follow-up letter, 582, 583
interviews, 580–82
job search techniques, 568–69
résumés for. *See* Résumés
Journal citations, 367–68, 370–71

Kapor, Mitch, 330
Key words, 307, 311
Knowledge Index, 311

Labels, 88, 163, 272. *See also* Headings
  accuracy of, 477
  for drawings, 150, 152–53, 163
  for photographs, 150, 161, 163
Laptop computers, 319
Layer graphs, 146, 334–35
Legal implications of documents, 121
Legal periodicals index, 309
Letters, 555–61. *See also* Inquiry letters
  cover, 577–80
  elements of, 560–61
  interview follow-up, 582, 583
  job application, 577–80
  types of, 556–60
  writing process for, 561–65
Libraries, 24, 116–17, 305–12, 379–80
Life on the Mississippi (Twain), 229
Line graphs, 136, 144–46, 149–50, 169–70
  spreadsheet-based, 160, 161, 330, 333–35
Lines, 187–88
Linguistic derivation, 208
Lists, 68–69, 178, 186–87
  parts, 474
Literature reviews, 377–93
  citation of sources in, 379, 380, 384–85
  completeness of, 323–24, 380
  examples of, 386–92
  form of, 378–79
  information collecting for, 379–80
  organization of first draft, 380–84
  publishing, 385–86
  quality review of, 385
  in research reports, 400
  situation, 378
  source material summaries for, 380
  writing principles for, 384–85
Logical argument, 281–82, 284–86, 293
Lotus 1–2–3, 330
Lutz, William, 166

Macros, 127
Management. *See* Executives
*Manual for authors of mathematical papers,
  A* (American Mathematical Society),
  362
Marginal glosses, 251–53
  design of, 177, 185, 187, 259
  writing of, 256–57, 258
Margins, 177
Media integration, 515
Meeting notes, 345–47
Memorandums
  elements of, 554–55
  types of, 553
  writing process for, 561–65
Miller, George, 272
Miscue analysis, 93–94
Mixed audiences, 20, 44, 48
MLA handbook for writers of research papers
  (Gibaldi & Achtert), 363

Multiple bar graphs, 142, 148
Multiple line graphs, 145–46, 149, 169–70

Negative sentences, 65
Nelson, Theodor, 521
Noise in argumentation, 283
Non sequitur, 285
Norman, Rose, 124
Note cards, 538
Noun strings, 65
Novices
  adapting description to, 48, 49–51, 222
  as audience, 19–20, 43
  computer tutorials for, 499
  instructions for, 471
Numbers tables, 138–40

Objections, anticipation of, 287, 288, 290
Objective of research reports, 399
Observations, 24, 231–32, 319–20
  data sufficiency of, 336
  description of, 235
  as evidence, 286
  verification of, 446
Occupational Safety and Health
  Administration, 478
On-line information, 116–17, 305
  catalogs, 116, 306–8
  for computer help, 499, 507–10
  databases, 311–12
  on grants, 413
  indexes. *See* Indexes
  search strategies, 307–8, 310–12
On-site visits, 319
Operational feasibility, 433–34, 439–40
Oral presentations, 531–51
  customer presentations, 533–34
  evaluation of, 549
  graphics for, 538–40, 543–44
  location for, 535
  note cards for, 538
  organization of, 535–37, 539
  practicing, 545–46
  principles of, 546–49
  professional meetings, 534–35
  prompts for, 537–38
  types of, 532–35
  visual aids for, 535, 538–45
  work group reports, 532–33
Order-of-importance organization of
  document, 341–42
Organizational qualifications, 419
Organization chart, 137, 153–54
Organization of document, 25, 66–70,
  338–53. *See also* Format
  clarity of, 47, 48
  content-driven, 339–44
  correspondence, 564–65
  feasibility studies, 341, 342, 442–45
  format-driven, 25, 345–48

instructions, 67–68, 339–40, 479–81
literature review, 380–84
part-by-part description, 218
progress reports, 341, 342, 459–61
proposals, 341–44, 422–26
quality review for, 82, 90
research report, 341, 405–6
subdivisions, 223
summary as organizer, 248–49
technical description, 339–41
technical process description, 233–34, 339–40
templates for, 348–50
Organization of oral presentations, 535–37
Outlines
of instructions, 479
of literature reviews, 382–84
of research reports, 397
for task management, 105
templates, 123–24, 348–50

Page layout, 175–79
of computer manuals, 501, 506
of instructions, 480–81
of résumés, 576
Paragraphs, 66–67
Parallel structure, 69
Paraphrase, 357–59
Part-by-part description, 218, 225–26
Particle-wave-field questions, 299
Partitions
example of, 275
role of, 265–66, 267
writing of, 269–73
Parts list, 474
Performance tests, 95–96
Periodicals
citations of, 367–68, 370–71
indexes of, 306
as literature review source, 379
Persuasive writing, 281
PERT chart, 154
Photographs, 150, 152
computer scanners to mark, 160–61
in descriptions, 219–20
production of, 158
in research reports, 407
Phrase definitions, 207, 212
Pie graphs, 136, 147–48, 149–50
spreadsheet-based, 160, 330–33
Plagiarism, 122, 361
Positive sentences, 65
Posters, 544
Povlacs, Joyce, 20
Pre-proposals, 427
Primary sources, 24, 302–3
Printing. See Publishing
Problem-cause-solution organization of document, 344

Procedures
checking, 335–36
research report section, 400–401, 406
Process description. See Technical process description
Professional libraries, 305
Professional meetings, 534–35
Professional style manuals, 56–57
Progress reports, 229, 453–69
definition, 454
example of, 462–68
form of, 455–59
information collecting for, 459
oral presentation of, 533–34
organization of, 341, 342, 459–61
publishing, 462
quality review of, 461–62
reference numbers of, 456
schedule updates in, 460–61
situation, 454–55
Project identification, 457
Prompts, 537–38
Proofreading
of letters of applications, 578–80
of résumés, 576–77
of scavenged documents, 123
Proposals, 229, 410–31
arguments in, 281
audience for, 414–15
example of, 428–31
form of, 346–48, 415–21, 423–26
group writing and, 102
information collecting for, 421–22
internal, 413
oral presentation of, 533
organization of, 341–44, 422–26
pre-proposals, 427
publishing, 427
quality review of, 426–27
scavenging to write, 122
schedule for writing, 421
situation, 411–15
solicited, 411–12
unsolicited, 412–13
Psychologist's Companion: A Guide to Scientific Writing for Students and Researchers, The (Sternberg), 283
Publication Manual of the American Psychological Association, 362–64, 366–68
Public libraries, 305
Publishing, 32–33
feasibility studies, 446
illustrations, 485
instructions, 485–86
letters of applications, 578
literature reviews, 385–86
quality review and, 78
research reports, 406–7
résumés, 575, 577

Purpose of oral presentation, 536
Purpose question, 300–301
Purposes of document, 19, 36–42
    adapting to audience to achieve, 47–52
    checklist for, 52
    explicit, 37–41
    implicit, 41–42

Quality reviews, 74–100. *See also*
        Colleague-based reviews; User testing
    of feasibility studies, 446
    of instructions, 482–85
    of literature reviews, 385
    problems of, 78–80
    of progress reports, 461–62
    of proposals, 426–27
    reasons for, 78
    of research reports, 406
    writer-based, 30, 80–89
Questionnaires, 315–19
Question sets, 298–302, 324–25
Quotations, 256–57, 355–57, 359–60

Ratio, 300–301
Readability, 85–86
Read-aloud protocols, 94
Readers. *See* Audiences
Readers' Guide to Periodical Literature, 309
Recommendations, 437, 443
*Recommended practice for style of references
    in engineering publications*, 363
Redundancy, 63
Reference books, 306
Reference guides (computer), 502–7
Reference lists, 362–63, 366–71
    example of, 390–92
    in literature reviews, 379, 386, 390–92
Reference numbers of reports, 456
References (job), 575
Reminder memos, 553
Repair manuals, 229
Repetition, 536
Reports. *See also* Progress reports; Research
        reports
    example of summary in, 277–78
    group writing and, 102
Requests for proposals (RFPs), 411–12,
        423–25, 427, 439
Research. *See* Information
Research reports, 394–409
    arguments in, 281
    example of, 407, 408–9
    form of, 345, 397–404
    graphics in, 401–2, 407
    information collecting for, 404–5
    organization of, 341, 405–6
    process description in, 229
    publishing, 406–7
    quality review of, 406
    situation, 395–97

Resources. *See* Information sources
Results in research reports, 401–3, 405,
        406–7
Résumés, 569–77
    elements of, 573–75
    guidelines for, 569–70
    printing of, 575, 577
    types of, 570–73
Reviewing/revising, 30–31. *See also*
        Quality reviews
Review/revision, 122–23, 128
RFPs. *See* Requests for proposals (RFPs)
Rogers, Carl, 287
Rosenbaum, Stephanie, 36
Running head, 175–76, 187, 189

Sachs, Jonathan, 330
Safety issues in documents, 121, 478–79,
        482. *See also* Warnings in instructions
Salutation, 560
Scanners, computer, 160–61, 162
Scavenging as writing technique, 122–23
Scene, 300–301
Schedules, 22
    for correspondence, 564
    for progress reports, 454–55
    progress report updates of, 460–61
    for proposals, 421
Searches, computer
    on-line information, 307–8, 310–12
    search-and-replace, 123, 127–28
Secondary sources, 24, 232–33, 302–3
Sentence definitions, 207
Sentences
    clarity of, 62–66
    topic, 66–67
Sexist language, 59
Shortcuts in writing, 120–22, 128
Side glosses. *See* Marginal glosses
Signature, 561
Similes, 222–23
Single line graphs, 145
Single-word definitions, 206, 212
Slang, 61–62
Slides, 540–41
Social Science Index, 309
Software
    for graphics, 118, 158–59
    for group writing, 108, 110, 112, 120
    for presentation graphics, 543–44
    standards, 110
Source graphics, 157–58
Source material. *See* Information sources
Space, 179
Spatial organization, 218, 340–41
Specific-to-general organization of
        document, 342–44
Spell checking, 83–84, 195
Spreadsheets, 118, 326–30
    graphics based on, 160, 330–35

Stacked bar graphs, 143–44, 148
Statistical significance, 325
Sternberg, Robert, 283
Stipulative definitions, 205–6
Structured document processors, 124–25
Style, 26–27. *See also* Style guides;
    Technical style
  conventions, 55
  definition, 55
  management of, 108–9, 112
  of oral presentations, 548
  quality review for, 85–88
  of type, 183
Stylebook/editorial manual (American
    Medical Association), 362
Style guides, 21
  company, 56
  for group writing, 109
  professional, 56–57
  source citation formats in, 362–71
*Style manual for guidance in the preparation
    of papers* (American Institute of
    Physics), 362
Subdivisions of documents, 223
Summaries. *See also* Technical summary
  citation of sources in, 372
  for oral presentations, 540
  in progress reports, 457
  in proposals, 416–17
  of source material, 357–60, 380
Summary chart, 445

Table of contents, 507
Tables, 135–36, 138–40, 187. *See also*
    Spreadsheets
  in research reports, 401, 402
  summary chart, 445
  word processor production of, 158–59
Tabs, 158–59
Tagmemics, 324
Tape recorders, 314, 319
Task analysis, 104, 111
Task management, 105–6, 112
Taste, matters of, 288
Team building, 103–4, 111–12
Technical analysis, 25, 264–79
  examples of, 273–74
  graphics used in, 269, 271
  guidelines for, 272–73
  role of, 265
  types of, 265–67
  when to use, 265
  writing process, 267–71
Technical argumentation, 280–94
  anticipation of objections in, 287, 288, 290
  background information in, 284
  conclusion in, 287, 293
  evidence in, 286
  examples of, 288–93
  guidelines for, 287–88

logic in, 281–82, 284–86, 293
noise in, 283
presentation of, 282–87
in proposals, 418–19
role of, 281
thesis statement in, 283–84
types of, 281–82
when to use, 281
Technical definition, 201–14
  in descriptions, 217
  examples of, 212–14
  expanded, 207–9
  formal, 205
  informal, 204–5
  phrase, 207, 212
  placement in document, 210–12
  role of, 202
  sentence, 207
  single-word, 206, 212
  stipulative, 205–6
  when to use, 203–4
Technical description, 215–27. *See also*
    Technical process description
  examples of, 223–26
  guidelines, 222–23
  in instructions, 479
  introduction in, 217, 224–26
  organization of, 339–41
  part-by-part, 218, 225–26
  in progress reports, 458
  in proposals, 418
  role of, 216
  types of, 216–17
  when to use, 216
Technical feasibility, 433, 434, 440–41
Technical process description, 228–45
  details in, 234, 236, 239, 240
  of equipment, 233, 235–37
  example of, 240–43
  graphics in, 237–39
  guidelines for, 239–40
  information sources for, 231–33
  of observation methods, 235
  organization of, 233–34, 339–40
  of purpose of process, 234–35
  role of, 229
  time information in, 239–40
  types of, 229–31
  when to use, 229
Technical style, 26–27, 54–73
  checklist, 70–71
  computer tests of, 85–87
  conventions, 55–71
  definition, 55
  guides to, 56–57
  organization, 66–70
  of proposals, 427
  quality review for, 85–88
  reader involvement, 57–59
  sentence style, 62–66

Technical style *Continued*
    word choice, 59–62
Technical summary, 246–63
    examples of, 259–62
    guidelines for, 257–59
    role of, 247
    types of, 249–53
    when to use, 247–49
    writing methods, 253–57
Technical writing overview, 17–34
    conducting research, 23–26
    evaluating writing situation, 18–23
    preparing document, 26–29
    publishing, 32–33
    reviewing/revision, 30–31
Technicians, 19, 43, 48, 50–51, 471–72
Technology management, 110, 112
Templates, 123–24, 348–50
Tests. *See also* User testing
    of documents, 30–31, 78, 80
    in instructions, 481, 486
    time, 95
Text
    blocking, 179–81
    condensation, 135–36
    definitions in, 210
    graphics integrated with, 162–63
Thesis statement, 283–84
Time constraints, 22, 79
    for correspondence, 563
    group writing and, 102, 106–7, 112, 421
    for oral presentations, 545
    for proposals, 415, 421
Time information, 239–40
Time tests, 95
Titles
    in graphics, 163
    of progress reports, 456
    of research reports, 398
Topics (Aristotle), 298
Topic sentence, 66–67
Transitions, 68, 88
Transparencies, 542–43
Trouble-shooting guide, 475–77
Tufte, Edward, 165
Tutorials, 498–502
Twain, Mark, 229
Type size, font and style, 181–83, 386

United States Government Printing Office
    style manual, 363

Users. *See* Audiences
User testing, 31, 75–78
    comprehension tests, 92–95
    of computer tutorials, 501
    evaluation surveys, 96–98
    guidelines for, 98
    of instructions, 482
    performance tests, 95–96
    of proposals, 427

Vertical bar graphs, 141–42
Videotapes, 541–42
Visual aids for oral presentations, 535,
    538–45
Visual aspects of document. *See* Document
    design
Visual coherence, 175
*Visual Display of Quantitative Information,
    The* (Tufte), 165
Vocabulary. *See* Word choice

Warnings in instructions
    adequacy of, 482
    design of, 187–88, 190, 475
    example, 486, 487
Wave questions, 299
Wilsondisc, 308, 310
Word choice, 59–62. *See also* Technical
    definition
    in computer tutorials, 498–99, 501
    in proposals, 427
    in résumés, 576
    similes, 222–23
Word processing, 27, 119. *See also*
    Computer writing techniques
    document design using, 192–94
    graphics produced by, 118, 158–59
    greeking text with, 179–80
    for group writing, 108, 110, 112, 120
    macros for, 127
    search-and-replace function, 123, 127–28
    spell checking, 83–84, 195
    template creation using, 349–50
    type size choice in, 181
Work group reports, 532–33
Writer-based reviews, 30, 80–89
Writing to Learn (Zinsser), 283

Zinsser, William, 283